Recent Advances in Photovoltaics

Edited by
Meera Ramrakhiani

The ever growing demand for clean energy potentially can be met by solar-to-electrical energy conversion. This book on "Recent Advances in Photovoltaics" presents a detailed overview of recent research and developments in the field of photovoltaic and solar cells. It starts with the basic theory and gradual progress in the field of photovoltaic and various generations of solar cells. The search for new materials and/or new structures such as multi-junctions, nanostructures, photoelectrochemical cells, organic solar cells etc. for improved performance is discussed. The experimental investigations on certain materials and modelling for better results are also described in the book.

Recent Advances in Photovoltaics

Edited by

Meera Ramrakhiani

Department of Physics and Electronics, Rani Durgavti University Jabalpur (M.P.), India

Published by **Materials Research Forum LLC**
Millersville, PA 17551, USA

Published as part of the book series
Materials Research Foundations
Volume 17 (2017)
ISSN 2471-8890 (Print)
ISSN 2471-8904 (Online)

Print ISBN 978-1-945291-36-4
ePDF ISBN 978-1-945291-37-1

Distributed worldwide by

Materials Research Forum LLC
105 Springdale Lane
Millersville, PA 17551
USA
http://www.mrforum.com

Manufactured in the United States of America
10 9 8 7 6 5 4 3 2 1

Table of Contents

Preface

Chapter 1   The rise of solar cells 1

Chapter 2   Basic principles and theory of the photovoltaic effect 37

Chapter 3   Nanostructured solar cells 58

Chapter 4   Quantum dot as light harvester nanocrystals for solar cell
             applications 117

Chapter 5   Photoelectrochemical solar cells using nanocrystalline copper
             selenide photo electrode 134

Chapter 6   Photovoltaic response of nanocrystalline cadmium telluride in
             photoelectrochemical cells 185

Chapter 7   Studies on the photovoltaic effect of CdSe based
             nanocrystalline multilayered photoelectrodes in
             photoelectrochemical solar cells 241

Chapter 8   Status and potential of organic solar cells 269

Chapter 9   Efficiency rise in PCDTBT:PC$_{70}$BM organic solar cell using
             interface additive 291

Chapter 10   Recent advances in polymer solar cells 299

Chapter 11   Advancement in simulation and modeling of
             organic solar cells 310

Chapter 12   Fill factor analysis of organic solar cell 334

Keywords 345

About the editors 347

Preface

The conventional energy sources are non-renewable, finite and are in danger of depletion to various degrees in the future. Everyday energy resources have to be modified from petroleum and nuclear energies to renewable forms of energy, which will play a major role in the world's future energy supply. Among new energy sources, solar energy is extremely promising since it is practically inexhaustible. The task of utilizing solar energy is becoming more and more urgent as a result of the gradual exhaustion of fossil fuel resources and increasing problem of environmental pollution. Photovoltaic (PV) is a method of converting solar energy directly into electricity using semiconducting materials that exhibit the photovoltaic effect. The solar radiations reaching the surface of earth are much more than the energy requirement. The mechanism of converting optical energy into electrical energy is known as photovoltaic. Photovoltaic systems are being used for fifty years in specialized applications, and grid-connected PV systems have been in use for over twenty years. Silicon based solar cells have achieved power conversion efficiency of around 24%; however high efficiency is offset by high cost. Much effort has recently been directed towards developing new and better solar energy conversion systems. Attempts are being made to reduce the cost and increase the efficiency so as to become competitive with presently available technology by using new structures based on different architectural design (like multiple-junctions) and by developing new materials (like nanocomposites) to serve as light absorbers and charge carrier separator with better efficiency. Various types of solar cells have been developed such as amorphous silicon solar cells, thin film solar cells, photoelectrochemical solar cells, multi-junction solar cells, nanocrystal solar cells, polymer solar cells etc. This special volume on "Recent Advances in Photovoltaics" published by Material Research Forum LLC presents a detailed overview of recent research and developments in the field of photovoltaics and solar cells.

This book consists of **12** Chapters written by experts and researchers in the area of solar cells. The **1ˢᵗ Chapter** discusses the chronological history and scientific advancements in research and development activities related to solar cell technology since 1954 to present date. Different generations of solar cells and various types of solar cells are also described in this chapter. The **2ⁿᵈ Chapter** deals with the theoretical background and basic physical principles of the photovoltaic effect. Various characteristics and parameters of solar cells are described and suitable materials for solar cells are discussed.

During the last decade, the development of the photovoltaic device theory and nanofabrication technology enables studies of more complex nanostructured solar cells with higher conversion efficiency and lower production cost. The 3ʳᵈ and 4ᵗʰ chapters are devoted to solar cells based on nanotechnology. The use of nanostructures in solar cells may enhance the efficiency and performance. The fundamental principles and important features of these advanced solar cell designs are systematically presented and summarized in **Chapter 3.** The function and role of nanostructures and the key factors affecting device performance have been discussed with special attention on quantum effect. The **4th Chapter** reviews use of quantum dot in solar cells. Quantum dot solar cell reduces heat waste by multiple electron generation (MEG) and produces up to three electrons per photon. The tuning of band energies by size change in quantum dots gives new ways to control the response and efficiency of the solar cell.

The chapters 5 to 7 are based on experimental observations and results using different nano-crystalline materials in photoelectrochemical cells. **Chapter 5** describes synthesis and characterisation of nanocrystalline copper selenide thin films and their photovoltaic response as photolectrode in photoelectrochemical cells. **Chapter 6** deals with similar studies on cadmium telluride thin films. Better performance has been obtained by using nanocrystalline photoelectrodes. **Chapter 7** is based on investigation of multi-layered cadmium selenide nanocrystalline photoelectrodes in photoelectrochemical cells. Enhanced performance is observed in case of multilayered photoelectrode due tuning the band gap by change in size of nanocrystals in different layers.

The chapters 8 to 12 are related to organic solar cells (OSC). **Chapter 8** reviews the current status of the field of organic solar cells and discusses different production technologies as well as studies the important parameters to improve their performance. Organic solar cell research has developed during the past 30 years, but especially in the last decade it has attracted scientific and economic interest triggered by a rapid increase in power conversion efficiencies. This was achieved by the introduction of new materials, improved materials-engineering, and more sophisticated device structures. **Chapter 9** is a paper based on experiments observations on the use of a low-bandgap PCDTBT:PC$_{70}$BM-based PV layer that incorporates a PTE surfactant, which was used to the BHJ interfaces in OSCs. It is shown that a combination of PTE interface additives and high-performance low-band gap PV materials holds great potential for the development of a new generation of highly efficient OSCs.

Polymer solar cells belongs to a promising class of next-generation photovoltaic, because they hold promise for the realization of mechanically flexible, lightweight, large-area devices that can be fabricated by room-temperature solution processing. **Chapter 10** covers the scientific origins and basic properties of polymer solar cell technology, material requirements and device operation mechanisms. Potential future developments and the applications of this technology are also briefly discussed.

Simulations/modelling are powerful tools for optimization of OSCs, reveal new insights, and predict the behaviour, performance, limitations, stability, dependency of OSCs and maximum attainable efficiency. **Chapter 11** provides an overview of simulation models (optical/electrical) for modelling state of the art devices, corresponding development in recent years on the basis of device physics and working principle. **Chapter 12** is based on the analysis of parameters that affect the fill factor of organic solar cell using MATLAB.

I hope this volume will be useful for not only scholars and workers in the field of solar cells, optical properties of materials, non-conventional energy sources etc. but also for physics and chemistry students and the persons concerned about the present knowledge on photovoltaic to overcome energy crisis.

It is my pleasant duty to offer my sincerely thanks to all the authors for their valuable contributions and the reviewers for improving the quality of the articles. I am specially thankful to Prof. S.K.Pandey for his help in editing work. Material Science Forum LLC, especially Thomas Wohlbier deserves appreciation for patience and acceptance of delayed submissions, and publication work of this special volume.

Meera Ramrakhiani

Editor of the Special Volume

Chapter 1

The rise of solar cells

Sonal P. Ghawade[1], Abhay D. Deshmukh*[1], Kavita Abhay Deshmukh[2] and S. J. Dhoble[1]

[1]Dept. of Physics, Energy Materials & Devices Laboratory, RTM University, Nagpur, India

[2]Dept. of MME, Visvesvaraya National Institute of Technology, Nagpur, India

abhay.d07@gmail.com

Abstract

Solar power is a significant source of green and renewable type of energy. A photovoltaic cell is an electronic device which directly converts sunlight into electricity. Light falling on the solar cell generates both current and voltage to produce electric power. This process needs a material in which the absorption of light raises an electron to a higher energy state, and the movement of this higher energy electron from the solar cell into an external circuit gives rise to electric current. The electron further dissipates its energy in the external circuit and returns to the solar cell. In this chapter we will discuss the chronological history and scientific advancements in research and development activities related to solar cell technology since 1954 to present. We will also show various types of solar cells and their applications.

Keywords

Photovoltaic Cells, History of Solar Cells, Different Generation of Solar Cells, Types of Solar Cells

Contents

1.   Introduction..2

1.1 History of photovoltaic cells...4

1.2 What is photovoltaics? 5

1.3 Why solar cell?...7

2.   Different generations of solar cells.....................................8

2.1   First generation solar cells 9

2.2   Second generation solar cells 9

2.3   Third generation solar cells 10

2.4   Fourth generations of solar cells 11

3.   Types of solar cells and their applications 12

3.1   Monocrystalline silicon solar cells 13

3.2   Polycrystalline silicon solar cells 13

3.3   Thin film solar cells 15

3.3.1 CdTe solar cell 15

3.3.2 Copper indium gallium selenide (CIGS) 17

3.3.3 Copper zinc tin sulphide (CZTS) 18

3.4   Quantum dots solar cells 20

3.5   Multijunction amorphous nanotechnology-based solar cells 22

3.6   Dye sensitized solar cells 24

3.7   Gallium arsenide solar cells 27

3.8   Perovskite solar cells 28

References 30

1.   Introduction

The most significant characteristic of our current time is the fastest growing technological development in all areas. Today it takes only a few months to achieve the same number of important inventions and discoveries that took decades, if not centuries, in the past. **The question is:** *"Can these technological developments fulfil the need to safeguard the environment by creating a world powered by the Sun's Energy?"* Everyday energy resources have modified from petroleum and nuclear energies to renewable forms of energy, thanks to the worldwide essential environmental and economic demands. One hundred years ago, an Italian chemist predicted the use of solar energy in 'the future'. He asked a question "So far, human civilization has made use almost exclusively of fossil energy. *Would it not be advantageous to make better use of radiant energy?"* [1]. Solar energy is viewed as one promising solution to the global energy crisis. Among a variety of means for generating energy from the sun, solar cell is one of the effective approach to convert solar energy into electrical energy. In 2009, the global manufacture of photovoltaic cells and modules was 12.3 GW [2], and it enhanced to over 20 GW one year later [3].

This chapter explains the sequential history and technical advances related to research field and progress activities concerning to solar cell technology from 1954 to present. Material scientists have identified III-V semiconductor compounds, quantum dots, organic materials and nanotechnology that can be applied in designing and progress of solar cells, also known as photovoltaic (PV) cells. It is significant to point out that photovoltaic cells are not only environmentally friendly but also offer clean, efficient, reliable and continuous source of electrical energy. *Why is there a demand for solar technology? How is it so important?* Some of the reasons include adverse political environments, severe climate situation, global greenhouse. This has required energy planners to look for optional sources to diminish reliance on fossil fuels such as oil, coal and gas to change to other clean forms of energy desirable to protect our planet [4]. Photovoltaic is a gift to technology and is able to provide electricity to various markets around the world and out of the world (in outer space). It also provides us to do what we already do (produce electricity, which is spread over the transmission grid), only to do it in a sustainable, pollution-free, equitable fashion. *Why photovoltaics are equitable?* Since nearly everyone has access to sunlight! Electricity is the most common form of energy in our day to day life. It is what allows citizens of the evolved countries to have about hygiene, universal lighting on demand, refrigeration, and interior climate control in their homes, schools, and businesses and widespread access to different electronic and electromagnetic media. Access to and consumption of electricity is nearly correlated with quality of life. Human Development Index (HDI) usually depends on the annual per capita electricity used. The HDI is collected by the UN and calculated on the basis of life expectancy, educational achievement, and per capita Gross Domestic Product. To progress the quality of life in lots of countries, as measured by their HDI, will require increasing their electricity consumption by factors of 10 or more, from a few hundred to a few thousand kilowatt-hrs (kWh) per year. How will we do it? Our selections are to continue utilizing the answers of the last century such as burning more fossil fuels (and releasing megatons of CO_2, SO_2, and NO_2) or building more nuclear plants (despite having no method of safely disposing of the high-level radioactive waste) or to apply the new millennium's answer of renewable, sustainable, nonpolluting, widely available clean energy like photovoltaics and wind. (Wind presently generates over a thousand times more electricity than photovoltaics but it is very site-specific, whereas photovoltaic is generally applicable to most locations.)

A large amount of carbon dioxide is emitted when electrical power is produced from coal, gas and wood. The most evident solution to the problem is to substantially reduce emissions of greenhouse gases, in particular CO_2. In order to achieve this we must develop clean, efficient, renewable sources of energy. Photovoltaic (PV) cells which convert sunlight directly into electricity are presently one of the leading potential candidates to supply this energy. This chapter aims to review of various types of solar cells and their applications.

1.1   History of photovoltaic cells

Photovoltaic effect (PV) was first discovered by French physicist Alexandre-Edmond Becquerel in 1839. His experiment was finished by illuminating two electrodes with different types of light. The electrodes were covered by light sensitive materials AgBr or AgCl, and carried out in a black body surrounded by an acid solution. The electricity increased when the light intensity increased [5]. In 1877, Willoughby Smith, an English Engineer discovered PV effect based on selenium and lead to the first selenium solar cell construction which was built in 1877. Explanation of first solar cells made from selenium wafer was given by Charles Fritts in 1883 [6]. In 1905, Albert Einstein published in his paper about the photoelectric effect. There he claimed that light consist of number of 'packets' or 'quanta' of energy, called as photons. This energy changes only with its frequency (electromagnetic waves or color of the light). This theory was extremely simple, innovative and explained very well the absorption of photons related to the frequency of the light [7]. The PV effect based on cadmium-selenide was observed in 1932. Nowadays, CdSe belongs to one of the significant material for solar cell production. First germanium solar cells were made in 1951. Three researchers Gerald Pearson, Daryl Chapin and Calvin Fuller at Bell Laboratories invented a silicon solar cell in 1954, which was the first material to directly convert enough sunlight into electricity to run electrical devices. The efficiency of the silicon solar cell developed in Bell laboratory was nearly 4%, which later increased to 11%. The cells were made by hand and cost more than $1000 per watt [8]. In 1955 Hoffman electronics invented the first PV product with 2% efficiency for US $25 per cell with 14 mW peak power. In 1960 another solar cell with 14% efficiency was introduced by Hoffman Electronics. In 1965 first Sun-powered automobile (SUNMOBILE) was demonstrated in Chicago. The first working photovoltaic module from silicon solar cells was developed by Sharp Corporation in 1963. The journey towards the large scale application of photovoltaic started in 1970 with the CdTe- based and amorphous Silicon thin film solar cell modules. During the period of

1970 to 1979 large number of photovoltaic companies has been recognized. These consist of Solar Power Corporation in 1972, Solarex Corporation in 1973, Japanese Sunshine Project in 1974, Solac International 1975, Solar Technology International in 1975 etc. In 1984 the first amorphous Silicon PV modules came into the market. During the period of 2000 to 2012 there is an expansion in the PV market. The majority of the PV challenges for PV system and components are overcome. The PV modules become very reliable (warranted for 25 years). The market growth rate is increased by 30% per year. On 24th September 2013 German Fraunhofer institude for solar energy system, Soitec CEA Leti and the Helmholtz Centre Berlin announced that they made up of four solar sub-cells based on III-V compound semiconductors for concentrator photovoltaic (CPV) with conversion efficiency 44.7% which is a world record [9].

1.2   What is photovoltaics?

What is the basis principle of solar cells operation? Semiconductor materials are used to make up solar cells, which have weakly adhered electrons in a band of energy called valence band. When energy above a certain threshold, called band gap energy, is employed to valence band electron, the bonds are broken and the electron is rather "free" to travel around in a new energy band called conduction band whereas it is able to "conduct" electricity. Therefore, the numbers of free electrons in the conduction band are separated from the valence band by the band gap (band gap measured in electron volts or eV). Such energy required to free the electron can be provided by the photons, which are particles of light. Figure 1 illustrates the idealized relation between band gap energy and the spatial boundaries. When the solar cell is exposed to sunlight, bundles of photons strike the valence electrons, due to breaking the bonds and jumping them to the conduction band. Then a particularly created selective contact gathered conduction band electrons- such type of electrons moves to the external circuit. The electrons lose their energy by moving in the external circuit such as pumping water, powering a sewing machine motor, a light bulb, spinning a fan or a computer. They are restored to the solar cell through the return loop of the circuit by a second selective contact, which comebacks to the valence band with the same energy they started with. The flow of electrons in the external circuit and contacts is called the electric current.

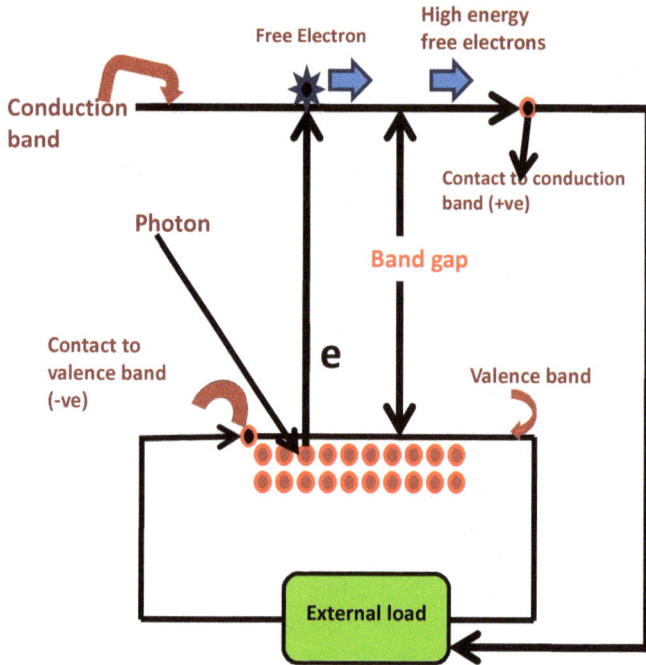

Figure 1: *Schematic of a Solar Cell.*

Electrons are pumped by photons from the valence band to the conduction band. There they are extracted by a contact selective to the conduction band (a n-doped semiconductor) at a higher (free) energy and delivered to the outside world via wires, where they do some useful work, then are returned to the valence band at a lower (free) energy by a contact selective to the valence band (a p-type semiconductor) [10]. The potential at which the electrons are presented to the external world is slightly less than the threshold energy that excited the electrons; called as the band gap. As a result, if a material band gap is 1 eV, then electrons excited by a 2 eV photon or 3 eV photon will both have a potential of slightly less than 1 V (i.e. the electrons are delivered with an energy of 1 eV). The electric power created is the product of the current times the voltage; that is, power is the number of free electrons times their potential.

As practical applications, a huge number of solar cells are interconnected and encapsulated into units called PV modules, which is the product typically sold to the customer. They give DC current that is usually transformed into further useful AC current by an electronic device called an inverter. The inverter, the rechargeable batteries (when storage is needed), the mechanical structure to mount and aim (when aiming is necessary) the modules, and any other elements necessary to build a PV system are called the balance of the system (BOS).

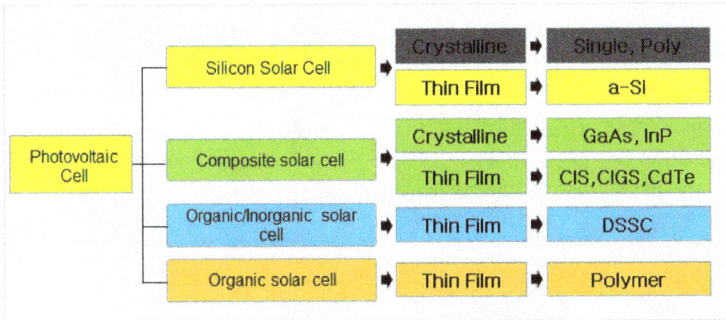

Figure 2: *The Different Types of Solar Cell.*

1.3   Why solar cell?

The demand for energy has forever been the primary driving force in the progress of industrial capability. The discovery of the steam engine sparked the industrial revolution and the resulting development of an energy economy based on wood and coal. At that time the continuous development of the energy economy has focused on different sources of energy, such as oil, water, wind, nuclear and gas. Nuclear energy is very expensive and poses radiation hazards and nuclear waste problems. Electrical energy sources using gas, wood, coal, and oil produce huge amounts of pollution or carbon dioxide emissions, so posing health risks. These electrical energy power sources involve large funds and scheduled maintenance. In case of coal-fired power plants, coal transportation cost, high capital investment, and delivery delay under adverse climatic conditions could pose dangerous problems. Alternatively, a solar energy source supplies reliable, pollution-free, quiet, long-term, maintenance-free, self-contained, and year-round continuous and unlimited operation at moderate costs. Even though all these profit of solar cells, almost 55 years after their invention, PV solar cells are giving only 0.04 % of the world's on-grid electricity because of the high cost of solar cells, which is outside the reach of the common consumer. Based on the 2007 statistical review of world energy consumption, 30% of the electrical energy is from coal, 16 % from natural gas, 15% from water generators, 9 % from oil, 4 % from nuclear reactors, and only 1 % from solar cells. In the United States, solar energy of all kinds fulfils less than 0.1 % of the electrical demand.

All industrial and Western countries such as the United States, Japan, Spain, Brazil, Italy, Germany, and other European countries are turning to electrical power generation from solar cells, for the reason that of the high capital investments, radiation, and carbon dioxide emissions related with coal-based, gas-based, nuclear-based and oil-based power plants.

2.   Different generations of solar cells

It is true that PV is becoming a part of the solution to the growing energy challenge and an essential component of future global energy production. Solar cells are usually divided into four main categories called generations [11-16]. Different generations of PV technologies, their cost effectiveness and efficiency limits are shown in Figure 3.

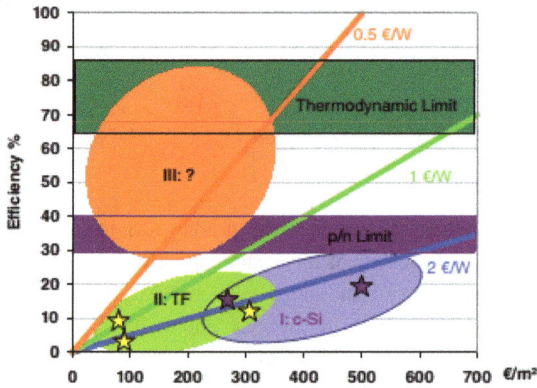

Figure 3: *Relation of the cost per watt of solar energy to the surface cost of manufacturing solar devices (modules) and the device efficiency.*

The blue, green, and orange oval regions represent the ranges found for crystalline silicon (first-), thin-film (second-), and third-generation solar technologies, respectively. The white zone marks the anticipated range for very high efficiency devices. For comparison, limiting efficiencies derived from thermodynamic constraints are also indicated as horizontal bands (low range: no concentration, high range: maximal concentration). Stars indicate industrial production costs as they could be estimated in 2010 from available data: filled stars for c-Si modules and hollow stars for thin-film technologies [17]. The first generation contains solar cells that are relatively expensive to produce and have a high efficiency. The second generation contains types of solar cells that have a lower efficiency, but are cheaper to produce, such that the cost per watt is lower than first generation cells. The term third generation is used for cells that are generally newer compared to other second generation cells. Most technologies in this generation are not yet commercial, but there is a lot of research going on in this area. The goal is to make third generation solar cells cheap to produce.

2.1   First generation solar cells

First generation solar cells are made of single crystal silicon wafers (c-Si). A low band gap of 1.1eV for silicon provides broad spectral absorption range. Laboratory scale cells have already achieved 25% efficiency. The greatest disadvantage of such cells is the expensive and energy intensive manufacturing technology which increase the generated power cost to ~4$/W.

Figure 4: First generation of single crystal wafers based Solar Cell.

2.2   Second generation solar cells

Second generation solar cells are based on the use of thin film semiconductor like cadmium telluride (CdTe), amorphous silicon (a-Si), polycrystalline silicon(p-Si), copper indium gallium diselenide (CIGS) etc. Use of thin films reduces the amount of essential materials for cell design, which contributes greatly to reduced costs for thin film solar cells. Various second generation technologies or semiconductor materials are presently under investigation for mass production. For an applied power rating, they show improved performance at low light levels than crystalline panels. The Polycrystalline Si consists, exclusively of crystalline Silicon grains (1mm), divided by grain boundaries and has low band gap ~1.1eV. Charge carrier mobility in p-Si is one order of magnitude higher than a-Si, hence shows greater stability under electric field and light-induced stress. Maximum efficiency of p-Si cells is ~20%. However p-Si cells can be fragile and the cost is still too high for widespread use. CdTe has a band gap of ~1.58 eV and laboratory scale efficiency of ~18% has already been achieved. CIGS alloy has an efficiency of ~ 20%, but CIGS cells have a complex heterojunction structure and are very difficult to prepare. The greatest disadvantage of a-Si panels is that they are much less efficient per unit area (~ 10%) and are generally not suitable for roof installations. One would typically need nearly double the panel area for the same power output. Long term stability of amorphous silicon cells is also a matter of concern.

Figure 5: Second Generation of Thin Film Solar Cell.

2.3   Third generation solar cells

The third generation comprises of two categories viz. IIIa and IIIb. Category IIIa consist of novel approaches that strive to very high efficiencies. The IIIa technologies have theoretical maximum efficiencies well above the 31% limit (Shockley-Queisser limit) for single-junction devices. Hence, these high efficiencies cells can afford higher costs and still show a favourable cost/watt balance. In the second type of third generation device (IIIb), the goal is exactly opposite. A low cost/watt balance would be achieved via moderate efficiencies (15-20%), but at very low production cost. This will require inexpensive semiconductor materials, packing solutions, production processes (low-temperature atmospheric routes), high fabrication throughout, low investment into the production facility and a production-on-demand scenario. Various promising technologies for these third generation low cost PVs are currently available and grouped under the appellation organic PVs (OPVs). They all have in common that at least one of the key functionalities for PV energy transition is handled through organic semiconductor or conductor. Dye sensitized solar cells (DSSC) have achieved ~12% efficiency. However, in the last 15 years no significant increase in efficiency has been achieved. Moreover sealing of a liquid electrode is still a challenge. Many small molecule organic CuPc (Copper phthalocyanine), ZnPc(Zinc phthalocyanine) etc. are also in contention to meet future energy demands. Use of conjugated polymers as the donor materials is another emerging technology where steady increase in efficiency has been achieved in the last decade. Recently over 10% efficiency has been achieved by using a low band-gap polymer as donor and fullerene derivative as acceptor material. Due to the inherent differences in materials properties of polymer and their inorganic counterpart, the device structure of polymer solar cell (PSC) is significantly different than that of conventional silicon solar cells. It is therefore, necessary to understand the differences between the materials properties and the resultant changes required in polymer based solar cells. Use of Quantum Dots (QDs) is another way of enhancing the efficiency of solar cells. These quantum dots are given so much attention because when a photon hits a quantum dot made of the same materials, there may be multiple electron–hole pair generation (typically 2-3); whereas that normally one photon excites one electron creating one electron-hole pair (EHP) in bulk material. Another way to increase the efficiency is to use several layers solar cells with different band gaps in a stack. Each layer will utilize light with different wavelength and can get higher efficiency [18].

2.4 Fourth generations of solar cells

The fourth generation (4G) solar cells are based on 'inorganics-in-organics' which offers improved power conversion efficiency compare to current third generation solar cells, while maintaining their low manufacturing cost. Within a single layer, the fourth generation solar cells consist of a low cost and flexibility of conducting polymer films (organic) with the lifetime stability of novel nanostructures (inorganic) material such as carbon nanotubes (CNTs) or metal nanoparticles (mNPs), graphene, metal oxides and nano-hybrid materials as shown in Figure 6.

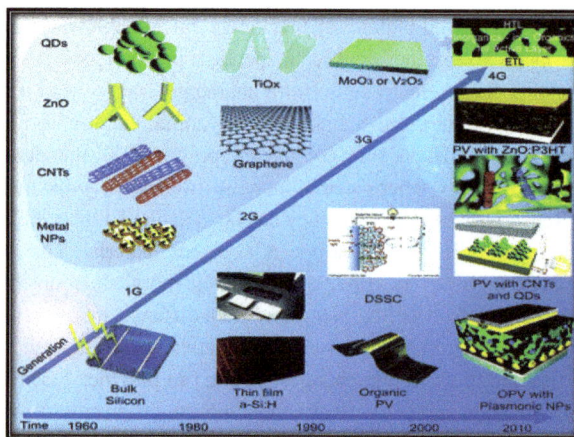

Figure 6: *Timeline of the four generations of photovoltaic devices, illustrating the changes from first generation (1G) to fourth generation (4G) with associated nanomaterial components that comprise half of the 4G devices [19].*

Leverage on the properties of these new, hybrid active materials can show their performance beyond that of third generation devices. Incorporation of active inorganic nanomaterials improves the harvesting of solar energy and the manipulation of electrical charges within these fourth generation solar cells, enhancing efficiency and lifetime stability. New fourth generation solar cells are being developed in the recently commenced Euro 11.6 M European Union FP7 SMARTOICS program, led by the Aristotle University of Thessaloniki, Greece. The SMARTONICS project aims to develop smart machines, tools and processes for large area production of fourth generation solar cells engineered at the nanoscale, using roll-to roll printing technology for high throughput and cost-efficient fabrication. It is believed that the fourth generation solar cells will be indeed the technology for future PV energy sources.

3.   Types of solar cells and their applications

Solar cells are usually named following the semiconducting material they are made of. These materials must have definite characteristics in order to absorb sunlight. Some cells are considered to handle sunlight that passes the atmosphere to Earth's surface, when others are optimized for use in space. Solar cells can be built of only one single layer of light-absorbing material (single-junction) or use multiple physical configurations (multi-junctions) to take advantage of various absorption and charge separation mechanisms.

3.1 Monocrystalline silicon solar cells

Monocrystalline silicon (or "single-crystal Si", "mono c-Si", or just mono-Si) is the base material for silicon chips widely employed in practically all electronic equipment today. Monocrystalline silicon also shows photovoltaic effect, and used as light absorbing material in the fabrication of solar cells. It comprises of solid silicon with continuous crystal lattice, unbroken to its edges, and free of any grain boundaries. Mono-Si is capable to be prepared intrinsic, consisting but of extremely pure silicon, or doped, containing very little quantities of other elements added to vary its semiconducting properties. Mainly silicon monocrystals are developed by the Czochralski process into ingots of up to 2 meters in length and weighing various hundred kilograms. These cylinders are then sliced into thin wafers of a few hundred microns for advance working. Monocrystalline silicon is different from other allotropic forms, such as amorphous silicon applied in thin film solar cells, and polycrystalline silicon, that consists of small crystals, also known as crystallites. Monocrystalline silicon is used in the production of high performance solar cells, yet because of solar cells are less challenging than microelectronics concerning structural imperfections, monocrystaline solar grade (Sog-Si) is often employed. Lab efficiencies of 25% for monocrystalline Si solar cells are the highest in the commercial PV market, ahead of polycrystalline silicon solar cell with 20.4 % and all established thin-film technologies namely, CIGS cells (19.8%), CdTe cells (19.6%), and a-Si cells (13.4%). These high efficiencies can be combined with other technologies, such as multi-layer solar cells in applications where space and weight turn into an issue such as powering satellites. Single-crystal silicon is possibly the most significant technological material of the last few decades- the "silicon era" [20]. Since its availability at an inexpensive cost has been required for the growth of the electronic devices on which the present day electronic and informatics revolution is based [21].

3.2 Polycrystalline silicon solar cells

Polycrystalline silicon is also called polysilicon or poly-Si. Polycrystalline and paracrystalline phases are made up of a number of smaller crystals or crystallites. Polycrystalline silicon (or semi-crystalline silicon, polysilicon, poly-Si, or simply "poly") is a material consisting of multiple small silicon crystals. Polycrystalline cells can be familiar by a visible grain, a "metal flake effect". Polycrystalline silicon is developed from metallurgical grade silicon through a chemical purification process, called Siemens process. This process takes distillation of volatile silicon compounds, and their decomposition into silicon at high temperatures. It is a high purity material. Polycrystalline silicon form of silicon is used as a raw material by the solar photovoltaic and electronics industry. Photovoltaic industry also creates advanced metallurgical- grade silicon (UMG-Si), utilizing metallurgical instead of chemical purification processes. When developed for the electronics industry, polysilicon contains impurity levels of less than one part per billion (ppb), although polycrystalline solar grade silicon (SoG-Si) is generally less pure. A few companies from China, Germany, Japan, Korea and the United States, such as GCL-Poly, Wacker Chemie, OCI, and Hemlock Semiconductor, as well as the Norwegian headquartered REC, accounted for most of the worldwide production of about 230,000 tons in 2013 [22]. Polysilicon is composed of small crystals, also well-known as crystallites. Although polysilicon and multisilicon are frequently used as synonyms, multicrystalline usually refers to crystals larger than 1 mm. Multicrystalline solar cells are the most common type of solar cells in the fast-growing PV market and consume most of the worldwide produced polysilicon. About 5 tons of polysilicon is required to manufacture 1 megawatt (MW) of conventional solar modules [23]. Polysilicon is different from monocrystalline silicon and amorphous silicon. Polycrystalline silicon is the key feedstock in the crystalline silicon based photovoltaic industry and employed for the manufacture of conventional solar cells. For the first time, in 2006, over half of the world's supply of polysilicon was being used by PV manufacturers [24]. The solar industry was severely hindered by a shortage in supply of polysilicon feedstock and was forced to idle about a quarter of its cell and module manufacturing capacity in 2007 [25]. Only twelve factories were known to produce solar-grade polysilicon in 2008; however by 2013 the number increased to over 100 manufacturers [26]. The use of polycrystalline silicon in the manufacturing of solar cells requires less material and therefore supplies higher profits and increased manufacturing throughput. Polycrystalline silicon does not need to be deposited on a silicon wafer to form a solar cell; rather it can be deposited on other cheaper materials, thus reducing the cost. Not requiring a silicon wafer alleviates the silicon shortages occasionally faced by the microelectronics industry [27].

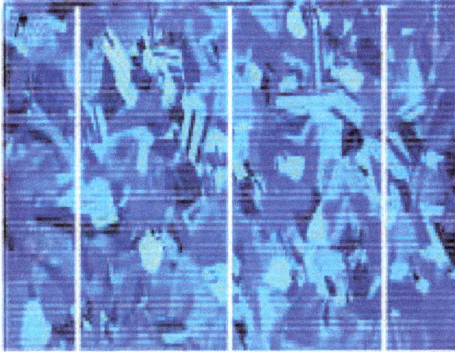

Figure7: Polycrystalline Solar Cell.

3.3 Thin film solar cells

In addition to the low materials consumption in thin film, one of the advantages of thin film solar cells is that they can be easily connected in series in an integral manner on large-scale substrates; therefore, whole modules are fabricated by the deposition process. This is advantageous economically, however, it is also very demanding for the process technology, because large areas have to be processed without defects. One problem is that a number of different materials are being pursued and it is not all clear which one is the best choice. The most significant materials will be described below.

3.3.1 CdTe solar cell

Thin-film technology of cadmium telluride (CdTe) is considered as one of the leading material for the development of cost-effective photovoltaics (PV), and first PV technology with the price for Wp below $1 ($0.85) [28]. A photovoltaic (PV) technology is based on the use of cadmium telluride, a thin semiconductor layer designed to absorb and convert sunlight into electricity. Cadmium telluride (CdTe) is a crystalline compound formed from cadmium and tellurium. It is used as a solar cell material. It is usually sandwiched with cadmium and sulphide to from a p-n-junction photovoltaic cell. CdTe is one of the thin film photovoltaic technologies with lower costs than conventional crystalline silicon solar cell. CdTe has a direct band gap of 1.5 eV, which is close to the ideal value for photovoltaic conversion efficiency. Meanwhile, high optical absorption coefficient and high chemical stability also appear in CdTe. These properties make CdTe a very attractive material for thin film solar cells. The theoretical efficiency of CdTe thin-film solar cells is expected to be 28%-30% [29-30].

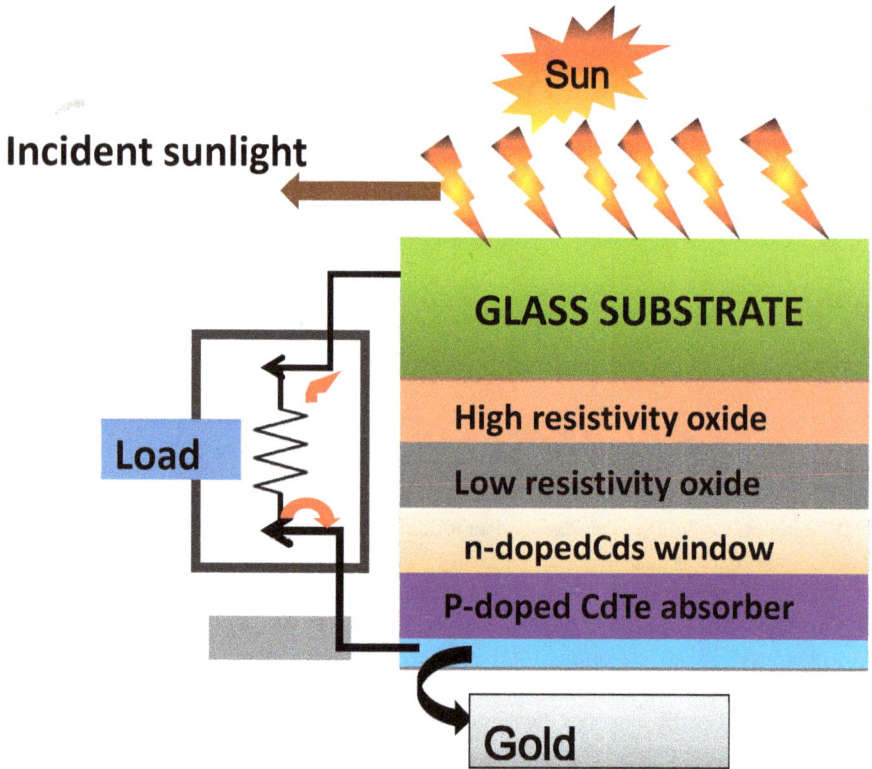

Figure 8: *The Five Layers of CdTe Solar Cell [34].*

A simple heterojunction design has been developed in which p-type CdTe was matched with n-type cadmium sulfide (CdS). The cell was finished by adding top and bottom contacts. According to lifecycle basis, CdTe PV has the smallest carbon footprint, lowest water use and shortest energy payback time of all solar technologies [31]. CdTe solar cells energy payback time of lesser than a year changes for faster carbon reductions with no short-term energy deficits. CdTe is rising quickly in acceptance and second most applied solar cell material in the world. A large fabrication of CdTe thin film solar cell is done by company named Tempe and Arizon. CdTe photovoltaics cell is applied in some of the world's main photovoltaic power stations, such as the Topaz Solar Farm. Among a share of 5.1% of worldwide PV manufacture, CdTe technology accounted for more than half of the thin film market in 2013 [32].

CdTe thin film solar cells are being used in calculators for more than two decades. The maximum conversion efficiency of CdTe thin film solar cells at laboratory scale is near about 16%, but commercial module efficiencies are in the range of 6-8%. At present, First Solar has declared a new world record this year for CdTe PV solar cell efficiency of 17.3% with a test of cell concept using commercial-scale manufacturing equipment and materials, and its average efficiency of modules formed in the first quarter of 2011 was 11.7% [33].

3.3.2 Copper indium gallium selenide (CIGS)

Copper-Indium-Gallium-Selenide (CIGS) solar cell is one of the thin film types of solar cell. Copper Indium gallium diselenide (CIGS) is a I-III-VI2 compound semiconductor material composed of copper, indium, gallium and selenium. It is a direct band gap semiconductor. The band gap varies from 1.0 eV (CIS) about to 1.7 eV (CGS) depending on the Ga/(In+Ga) ratio [35]. It is a tetrahedrally bonded semiconductor and chalcopyrite crystal structure. A copper indium gallium selenide solar cell (or CIGS cell, sometimes CI(G)S or CIS cell) is thin film solar cell used to convert sunlight into electric power. CIGS has very high absorption coefficient of more than 10^5/cm for 1.5 eV and higher energy photons [36]. It is used as light absorber material for thin film solar cell. A low-cost and thin substrate together with a thin and flexible encapsulant, glass-free, layered would combine the advantages of flexible solar cells and also cost-effective production of CIGS solar cell.

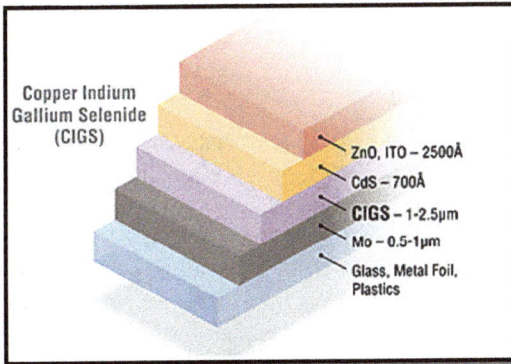

Copper Indium
Gallium Selenide
(CIGS)

ZnO, ITO – 2500Å
CdS – 700Å
CIGS – 1-2.5µm
Mo – 0.5-1µm
Glass, Metal Foil,
Plastics

Figure 9: *Graphic showing the five layers that consist of CIGS Solar Cells.*

CIGS cells continue being developed, as they promise to reach silicon like efficiencies, while maintaining their low costs, as is typical for thin-film technology [37]. Prominent production of CIGS photovoltaics was done by now bankrupt companies i.e. Nanosolar and Solyndra. Current market leader is the Japanese company Solar Frontier, making solar modules free of any heavy metals such as cadmium or lead [38]. CIGS solar cells show good stability from electron and proton irradiation; whereas, for example, high-efficiency Si or GaAs solar cells are more prone to radiation damage [39]. Therein, CIGS has one of the major advantages of being able to be situated on flexible substrate materials, producing lightweight and highly flexible solar panels. Advancements in efficiency have made CIGS a well-known technology among alternative cell materials. The most common device construction for CIGS solar cell is shown in Figure 9.

Glass is commonly used as a substrate, still lots of companies are also looking at lighter and more flexible substrate such as polyamide or metal foils. A molybdenum (M_O) layer is situated (commonly by sputtering method) which serves as the back contact and to reflect most unabsorbed light back into the absorber. Next M_o deposition, a P-type CIGS absorber layer is developed by one of the various method. A thin n-type buffer layer is added on top of the absorber. The buffer layer is typically CdS deposited via chemical bath deposition. The buffer is covered with a thin, intrinsic ZnO layer which is capped by a thicker, Al doped ZnO layer. Although increasing the series resistance, the intrinsic ZnO layer is valuable to cell performance. The Al doped ZnO serves as a transparent conducting oxide to collect and move electrons out of the cell while absorbing very small amount of light. There is also environmental concern for the hazards during the operation of CIGS modules with one potential risk being the leaching of critical materials into rain water. This only happens if a module is broken or crushed, so the normally well-encapsulated active layers are exposed. The main hazard during the active life of the CIGS modules is related to the accidents.

3.3.3 Copper zinc tin sulphide (CZTS)

Copper Zinc Tin Sulfide (CZTS) is a 4th semiconducting compound which has obtained enhancing interest as in the late 2000s for applications in solar cells. The class of concerned materials includes other such I2-III-IV-VI4 compounds as copper zinc tin selenide (CZTSe) and the sulphur selenium alloy CZTSSe. The CZTS provides favourable optical and electronic properties related to copper indium gallium selenide (CIGS) and make it well suited for use as a thin-film solar cell absorber layer. The CZTS is made up of only abundant and non-toxic elements. Concerns with the price and accessibility of indium in CIGS and tellurium in CdTe and toxicity of cadmium have been a big motivator to explore for option thin film solar cell materials. Latest materials development for CZTS has improved efficiency to 12.6% in laboratory cells, however more work is desired for their commercialization [40]. The typical CZTS solar cell structure is indicated in Figure 10. It consist of a molybdenum (Mo) covered soda lime glass (SLG) as the electrical contact, a thin CZTS light absorbent layer which is in contact with an n-type CdS layer to produce a p-n junction and a thin i-ZnO/Al: ZnO layer on top of the CdS layer playing the role of a window layer with an electrical contact. The CZTS was first created in 1967 [41] and was later shown to exhibit the photovoltaic effect in 1988 [42]. The CZTS solar cells with efficiency up to 2.3% were first reported in 1997 [43]. The solar cell efficiency in CZTS was improved to 5.7% in 2005 by optimizing the deposition process [44]. A bifacial photo electrochemical device, utilizing CZTS absorber material and transparent conducting back contact was reported in 2011 [45] which can create photocurrent on either side of illumination. Lately, it has been demonstrated that sodium has an raising effect on the structural and electrical properties of CZTS absorber layers [46]. These advances, on side the beginning of CIGS construction on commercial scale in the mid-2000s catalysed research interest in CZTS and connected compounds. Since 1988 CZTS was considered as an option to CIGS for commercial solar cell systems. The advantage of CZTS is the deficiency of the quite rare and expensive element indium. In 2010, a solar energy conversion efficiency of about 10% was achieved in a CZTS device [47]. The CZTS technology is now being produced by various private companies. In August 2012, IBM announced that they had produced CZTS solar cells able to convert 11.1% of solar energy to electricity [48]. In November 2013, the Japanese thin-film solar company Solar Frontier announced that in joint research with IBM and Tokyo Ohka Kogyo (TOK), they have developed a world-record setting CZTS solar cell with 12.6% energy conversion efficiency [49].

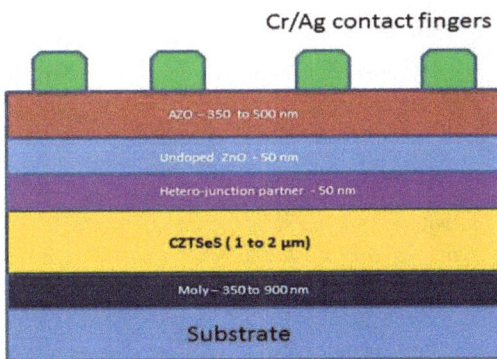

Figure 10: *Schematic Diagram of a Cu$_2$ZnSnS$_4$ Solar Cell.*

3.4   Quantum dots solar cells

Scientists at the University of Minnesota have planned photovoltaic solar cells by nanowires, nanocrystals, and quantum dots. Semiconductor nanocrystals are also known as quantum dots. A quantum dot solar cell is a device that uses quantum dots as the absorbing photovoltaic material. It is important to point out that quantum dots have advantages of improvement by use of photosensitive dyes in the manufacturing of solar cells. Moreover, quantum dots have a superior ability to match the solar spectrum, because their absorption spectrum can be tuned with the particle or dot size. Quantum dots have demonstrated a capability to generate multiple electron-hole pairs per individual photon, which could result in enhanced conversion efficiency for a solar cell. Research scientists consider that preparation of quantum dot technology in fabrication of solar cells will meet both critical requirements, namely, low fabrication cost and higher conversion efficiency. Quantum dots have band gaps that are tunable over a broad range of energy levels by changing the dots' size. In bulk materials the band gap is fixed via the choice of material(s). Such property makes quantum dots attractive for multi-junction solar cells, where a variety of materials are used to progress efficiency by harvesting multiple portions of the solar spectrum. Quantum dots are semiconducting particles that have been reduced to the size of the Exciton Bohr radius and due to quantum mechanics considerations, the electron energies that can exist within them become finite, much alike energies in an atom. Quantum dots have been termed as "artificial atoms" in which the energy levels are tuneable by varying their size, which in turn defines the band gap. The dots can be acquired or grown over a range of sizes, allowing them to express a variety of band gap without changing the underlying material or construction techniques [50]. In characteristic wet chemistry preparations, the tuning is achieved by changing the synthesis duration or temperature. The ability to tune the band gap makes quantum dots suitable for solar cells. Single junction implementations applying lead sulphide (PbS) CQDs have band gaps that can be tuned into the far infrared frequencies that are generally difficult to reach with traditional methods. Half of the solar energy contacting the Earth is in the infrared region. A quantum dot solar cell makes infrared energy as accessible as any other [51]. Furthermore, CQDs present easy synthesis and preparation method. Although suspended in a colloidal liquid form they can be simply handled throughout production, with a fume hood as the majority complex equipment required. CQDs are generally synthesized in small batches, however can be mass-produced. The dots can be spread on a substrate through spin coating, either as a result of hand or in an automated process. Large-scale production could apply spray-on or roll-printing systems, dramatically falling module structure costs. Early examples employed costly molecular beam epitaxy processes, simple less expensive manufacturing methods were developed later. These employ subsequent solution processing and wet chemistry (colloidal quantum dots - CQDs). Concentrated nanoparticle solutions are stabilized via long hydrocarbon legends that remain the nanocrystals suspended in solution. To produce a solid, the solutions are cast down and the long stabilizing legends are exchanged with short-chain cross-linkers. Chemically engineering the nanocrystal surface can better passivate the nanocrystals and reduce detrimental trap states that would curtail device performance by means of carrier recombination. This approach produces an efficiency of 7.0% [52]. In 2014 the use of iodide as a legend that does not bond to oxygen was introduced. This maintains stable n- and p-type layers, boosting the absorption efficiency, which produced power conversion efficiency up to 8% [53].

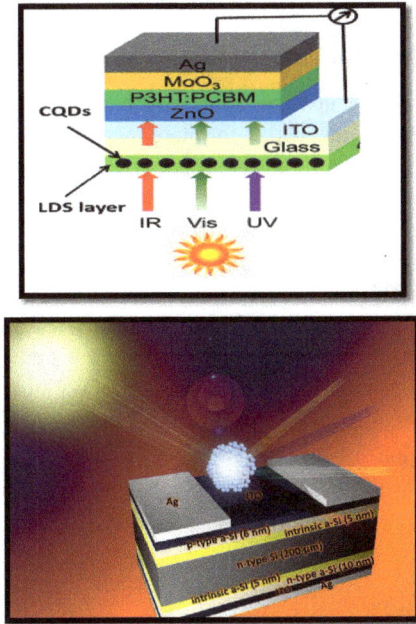

Figure 11: (a) Schematic architecture of the BHJ solar cell with LDS layer [54]. (b) Graphene based quantum dots solar cell [55].

21

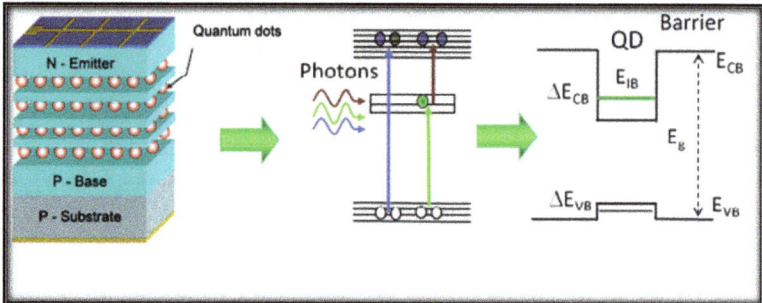

Figure 11: *(c) Schematic of operating principle and band energy diagram III(As-SB) of Quantum Dots.*

3.5   Multijunction amorphous nanotechnology-based solar cells

Amorphous silicon (a-Si) is the non-crystalline form of silicon. It is prepared by thin film technology and have been on the market for more than 15 years. It is generally used in pocket calculators; however it also powers some private homes, buildings, and remote area facilities. The latest research studies show that a better triple-junction amorphous solar cell consisting three different semiconductor layers with optimum thicknesses be able to achieve theoretical conversion efficiencies more than 35 percent by deploying solar concentrators [56]. This triple-junction amorphous nanotechnology-based solar cell, as shown in Figure 12, consists of three layers of semiconductor material, namely the gallium indium phosphide (GaInP) layer, best suitable for the short-wavelength region of the solar spectrum; the gallium indium arsenide (GaInAs) layer, best suitable for the middle region of the spectrum; and the germanium (Ge) layer, which gets most energy from the infrared region of the spectrum. The sub-cells, each defined by its semiconductor material layer are stacked one over another as shown in Figure12.

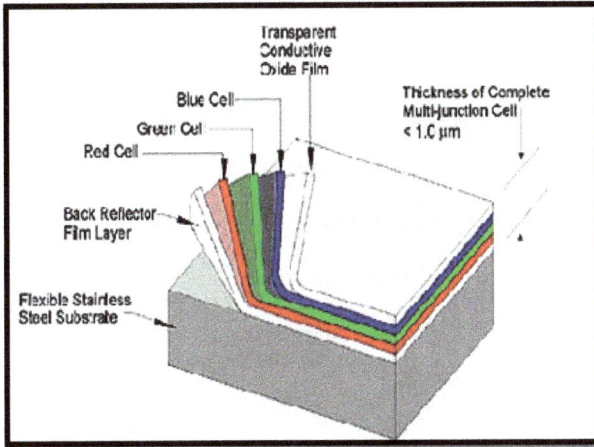

Figure 12: *Amorphous Silicon is Uni-Solar. They use a triple layer system [57].*

Typical characteristics of photon-detector materials are summarized in Table 1. It is significant to mention that major improvement in conversion efficiency of this triple-junction solar cell consisting three various semiconductor layers is a result of optimum absorption or extraction of the incident solar energy above the wide spectral region ranging from 400 nm to 1800 nm.

Table 1: *Typical Characteristics of Photon-Detector Materials*

Detector Material	Spectral Range (nm)	Responsivity (A/W)	Current Gain Ratio
Germanium (Ge)	40–1800	0.2 @ 400/0.8 @ 800 nm	200:1
		0.8 @ 1700/0.9 @ 1800 nm	
InGaAs	1000–1700	0.6 @ 1000/ 0.8 @ 1700 nm	10:1
GaAs	800–900	0.8 @ 850 nm	20:1

The structure of this particular solar cell is based on a metamorphic principle well-known as lattice mismatch, which allows for energy band gaps to be controlled so that each layer or sub-cell is further well matched to a specific spectral range most important to higher absorption rate. These multijunction amorphous solar cells were produced specially to provide electrical power for spacecraft or satellite applications. Research scientists consider that integration of solar concentrators will construct these devices more affordable and will be best suited mostly for terrestrial applications. The scientists further believe that a solar power system using such cells can deliver electrical power in excess of 35 kW, almost 50 % more than that which could be achieved with silicon-based concentrated solar cells. Research scientists at Spectrolab Inc. in Sylmar, California, a subsidiary of Boeing Co., are experimenting with four-junction [58] amorphous solar cells by different metamorphic materials, further cell architectures and exotic design approaches, which may well achieve conversion efficiencies close to 45 percent. At present, four junction cells with solar concentrators are viewing conversion efficiencies more than 36 percent and the company claims that such solar cells with efficiencies close to 42.1 percent will be available in near future. Amorphous silicon panels are produced by vapor-depositing a thin layer of silicon material - about 1 micrometer thick - on a substrate material such as metal or glass. Amorphous silicon can also be deposited at very low temperatures, as low as 75 degrees Celsius, which allows for deposition on plastic as well. In its simplest form, the cell design has a single sequence of p-i-n layers. But, single layer cells suffer from important degradation in their power output (in the range 15-35%) when showing to the sun. The mechanism of degradation is called the Staebler-Wronski Effect, following its discoverers. Superior stability requires the use of thinner layers in order to increase the electric field strength across the material. However, this reduces light absorption, hence cell efficiency. This has led the industry to develop tandem and even triple layer devices that contain p-i-n cells stacked one on top of the other. One of the pioneers of developing solar cells using amorphous silicon is Uni-Solar. They use a triple layer system (see illustration below) that is optimized to capture light from the full solar spectrum). As you can see from the illustration, the thickness of the solar cell is just 1 micron, or about 1/300th the size of mono-crystalline silicon solar cell. While crystalline silicon achieves a yield of about 18 percent, amorphous solar cells' yield remains at around 7 percent. The low efficiency rate is partly due to the Staebler-Wronski effect, which manifests itself in the first hours when the panels are exposed to sunlight, and results in a decrease in the energy yield of an amorphous silicon panel from 10 percent to around 7 percent. The principal advantage of amorphous silicon solar cells is their lower manufacturing costs, which makes these cells very cost competitive.

3.6   Dye sensitized solar cells

Dye-sensitized solar cells (DSSCs) are an inexpensive option to the universal silicon (Si) solar cell technology. The most predominant technology developed recently is DSSCs. Dye Sensitized solar cells (DSSC), also sometimes called as dye sensitized cells (DSC), are a third generation of photovoltaic (solar) cell that converts visible light into electrical energy. Dye Sensitized solar cells (DSSC) were first invented in 1991 by Professor Michael Graetzel and Dr Brian O'Regan at ÉcolePolytechniqueFédérale de Lausanne (EPFL), Switzerland and is often called as to as the Graetzel cell, then we call it G Cell [59]. The modern version of a dye solar cell is also known as Grätzel cell. A dye-sensitized solar cell (DSSC, DSC or DYSC) belongs to the group of thin film solar cells, and it is a low-cost solar cell [60]. Dye Sensitized solar cells are based on a semiconductor placed between a photo-sensitized anode and an electrolyte: a photoelectrochemical system (Figure 13). The DSSC has a number of attractive features. It is simple to construct by using conventional roll-printing techniques, is semi-flexible and semi-transparent which offers several uses which are not possible in glass-based systems, and the majority of the materials used are low-cost. In practice it has demonstrated that it is very hard to reduce a number of expensive materials, particularly platinum and ruthenium, and the liquid electrolyte exhibits a serious challenge for making a cell suitable as use in all weather. While its conversion efficiency is lower than the best thin film cells, in theory its price or performance ratio should be sufficient to allow them to compete with electrical generation from fossil fuel by achieving grid parity. Commercial applications, which were held up due to chemical stability problems [61], would be overcome and it is forecast in the European Union Photovoltaic Roadmap that these cells would significantly contribute to renewable electricity generation by 2020.

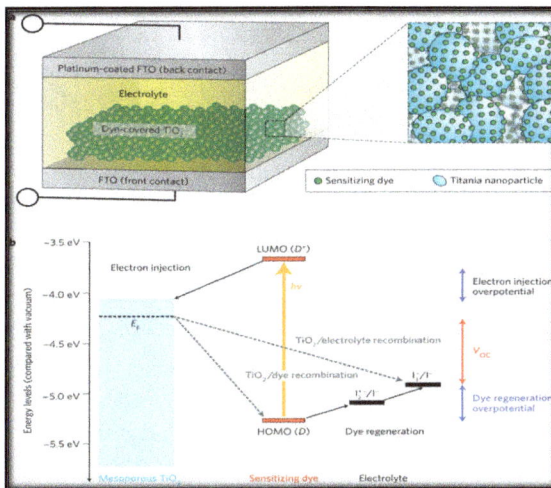

Figure 13: *Schematic Diagram of Dye Sensitized Solar Cell [63].*

The key difference of DSSCs from the silicon junction devices is that the contact phase with the semiconductor is exchanged by an electrolyte (liquid, gel and quasi-solid) in DSSC, which makes it a photochemical cell. Advancement in nanocrystalline materials manufacturing and characterization made it likely to attain systems with interpenetrating networks of mesoscopic semiconductors exhibiting higher conversion efficiencies which is one of the critical features for growing of the DSSC technology. Also second times, achieving optical absorption and charge separation processes using a combination of a sensitizer as a light-absorbing material with a nanocrystalline wide band gap semiconducting material is directed to DSSC fabrication. Multi-junction concentration solar cell has achieved a record high efficiency of 40.7% in the year 2010, as compared to 20-27% efficiency in crystalline solar cells. DSSCs, which have a simple fabrication process and shows potential applications on flexible substrates, are considered as third generation solar cells and their cost is predicted to be five times less than that of the second generation solar cells (silicon solar cells) as shown in Figure 14. The recorded efficiency of 11.1% for the DSSCs has been reported by Y. Chiba et al [62].

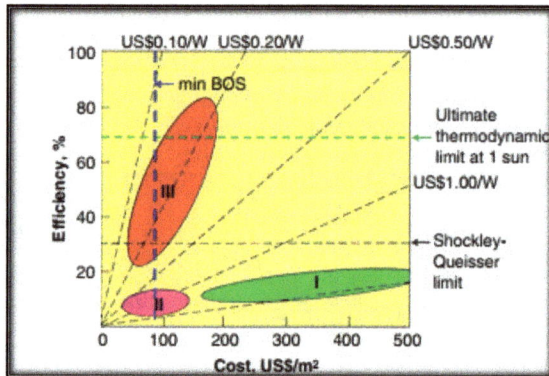

Figure 14: *The cost comparison of different solar cells.*

The DSSC technology is considered as a third generation photovoltaic technology, which is exhibiting great potential to become a innovative technology that has several indoor applications (window panels and doors) in low light environment. Additionally, the DSSC fabrication process is simple and cost effective, and does not need clean-room facility and highly pure materials. The efficiency of the DSSCs is around 12% based on today's technology, which is higher than the organic solar cells and lower than the traditional silicon-based solar cells. There are several challenges still to be solved related with DSSCs for commercialization, and one of those challenges is related to its longevity and stability. Titanium dioxide (TiO_2) suited the semiconductor of choice due to its several advantages for sensitized photochemistry and photoelectrochemistry. TiO_2 is a low cost, non-toxic, broadly available and biocompatible material.

3.7   Gallium arsenide solar cells

Gallium arsenide (GaAs) is made up of two base elements i.e. gallium and arsenic. It is a III-V direct band gap semiconductor. It has zinc blende crystal structure. GaAs is a very commonly used compound semiconductor in multijunction solar cells. Gallium arsenide is applicable to the production of various types of devices such as microwave frequency integrated circuits, monolithic microwave integrated circuits; infrared light emitting diodes, laser diodes, solar cells and optical windows [64]. GaAs based solar cells are also very significant for solar space applications. Besides the possibilities of constructing multijunction cells with GaAs and its derivative compounds, very high efficiency cell, over 40% [65] has been constructed using this material. There are number of other possible compounds such as InP, AlP, AlAs and GaP that are part of the family of III-V compounds. A few electronic properties of gallium arsenide are superior to silicon. Gallium arsenide has a higher saturated electron velocity and higher electron mobility, according transistors to function at frequencies in excess of 250 GHz. GaAs devices are comparatively insensitive to overheating due to their wider energy band gap and they also tend to create less noise in electronic circuits than silicon devices, especially at high frequencies. Gallium arsenide also has a higher breakdown voltage. Breakdown voltage is the minimum (reverse) voltage required to make a part of the component electrically conductive. Typical p/n GaAs solar cell is shown in Figure 15. Generally the junction is formed 0.5µm from the front window layer and a back surface field (BSF) is formed by the AlGaAs layer to make certain the diffusion gradient for minority holes is directed towards the junction [66]. This structure leads to almost ideal carrier collection in the base and emitter regions.

The solar cell is made up of three layers such as Indium Gallium Phosphate (InGaP), Gallium Arsenide (GaAs) and Germanium (Ge) as shown in Figure 15. Radiation degradation behaviour of this solar cell is more complex than that of conventional one, since it is multilayer structure. Solar cell is formed to do one of the predominate task. That is the creating of electricity by the absorption of photons. When light, i.e. the bundles of radiant energy from the sun, strikes the cell, a definite portion of it is absorbed inside the semiconductor material. In this process gallium arsenide is used as a semiconducting material, such that the energy of the absorbed light is transmitted to the semiconductor, in our case the gallium arsenide. The energy excites electrons, knocking them loose from their earlier bound state. This allows them to flow freely. Solar cell as well as photovoltaic cell also has one or more electric fields that act as an intermediate. This field forces the electrons liberated by light absorption, to flow in a definite direction. This flow of electrons, like many others, is a current. This current can be harnessed via placing metal contacts on the top and bottom of the cell. Among these newly placed contacts, the current can be drawn off to power just about any external application. There are many ways to create a GaAs solar or photovoltaic cell. Firstly the GaAs crystal must be produced. A number of methods to obtain GaAs crystals include Molecular Beam Epitaxy, Metalorganic Vapour Phase Epitaxy and Electrochemical Deposition (or Electroplating).

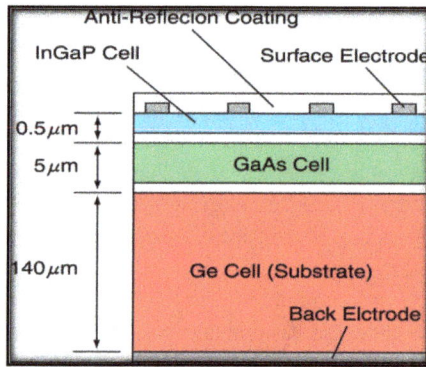

Figure 15: *A crossectional diagram of InGaP/GaAs/Ge triple junction Solar Cell.*

3.8 Perovskite solar cells

A perovskite solar cell includes a perovskite structured compound, almost commonly a hybrid organic-inorganic lead or tin halide - based material, as the light - harvesting active layer [67]. These perovskite materials have been familiar for many years, but the first incorporation into a solar cell was reported by Miyasakaet al. in 2009 [68].

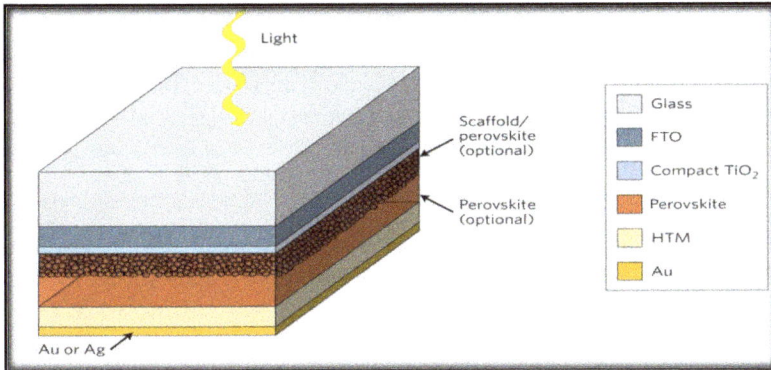

Figure 16: *Schematic Diagram of Perovskite Solar Cell.*

This was based on a dye-sensitized solar cell architecture, and generated only 3.8% power conversion efficiency (PCE) with a thin layer of perovskite on mesoporous TiO_2 a selectron-collector. The name 'perovskite solar cell' is derived from the ABX_3 crystal structure of the absorber materials, which is referred to as perovskite structure. Perovskite materials such as methylammonium lead halides ($CH_3NH_3PbX_3$, where X is a halogen ion such as Iodine, Bromine, Chlorine) has an optical bandgap between 2.3 eV and 1.6 eV depending on halide content. Formamidinum lead trihalide ($H_2NCHNH_2PbX_3$) has also shown a promising material, with bandgaps between 1.5 and 2.2 eV. The minimum band gap is nearer to the optimal for a single-junction cell than methylammonium lead trihalide, therefore it should be able to operate at of higher efficiencies [69]. A common concern is the inclusion of lead as constituent of the perovskite materials; solar cells based on tin-based perovskite absorbers such as $CH_3NH_3SnI_3$ have also been reported with lower power-conversion efficiencies [70-72]. Perovskite materials are cheap to construct simple manufacturing of solar cells. Solar cell efficiencies of devices applying these materials have increased from 3.8% in 2009 [68] to 21.0% in 2015 [73], this is the fastest-advancing solar technology to date [67]. With the potential of achieving even higher efficiencies and the very low production costs, perovskite solar cells have become commercially attractive, with start-up companies already promising modules on the market by 2017 [74-75]. Perovskite solar cells hold an improvement over traditional silicon solar cells in the simplicity of their manufacturing process. Traditional silicon cells require expensive, multistep processes, conducted at high temperatures (>1000 °C) in a high vacuum in special clean room facilities [76]. Meantime the organic-inorganic perovskite material can be fabricated with simpler wet chemistry techniques in a traditional lab environment. Almost particularly, methylammonium and formamidinium lead trihalides have been created using a variety of solvent techniques and vapor deposition techniques, both of which have the potential to be scaled up with relative feasibility [77]. In December 2015, a new record efficiency of 21.0% was achieved by researchers at EPFL [73].

Conclusion

Thus solar cell technology is a clean and green technology and provides a means of obtaining useful energy from solar radiation which is an inexhaustible source. There has been rapid progress in developing new materials and new device structure for harvesting solar energy in the recent past. It is expected that it will be possible to produce sufficient electricity from solar radiations to supply to the main grid and overcome energy crisis without further polluting our environment.

References

[1] Ciamician G (1912) Photochemistry of the Future, Science 36: 385-394. https://doi.org/10.1126/science.36.926.385

[2] W. P. Hirshman, "Surprise, surprise (cell production 2009: survey)," Photon International, pp. 176-199, 2010.

[3] A. Sharma, "PV demand database-quarterly," IMS Research, 2011.

[4] Dave Clark, R. Patel, et al. "Laser may bring the solar cell market," Photonic Spectra, June 2007, 54-58.

[5] J. J. Berger: The Business of Renewable energy and what it means of America (1998).

[6] C. C. Wang and K. J. Lin: Applied Mechanics and materials 130 (1989) 1286.

[7] A. Einstein: An Phys.17 (1905) 132 and Am. J. Phy. 33 (1965) 367 (in English).

[8] D. M. Chapin, C. S. Fuller and G.L. Pearson: Journal of Applied physics 25 (1954) 676. https://doi.org/10.1063/1.1721711

[9] A. W. Bett, S. P. Philipps , S. ESSig, S. Heckelmann, R. Kellenbenz, V. Klinger, M. Niemeyer, D. Lackner , F.Dimorth: 28 th European photovoltaic solar energy conference and exhibition, Paris, France pp. 1-6 (2013).

[10] Handbook of Photovoltaic Science and Engineering, Antonio Luque, Steven Hegedus, 2003 ISBN 0-471-49196-9).

[11] S. R. Wenham and M.A. Green: Prog. Photovolt: Res. Appl. 43 (1996).

[12] A. V. Shah, H. Schade, M. Vanecek, J. Meier, E. Vallat-Sauvain, N. Wyrsch, U. Koll, C. Droz and J. Bailat: Prog. Photovolt: Res. Appl. 12 113 (2004). https://doi.org/10.1002/pip.533

[13] K. L. Chopra, P. D. Paulson and V. Dutta: Prog. Photovolt: Res. Appl 12 69 (2004). https://doi.org/10.1002/pip.541

[14] C. J. Brabec, S. Gowrisanker, J. J. M. Halls, D. Laired, S. Jia and S. P. Williams: Adv. mater. 22 3839 (2010). https://doi.org/10.1002/adma.200903697

[15] Website: http://www.nrel.gov

[16] M. A. Green, K. Emery, Y. Hishikawa, W. Warta and E. D. Dunlop: Prog. photovolt: Res. Appl. 21 (2013) 1. https://doi.org/10.1002/pip.2352

[17] Solar Cell Materials Developing Technologies ISBN: 9780470065518 Gavin Conibeer, Arthur Willoughby.

[18] Website: http://org.ntnu.no/solarcells/pages/generatins.php.

[19] K. D. G. ImalkaJayawardena, Lynn J. Rozanski, Chris A. Mills, Michail J. Beliatis, N. AaminaNismy and S. Ravi P. Silva, Nanoscale, 2013, 5, 8411.

[20] Photovoltaics Report, Fraunhofer ISE, July 28, 2014 (http://www.ise.fraunhofer. de/en/downloads-englisch/pdf-files-englisch/photovoltaics-report-slides.pdf), pages 24, 25.

[21] W. Heywang, K. H. Zaininger, Silicon: the semiconductor material (http://books.google.com/books?id=ATFo8Pr67uIC&pg=PA25&dq=%22silicon+era%2& hl=en&ei=tGdFTMrmK8Oblgfi763tAw&sa=X&oi), in Silicon: evolution and future of a technology, P. Siffert, E. F. Krimmel eds., Springer Verlag, 2004.

[22] "Solar Insight, Research note - PV production 2013: an allAsianaffair" (PDF). Bloomber New Energy Finance. 16 April 2014. pp. 2-3. Archived from the original on 30 April 2015.

[23] "China: The new silicon valley - Polysilicon". 2 February 2015. Archived from the original on 30 April 2015.

[24] Photovoltaics: Getting Cheaper (http://www.nyecospaces. com/2007/09/photovoltaics-getting-cheaper. html).

[25] The Wall Street Journal, A Shortage Hits Solar Power. April 29, 2006.(http://online. wsj. com/article/SB114624912379938991. html).

[26] http://www.enfsolar.com/directory/material/polysilicon.

[27] Basore, P. A. (2006), "CSG-2: Expanding the production of a new polycrystalline silicon PV technology" (PDF), Proceedings of the 21st European Photovoltaic Solar Energy Conference.

[28] V. Avrutin, N.Izyumskaya, and H. Morko, "Semiconductor solar cells: recent progress in terrestrial applications," Superlattices and Microstructures, vol. 49, no. 4, pp. 337-364, 2011. https://doi.org/10.1016/j.spmi.2010.12.011

[29] S. M. Sze, "Physics of semiconductor devices (2nd edition), Wiley, Amsterdam (1981)," Microelectronics Journal, vol. 13, no. 4, p. 44, 1982.

[30] A. Bosio, N. Romeo, S. Mazzamuto, and V. Canevari, "Polycrystalline CdTe thin films for photovoltaic applications," Progress in Crystal Growth and Characterization of Materials, vol. 52, no. 4, pp. 247-279, 2006. https://doi.org/10.1016/j.pcrysgrow.2006.09.001

[31] Mohammad Bagher, Mirzaei Mahmoud AbadiVahid, Mirhabibi Mohsen, Types of Solar Cells and Application, American Journal of Optics and Photonics. Vol. 3, No. 5, 2015, pp. 94-113. doi: 10.11648/j.ajop.20150305.17. https://doi.org/10.11648/j.ajop.20150305.17

[32] Fraunhofer ISE Photovoltaic Report (http://www. ise. fraunhofer. de/en/downloads-englisch/pdf-files- englisch/photovoltaics-report-slides. pdf), July 28, 2014, pages 18, 19.

[33] Achievements and Challenges of CdS/CdTe Solar Cells Zhou Fang, Xiao Chen Wang, Hong Cai Wu, and Ce Zhou Zhao, Hindawi Publishing Corporation International Journal of Photoenergy, Volume 2011, Article ID 297350, 8pages. https://doi.org/10.1155/2011/297350

[34] Solar cell technology and their Applications. 2010 ISBN 978-1-4200-8177-0.

[35] Tinoco, T.; Rincón, C.; Quintero, M.; Pérez, G. Sánchez (1991). "Phase Diagram and Optical Energy Gaps for CuInyGa1−ySe2 Alloys" Physica Status Solidi (a) 124 (2): 427. Bibcode: 1991PSSAR. 124. 427T. doi: 10. 1002/pssa. 2211240206.

[36] Stanbery, B. J. (2002). "Copper Indium Selenides and RelatedMaterials for Photovoltaic Devices". Critical Reviews in Solid State and Materials Science doi:10. 1080/20014091104215.

[37] Andorka, Frank (2014-01-08). "CIGS Solar Cells, Simplified". http://www.solarpowerworldonline.Com. Solar Power World. Archived from the original on 16 August 2014. Retrieved 16 August 2014. External link in |website= (help).

[38] "CIS - Ecology". Solar Frontier. Retrieved July 2015.

[39] Jasenek et al., 2001; La Roche et al., 2000.

[40] M.T.Winkler, W.Wang.O.Gunawan, H.J.Hovel, T.K.Todorv and D.B.Mitzi: Energy Environ. Sci. 7 (2014) 1029. https://doi.org/10.1039/C3EE42541J

[41] R. Nitsche, D. F. Sargent, and P. Wild: Journal of Crystal Growth 1/1 (1967) 52. https://doi.org/10.1016/0022-0248(67)90009-7

[42] K. Ito and T. Nakazawa: Japanese Journal of Applied Physics 27 (1988) 2094. https://doi.org/10.1143/JJAP.27.2094

[43] T. M. Friendlmeier, N. Wieser, T. Walter , H. Dittrich, and H. W. Schock: Proceeding of the 14th European photovoltaic solar energy coneference (1997).

[44] H. Katagiri, K. Jimbo, W. S. Maw, K. Oishi, M. Yamazaki, H. Araki and A. Takeuchi: Thin Solid Films 517/7 (2009) 2455. https://doi.org/10.1016/j.tsf.2008.11.002

[45] P. K. Sarswat and M. L. Free: Physica Status Solidi A 208/12 (2011) 2861. https://doi.org/10.1002/pssa.201127216

[46] Tejas Prabhakar and J. Nagaraju: Solar Energy Materials and Solar cells 95/3 (2011) 1001. https://doi.org/10.1016/j.solmat.2010.12.012

[47] T. K. Todorov, K. B. Reuter and D. B. Mitzi: Advanced Materials 22/20 (2010) E 156.

[48] Website: http://ibmresearchnews.blogspot.in

[49] W. Wang, M. T. Winkler, O. Gunawan, T. Gokmen, T. K. Todorov, Y. Zhu and D. B. Mitzi: Advanced materials (2013) doi.10.1002/aenm.201301465. https://doi.org/10.1002/aenm.201301465

[50] Baskoutas, Sotirios; Terzis, Andreas F. (2006). "Sizedependent band gap of colloidal quantum dots". Journal ofApplied Physics 99: 013708. Bibcode:2006JAP....99a3708B. doi:10.1063/1.2158502. https://doi.org/10.1063/1.2158502

[51] H. Sargent, E. (2005). "Infrared Quantum Dots" (PDF). Advanced Materials 17 (5): 515-522. doi:10.1002/adma.200401552. https://doi.org/10.1002/adma.200401552

[52] Ip, Alexander H.; Thon, Susanna M.; Hoogland, Sjoerd; Voznyy, Oleksandr; Zhitomirsky, David; Debnath, Ratan; Levina, Larissa; Rollny, Lisa R.; Carey, Graham H.; Fischer, Armin; Kemp, Kyle W.; Kramer, Illan J.; Ning, Zhijun; Labelle, André J.; Chou, Kang Wei; Amassian, Aram; Sargent, Edward H. (2012). "Hybrid passivated colloidal quantum dot solids". Nature Nanotechnology 7 (9): 577–582. 2012. https://doi.org/10.1038/nnano.2012.127

[53] Mitchell, Marit (2014-06-09). "New nanoparticles bring cheaper, lighter solar cells outdoors". Rdmag.com. Retrieved 2014-08-24. https://doi.org/10.1039/C4TC00988F

[54] Carbon quantum dots: synthesis, properties and applications.Youfu Wang and Aiguo HuJ. Mater. Chem. C, 2014, 2, 6921.

[55] Meng-Lin Tsai,Wei-Chen Tu, Libin Tang, Tzu-Chiao Wei, Wan-Rou Wei, Shu Ping Lau,Lih-Juann Chen, and Jr-Hau He pubs.acs.org/NanoLett,DOI: 10.1021/acs.nanolett.5b03814. https://doi.org/10.1021/acs.nanolett.5b03814

[56] M. A. Greenwood, News Editor. "Solar Technology: Seeking its day in the sun," Photonics Spectra, July 2007, 42-50.

[57] Askari Mohammad Bagher, Mirzaei Mahmoud AbadiVahid, Mirhabibi Mohsen. Types of Solar Cells and Application. American Journal of Optics and Photonics. Vol. 3, No. 5, 2015, pp. 94-113. https://doi.org/10.11648/j.ajop.20150305.17

[58] "Oerlikon Divests Its Solar Business and the Fate of Amorphous Silicon PV". Grrentech Media. March 2, 2012.

[59] Brian O' Regan, Michael Grätzel (24 October 1991). "A lowcost, high efficiency solar cell based on dye-sensitized colloidal TiO2 films". Nature, 353, (6346): 737-740. https://doi.org/10.1038/353737a0

[60] "Dye-Sensitized vs. Thin Film Solar Cells", European Institute for Energy Research, 30 June 2006.

[61] Tributsch, H (2004). "Dye sensitization solar cells: a critical assessment of the learning curve". Coordination Chemistry Reviews 248 (13–14): 1511-1530Y. Chiba, A. Islam, Y. Watanabe, R. Komiya, N. Koide and L. Han, Dye sensitized solar cells with conversion efficiency of 11.1%, Jpn. J. Appl. Phys. Vol. 45, 638–640, (2006).

[62] The renaissance of dye-sensitized solar cells, Brian E. Hardin, Henry J. Snaith& Michael D. McGehee. Nature Photonics 6, 162-169 (2012). https://doi.org/10.1038/nphoton.2012.22

[63] Moss, S. J. and Ledwith, A. (1987). The Chemistry of the Semiconductor Industry. Springer. ISBN 0-216-92005-1.

[64] Solar Photovoltaics Fundamentals, Technologies and Applications chetan Singh Solanki 2013ISBN-978-203-4386-3.

[65] S. P. Tobin, S. M. Vernon, C. Bajgar, L. M. Geoffroy, C. J. Keavney, M. M. Sanfacon, and V. E. Haven. Device processing and analysis of high-efficiency GaAs Solar Cells. Solar Cells, 24(1-2):103-115, 1988. https://doi.org/10.1016/0379-6787(88)90040-3

[66] Collavini, S. , Völker, S. F. and Delgado, J. L. (2015). "Understanding the Outstanding Power Conversion Efficiency of Perovskite-Based Solar Cells". Angewandte Chemie International Edition 54 (34): 9757-9759.doi:10. 1002/anie. 201505321.

[67] Kojima, Akihiro; Teshima, Kenjiro; Shirai, Yasuo; Miyasaka, Tsutomu (May 6, 2009). "Organometal Halide Perovskites as Visible-Light Sensitizers for Photovoltaic Cells". Journal of the American Chemical Society 131(17): 6050–6051. doi:10. 1021/ja809598r. PMID 19366264.

[68] Eperon, Giles E.; Stranks, Samuel D.; Menelaou, Christopher; Johnston, Michael B.; Herz, Laura M.; Snaith, Henry J. (2014). "Formamidinium lead trihalide: a broadly tunable perovskite for efficient planar heterojunction solar cells". Energy & Environmental Science 7 (3): 982. doi:10. 1039/C3EE43822H.

[69] Noel, Nakita K.; Stranks, Samuel D.; Abate, Antonio; Wehrenfennig, Christian; Guarnera, Simone; Haghighirad, Amir-Abbas; Sadhanala, Aditya; Eperon, Giles E.; Pathak, Sandeep K.; Johnston, Michael B.; Petrozza, Annamaria; Herz, Laura M.; Snaith, Henry J. (May 1, 2014). "Lead-free organic-inorganic tin halide perovskites for photovoltaic applications". Energy & Environmental Science 7 (9): 3061. doi:10. 1039/C4EE01076K.

[70] Wilcox, Kevin (May 13, 2014). "Solar Researchers Find Promise in Tin Perovskite Line". Civil Engineering. Archived from the original on October 6, 2014.

[71] Meehan, Chris (May 5, 2014). "Getting the lead out of Perovskite Solar Cells". Solar Reviews.

[72] "NREL efficiency chart".

[73] Oxford Photovoltaics (June 10, 2013) Oxford PV reveals breakthrough in efficiency of new class of solar cell (http://www.oxfordpv.com/oxford-pv-news/oxford-pv-reveals-breakthrough-in-efficiency-of-new-class-of-solar- cell/).

[74] Wang, Ucilia (September 28, 2014). "Perovskite Offers Shot at Cheaper Solar Energy". The Wall Street Journal. Retrieved May 7, 2015. (regi st rat i on requi red)

[75] Is Perovskite the Future of Solar Cells? (http://www.engineering. com/Blogs/tabid/3207/ArticleID/6773/Is-Perovskite-the-Future-of-Solar-Cells.aspx). engineering.com. December 6, 2013.

[76] Saidaminov, Makhsud I.; Abdelhady, Ahmed L.; Murali, Banavoth; Alarousu, Erkki; Burlakov, Victor M.; Peng, Wei; Dursun, Ibrahim; Wang, Lingfei; He, Yao; MacUlan, Giacomo; Goriely, Alain; Wu, Tom; Mohammed, Omar F.; Bakr, Osman

M. (2015). "High-quality bulk hybrid perovskite single crystals within minutes by inverse temperature crystallization". Nature Communications 6: 7586. doi:10. 1038/ncomms8586. PMC 4544059. PMID 26145157.

[77] Snaith, Henry J. (2013). "Perovskites: The Emergence of a New Era for Low-Cost, High-Efficiency Solar Cells". The Journal of Physical Chemistry Letters 4 (21): 3623-3630. doi:10. 1021/jz4020162.

Chapter 2

Basic principles and theory of the photovoltaic effect

M. Ramrakhiani*, Swati Dubey, Hemraj Waxar, Kamal Kumar Kushwaha and Pranav Singh

Department of Postgraduate Studies and Research in Physics and Electronics, Rani Durgawati Vishwavidyalaya, Jabalpur (MP), India

*mramrakhiani@hotmail.com

Abstract

Among new energy sources, solar energy is extremely promising since it is practically inexhaustible. Photovoltaic (PV) is a method of converting solar energy directly into electricity using semiconducting materials. The number of solar energy power projects using photovoltaic technology is gathering momentum due to deeper understanding of the technology and commercial and government policies. Of course improved affordability due to the falling cost is another major factor. In order to improve the efficiency and reduce the cost further, a clear understanding of the basic scientific principles of photovoltaic is necessary. The basic theory and principles of the photovoltaic effect are described here giving expressions for voltage and current. Various characteristics and parameters of solar cells are described and suitable materials for solar cells are discussed.

Keywords

Photovoltaic Effect, Solar Cell Characteristics, Solar Cell Parameters, Solar Cell Materials

Contents

1. Introduction..38

2. Solar Cell 39

3. Photovoltaic Effect..41

4. Theory of Photovoltaic Effect...44

5. Characteristics and Parameters of Solar Cells..........................48

5.1   Open Circuit Voltage (V_{oc})...**49**

5.2   Short Circuit Current (I_{sc} or J_{sc}) **50**

5.3   Maximum Power Output (P_{max}) **50**

5.4   Fill Factor (FF) or Curve Factor...**50**

5.5   Efficiency (η) **51**

5.6   Shunt Resistance (R_{sh})...**51**

5.7   Series Resistance (R_s)...**51**

5.8   Spectral Response (SR) **52**

6.   Materials for Solar Cells **53**

7.   Conclusion **56**

References...**56**

1.   Introduction

Industrial development and improvement in the way of life require an abundant supply of energy. The total energy consumed annually in a country is a measure of the level of national economy. It has become apparent that non-renewable energy sources are finite and are in danger of depletion to various degrees in future. The renewable sources of energy will have to play a major role in the worlds future energy supply.

Among new energy sources, solar energy is extremely promising since it is practically inexhaustible. The sun is the world biggest natural source of energy. The task of utilizing solar energy is becoming more and more urgent as a result of the gradual exhaustion of fossil fuel resources and increasing problem of environmental pollution. The technological waste from the existing energy installations and the thermal pollution produced by them present a threat to our ecology.

In contrast, the sun provides ecologically pure energy. The sun sends down 8000 times the amount of the world's present energy requirement. In several developing countries in general, the use of solar energy appears to be a reasonable proposition as an alternative source of energy. Though life exists on earth only because of solar energy, today it may be used for practical purposes in two different ways - as thermal energy and as electrical energy. In the past few decades a large number of efforts have been made to utilize solar energy for heating, air conditioning, production of electric power, etc.

Photovoltaic (PV) is a method of converting solar energy directly into electricity using semiconducting materials that exhibit the photovoltaic effect. A photovoltaic system employs solar panels composed of a number of solar cells, to supply usable solar power. Power generation from solar PV has long been seen as a clean sustainable energy technology, which draws upon the planet's most plentiful and widely distributed renewable energy source - the Sun. The direct conversion of sunlight to electricity occurs without any moving parts or environmental emissions during operation. The photovoltaic systems are being used for fifty years in specialized applications, and grid-connected PV systems have been in use for over twenty years. These were first mass produced in the year 2000, when German environmentalists including Euro solar succeeded in obtaining government support for a 100,000 roofs program.

Solar PV is now, after hydro and wind power, the third most important renewable energy source in terms of globally installed capacity. More than 100 countries use solar PV. Installations may be ground-mounted (and sometimes integrated with farming and grazing) or built into the roof or walls of a building (either building-integrated photovoltaic or simply roof top).

There is a huge potential for solar energy across the entire world. Solar energy in India has started to take-off. It is expected that due to the various advantages that PV Solar offers, it will far surpass wind energy in the country. With a boost by the Government in 2009-10 under Jawaharlal Nehru Solar Mission, Solar energy is all set to take the centre stage for a long time to come.

The PV is already quite competitive and in certain cases PV Solar projects have become financially viable even in absence of subsidies. The solar energy power projects using photovoltaic technology is gathering momentum due to deeper understanding of the technology and commercial and government policies. Of course improved affordability due to the fall of cost is another major factor. In order to improve the efficiency and reduce the cost further, a clear understanding of the basic scientific principles of photovoltaic is necessary. The basic theory and principles of photovoltaic effect are described here.

2.   Solar Cell

A solar cell (also called photovoltaic cell) is a solid state device that converts artificial or natural light directly into electricity via the photovoltaic effect. Assemblies of cells are used to make solar modules, also known as solar panels. The energy generated from these solar modules, referred to as solar power, is an example of solar energy. Photovoltaic (PV) comprises the technology to convert sunlight directly into electricity. The term "photo" means light and "voltaic" means electricity. A photovoltaic (PV) cell, also known as "solar cell," is a semiconductor device that generates electricity when light falls on it. The photovoltaic effect can continue to provide voltage and current as long as light is available. This voltage or current can be used as a source of power in an electrical circuit, as in a solar power system or to measure the brightness of the incident light.

The basic idea of a solar cell is to convert light energy into electrical energy. The energy of light is transmitted by photons, small packets or quanta of light. When photons are absorbed by matter in the solar cell, their energy excites electrons to higher energy states where the electrons can move more freely. In an ordinary material, if the electrons are not given enough energy to escape, they would soon relax back to their ground states. In a solar cell however, the way it is put together prevents this from happening. The electrons are instead forced to one side of the solar cell, where the build-up of negative charge makes a current flow through an external circuit. The current ends up at the other side (or terminal) of the solar cell, where the electrons once again enter the ground state, as they have lost energy in the transit through the device circuit. Consequently, the end result is a current of electrons, better known as electricity. An illustration is shown in Figure 1.

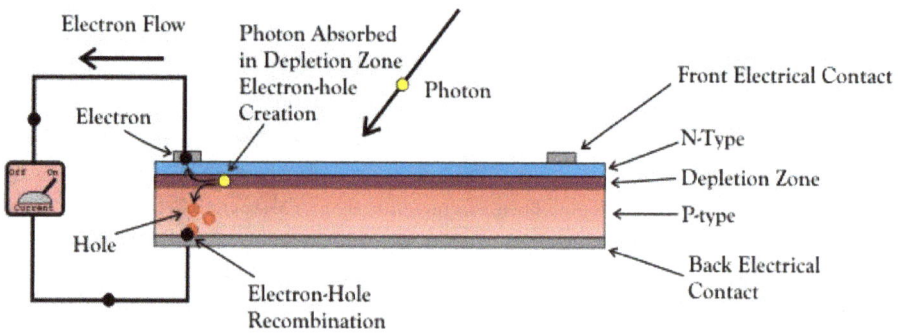

Figure 1: Structure of a Solar Cell.

3. Photovoltaic Effect

Photovoltaic is the field of technology and research related to the practical application of photovoltaic cells in producing electricity from light, though it is often used specifically to refer to the generation of electricity from sunlight. Cells are described as photovoltaic cells when the light source is not necessarily sunlight. These are used for detecting light or other electromagnetic radiation near the visible range, for example infrared detectors, or measurement of light intensity.

The photovoltaic effect was observed in 1839 by the French scientist Edmund Becquerel, while experimenting with an electrolytic cell made up of two metal electrodes. He discovered that when exposing

The French scientist Alexandre-Edmon Becquerel

certain materials to sunlight a weak electrical current is generated. He named this phenomenon the "photovoltaic effect". The diagram illustrates the operation of a basic photovoltaic cell, also called a solar cell. Solar cells are made of the some kinds of semiconductor materials, such as silicon, used in the microelectronics industry.

For solar cells, a thin semiconductor wafer forms an electric field, positive on one side and negative on the other due to different type of doping. When light energy strikes the solar cell, electrons are knocked loose from the atoms in the semiconductor material. If electrical conductors are attached to the positive and negative sides, forming an electrical circuit, the electrons can be captured in the form of an electric current that is, electricity. This electricity can then be used to power a load, such as a lamp or a device (Figure 2).

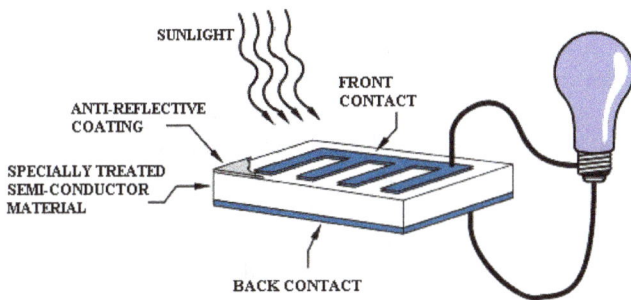

Figure 2: *Solar Cell Operation.*

When photons in the sun light collide with the silicon solar cell, one of three things can happen:

- The photon can be reflected at the surface of the silicon
- The photon can be absorbed by the silicon
- The photon can pass right through the silicon

The photovoltaic effect is because of absorption of photons, as the photons hit the atoms in the silicon, the energy is absorbed by the electrons and excited into a higher state of energy. When these free electrons flow through the material, electricity arises.

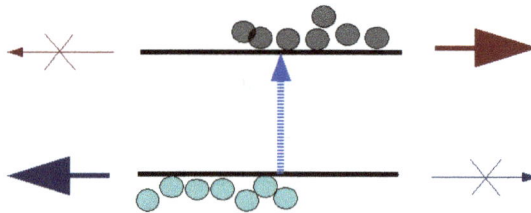

Figure 3: Schematic of charge separation in a PV device. The upper contact provides low resistance for electrons to flow but blocks holes, while the lower provides an easy path for holes but a barrier to electrons.

Photovoltaic energy conversion results from -

- Charge generation
- Charge separation
- Charge transport

Charge separation is caused by a driving force of charges which must be built into the device. There are many ways to provide a charge separation mechanism by variation in the electronic materials. Every material has its own band gap that describes how strong the electrons are bonded to the atoms. For semiconductors, such as silicon, the band gap refers to the energy difference between the valence band and the conduction band. When a negative electron is excited, it leaves behind a void which is called a positive hole. The presence of a missing covalent bond allows the bonded electrons of neighbouring atoms to jump into the hole, leaving another hole behind. Because of this, holes also move through the lattice. When photons are absorbed in the semiconductor, it can be said they create mobile electron-hole pairs. This is the case for semiconductors in general.

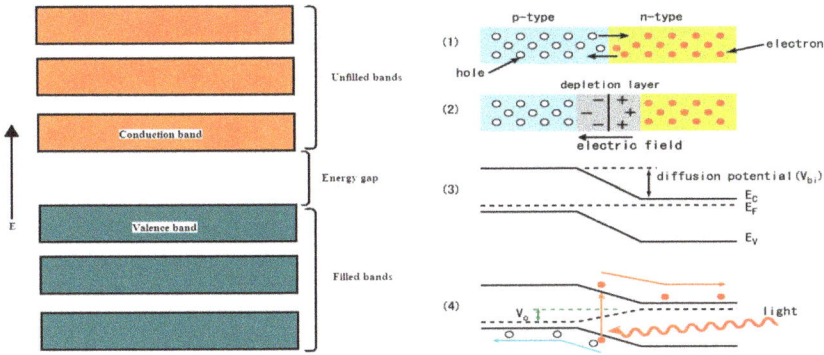

Figure 4: *Photovoltaic Process.*

When photons strike a PV cell, they may be reflected or absorbed, or they may pass right through. These photons contain different amounts of energy that correspond to the different wavelengths of the solar spectrum. The energy of a photon is transferred to an electron in an atom of the semiconductor device. With its newfound energy, the electron is able to escape from its normal position associated with a single atom in the semiconductor to become part of the current in an electrical circuit. A built-in electric field inside the PV cell provide the voltage needed to drive the current through an external load.

The photovoltaic effect is defined as creation of an electromotive force by the absorption of light in an inhomogeneous solid; most common is a p-n junction. The p-n junction is the most classical structure for solar cells. The semiconductor is doped in different regions with different elements, so that we have p- and n-type doping. An electric field is established because of the difference in a work function of n- and p-type material. When a p-n junction is formed between two semiconductor free carriers from one semiconductor diffuse to the other because of the concentration gradient. The diffusing holes and electrons are annihilated due to recombination such that n- and p- regions near the junction become depleted of their respective majority carriers, electrons and holes; this depletion layer opposes the flow of majority carriers. Hence a positive charge is built on the n-side of the junction and a negative charge is built on the p-side of the junction due to uncovered positive donor ions and covered negative acceptor ions. The P-side becomes negative and N-side becomes positive during diffusion of electrons towards the

P-side, and holes towards the n-side. Such type of barrier is called potential barrier or junction barrier. Only in-side the barrier, there is a positive charge on the n-side and negative charge on the p-side (Figure 5).

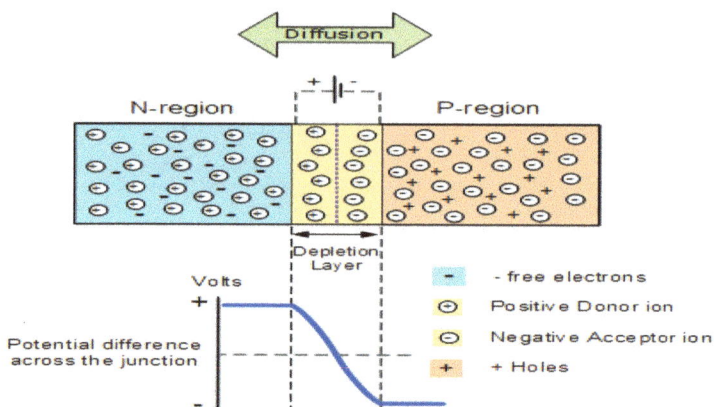

Figure 5: *Charge carriers & potential in p-n junction.*

When the unbiased p-n junction is exposed to light and photons possessing energy [$h\upsilon >$ E_g -band-gap energy] more than the required energy to break the bonds or excite electrons from the valence to the conduction band, electron-hole pairs are created. Photo-generation, or the photocurrent, is the most important generation in PV cells. It is the process where mobile electrons and holes are created due to absorption of electromagnetic radiation in the semiconductor. The photo-generation rate is determined by the number of incident photons.

4.   Theory of the Photovoltaic Effect

For photovoltaic effect, a mechanism is necessary to separate the electrons and hole pairs created by light, to ensure preferred direction of movement of charges, so that the current flows in an external circuit without the application of an external field. The potential barrier of p-n junction (or liquid semiconductor junction or certain discontinuity) provides this mechanism. A p-n junction has a depletion region with a potential barrier caused by immobile, ionized impurities. This potential barrier opposes the flow of majority carries and favours the flow of minority carriers across the junction. The minority carriers generated by the absorption of photons of energy hυ > Eg cross over the junction by diffusion in the negative direction of the concentration gradient.

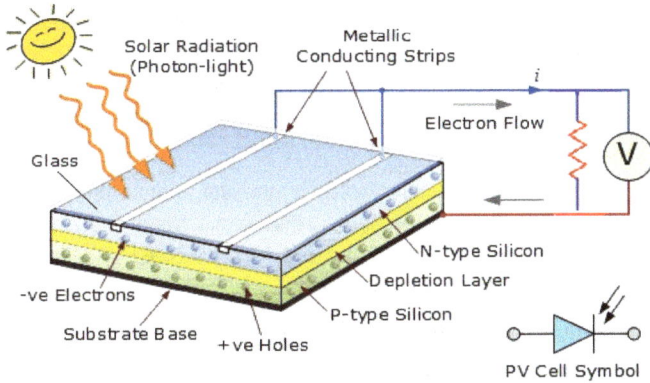

Figure 6a: *Schematic diagram of a p-n junction with radiation being absorbed and current flowing through external resistance.*

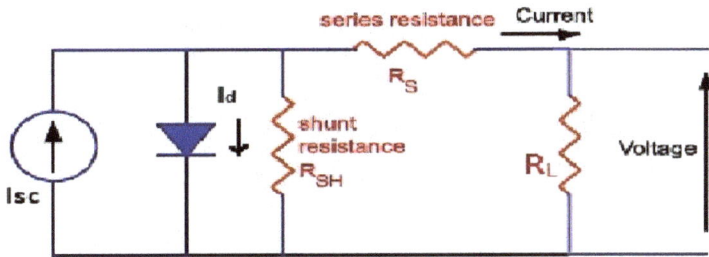

Figure 6b: *Equivalent circuit of p-n junction Solar Cell.*

The Electron-hole pairs are generated throughout the semiconductor but only those cross over the junction, which are within one diffusion length from the junction. The charge carriers which cross over the junction flow through external resistance R_L, which causes potential drop across the resistance and makes the junction forward biased. Flow of current is opposite in direction to the flow of radiation current. Figure 6a shows the schematic diagram of a p-n junction with radiation being absorbed and current flowing through external resistance and Figure 6b shows an equivalents circuit of the solar cell.

The energy band in n-region and p-region is shown in Figure 4. When a p-n junction is formed the conduction and valance bands of a depleted n-region bend upward while those in the depleted p-region bend downwards. The barrier height, gives the electrostatic field in the junction (Figure 5). Under illumination additional electron-hole pairs are generated in p-and n-regions. The minority carrier concentration drastically increases over its equilibrium value and flow across the junction.

The net photo-generated current results, which is the sum of hole flow from n- to p-region and electron flow from p- to n-region. This current has a direction opposite to forward bias current. The number (n) of charge carries produced by photons near the junction within the diffusion length is proportional to -

- the intensity (I) of light (or number of photons incident)

- absorption coefficient (α) (or number of photons absorbed)

- quantum yield (η) for generating electron-hole pair by a photon and

- diffusion length for electrons and holes (L_n & L_p)

$$n = \alpha\, I\, \eta\, (\, L_n + L_p) \tag{1}$$

Considering that all the minority carries within the diffusion length cross over the junction, the current density (J_r) due to radiations or light will be -

$$J_r = q\, \alpha\, I\, \eta\, (\, L_n + L_p) \tag{2}$$

where q is the electronic charge.

When no external contact is made between n and p parts, i.e. the cell is open circuited, a potential difference V_{oc} is obtained at the two ends lowering the built-in voltage by an amount V_{oc}. In case of an ideal diode without recombination, the V_{oc} would equal to the electrostatic potential barrier. In practice it may be lower.

When short circuited, the current flowing through the cell is -

$$J_{sc} = J_r \tag{3}$$

In order to use the generated power, a load is connected at the two ends of the cell. Under illumination, a part of photo-generated current flows through the external load from p-region to n-region. This results in lowering the potential difference between n and p regions. The net current flowing through the external load is photo-generated current from p to n region and diode current in opposite direction.

As a solar cell contains a p-n junction, it may be treated as a diode. For an ideal diode, the dark current density is given by -

$$J_{dark}(V) = J_0(e^{qV/k_B T} - 1) \tag{4}$$

Here J_0 is a reverse saturation current density, q is the electron charge, k_B is Boltzmann constant and V is the voltage between the terminals. The resulting current can be approximated as a superposition of the short circuit current and the dark current:

$$J = J_{sc} - J_0(e^{qV/k_B T} - 1) \tag{5}$$

Hence the net current density of an ideal Shockley p-n junction solar cell is given by

$$J_L = J_r - J_o [\, _{exp}\frac{(qV_L)}{KT} - 1] \tag{6}$$

where, J_L is load current density, J_r is photocurrent density, q is electronic charge, V_L is voltage across load resistance, K is Boltzmann constant, T is Absolute temperature, and J_O is reverse saturation current density.

Above equation is derived for ideal solar cell assuming photocurrent generation only in the top region. But photocurrent generation is not limited to the top region, it is also in the p and n type region and depletion region.

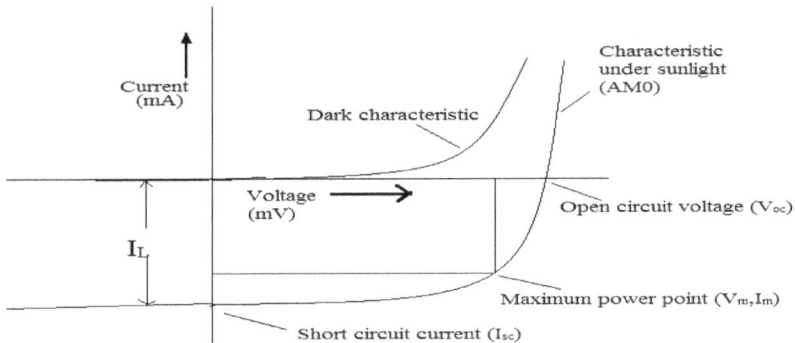

Figure 7: *I-V characteristics of p-n junction Solar Cell.*

Figure 7 shows the I-V characteristics of an illuminated p-n junction, in which voltage has positive value and current flows in a negative direction. The diode thus acts like a battery from which power can be obtained. This is the principle of photovoltaic solar cell.

It is seen that the I-V curve keeps the same shape but is offset along the negative current axis. As a result, an open circuit voltage appears on the positive voltage axis and short circuit current on the negative current axis. I-V characteristics show that at photovoltaic junction, the current flow in the reverse direction and voltage in the forward direction.

In a practical solar cell, when current flow through the external circuit, there is resistance due to semiconductor material as well as metallic contacts. Besides this leakage path across the p-n junction acts as shunt and reduces the current through load. Therefore an actual solar cell is represented by an equivalent circuit shown in Fig 6b. Here R_s represents the series resistance of the solar cell consisting of resistance of the front region, base region and the front and back contacts. The R_{sh}, known as shunt resistance, stands for the leakage resistance across the p-n junction. When these so called parasitic resistances are included, the current expression (6) becomes -

$$J_L = J_r - Jo \ \exp\frac{(qV_L + IR_S)}{nKT} - \frac{(V_L + IR_S)}{R_{sh}} \tag{7}$$

where n is diode ideality factor equal to one for p-n junction solar cell, R_{sh} is shunt resistance, R_s is series resistance.

The series resistance is caused by the fact that a solar cell is not a perfect conductor. The parallel resistance is caused by leakage of current from one terminal to the other due to poor insulation, for example on the edges of the cell. In an ideal solar cell, $R_s = 0$ and $R_{sh} = \infty$.

In practice, a solar cell has finite series and shunt resistance. The load current density is given by equation (7) which includes an empirical factor n, known as diode ideality factor. This depends on the quality of material used, junction placement and recombination of minority carries in the material. The value of n becoming greater than the unity leads to deterioration of the cell performance.

5. Characteristics and Parameters of Solar Cells

A solar cell under illumination generates photo-current, which passes through the external load resistance and a voltage is developed across the load. The voltage and current through the circuit depends on the load resistance. The most important characteristic of a solar cell is its I-V characteristics, which is obtained by measuring the current and voltage due to change in the load resistance. A typical I-V characteristic is shown in Figure 8. Theoretically it should be according to equation (7).

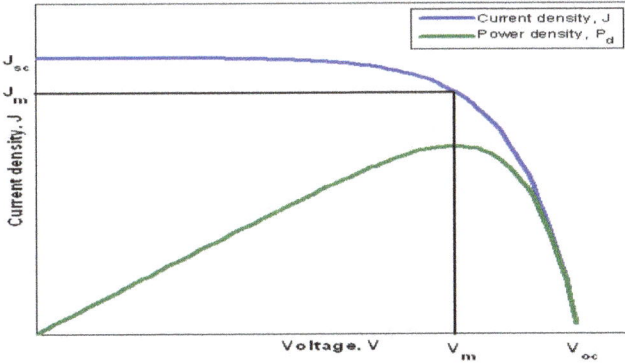

Figure 8: I-V Characteristics of an Ideal Solar Cell.

The solar cell parameters can be determined from the I-V curve. A solar cell is characterised on the basis of following parameters.

5.1   Open Circuit Voltage (V_{oc})

Open circuit voltage (V_{oc}) is the voltage between the terminals when no current is drawn (infinite load resistance). This is the voltage across the cell when load resistance R_L is infinite. It is obtained from the intercept of I-V curve on voltage axis at I=0. In general, V_{oc} increases linearly with the log of intensity of incident light. When a load is connected to the solar cell, the current decreases and a voltage develops as charge builds up at the terminals. The resulting current can be viewed as a superposition of the short circuit current, caused by the absorption of photons, and a dark current, which is caused by the potential built up over the load and flows in the opposite direction. To find an expression for the open circuit voltage, V_{oc}, we use equation (7) setting $J_L = 0$. This means that the two currents cancel out so that no current flows, which exactly is the case in an open circuit. The resulting expression is

$$V_{oc} = \frac{k_B T}{q} \ln\left(\frac{J_{sc}}{J_0} + 1\right)$$

$$(8)$$

5.2   Short Circuit Current (I_{sc} or J_{sc})

Short circuit current/current density (I_{sc} or J_{sc}) is the current when the terminals are connected to each other (zero load resistance). This is the current through the cell when load resistance R_L is zero. It is obtained from the intercept of I-V curve on current axis at V=0. The I_{sc} or J_{sc} increases linearly with the intensity of incident light. The short circuit current increases linearly with light intensity, as higher intensity mean more photons, which in turn mean more electrons. Since the short circuit current I_{sc} is roughly proportional to the area of the solar cell, the short circuit current density,

$$J_r = J_{sc} = I_{sc}/A \qquad (9)$$

is often used to compare solar cells.

5.3   Maximum Power Output (P_{max})

In general, the power delivered from a power source is $P = IV$, i.e. the product of voltage and current. If the current density J is used, we get the power density

$$P_d = JV \qquad (10)$$

The maximum power density occurs somewhere between $V = 0$ (short circuit) and $V = V_{oc}$ (open circuit) at a voltage V_m. The corresponding current density is called J_m, and thus the maximum power density is $P_{max} = J_m\,V_m$.

This is the maximum power which can be delivered to the Load. This is maximum product of current and voltage which is obtained at the knee of the I-V curve.

5.4   Fill Factor (FF) or Curve Factor

The fill factor, FF, is another quantity which is used to characterize a solar cell. It is defined as

$$FF = \frac{J_m V_m}{J_{sc} V_{oc}} \qquad (11)$$

and gives a measure of how much of the open circuit voltage and short circuit current is "utilized" at maximum power. It is the ratio of maximum output power and product of V_{oc} and I_{sc}.

$$F.F. = P_{max}/[V_{oc} \times I_{sc}] \qquad (12)$$

This is the area of largest rectangle within the I-V curve and exes.

5.5   Efficiency (η)

The efficiency of a solar cell is defined as the power (density) output divided by the power (density) input. If the incoming light has a power density Ps, the efficiency will be

$$\eta = \frac{J_m V_m}{P_s} \tag{13}$$

This is the ratio of maximum output power and input power.

$\eta = P_{max}/$ [area x intensity of incident light] x 100

Using FF we can express the efficiency as

$$\eta = \frac{J_{sc} V_{oc} FF}{P_s} \tag{14}$$

The four quantities J_{sc}, V_{oc}, FF and η are frequently used to characterize the performance of a solar cell.

5.6   Shunt Resistance (R_{sh})

This stands for the leakage resistance across the p-n junction. This is estimated from the slope of the I-V curve at short circuit point i.e. at V=0. $R_{sh} \gg 20$ kilo-ohm for good quality solar cells,. The shunt resistance is attributed to various leakage paths along the edges of the cell. These leakages can be caused by diffusion spikes along dislocations or grain boundaries, or possibly by fine metallic bridges along micro-cracks after the metal contacts are made. Low value of shunt resistance has adverse effect on the performance of solar cell.

5.7   Series Resistance (R_s)

This represents the internal resistance of the solar cell due to the front region, base region and the front and back contacts. This is estimated from the slope of the I-V curve at open circuit point i.e. at I=0. A higher value of series resistance drastically reduces the fill factor or curve factor.

It means that low value of shunt resistance and high value of series resistance significantly degrade the performance of the cell. Practical good quality solar cells are associated with shunt resistance of the order of 10 kilo ohm and series resistance about 0.01 ohms.

5.8 Spectral Response (SR)

The spectral response of the solar cell is given by the variation of short circuit current as a function of wave length incident light. The solar radiations are associated with photons of energy $E = hc/\lambda$ and for different values of wave length (λ) the response of the cell is different for different wavelength. The spectral response $SR(\lambda)$ at a wave length is the number of carriers collected per incident photons at wavelength λ.

$$SR(\lambda) = \text{Current} / [q \times N_{ph}(\lambda)] \tag{15}$$

where $N_{ph}(\lambda)$ is the number of photons of wave length λ incident on the cell.

The total current due to all the wavelengths from the incident light is given by-

$$J_r = q \int N_{ph}(\lambda) \times SR(\lambda)\, d\lambda \tag{16}$$

Figure 9 shows the typical spectral response of a solar cell. Spectral response often depends on intensity of light.

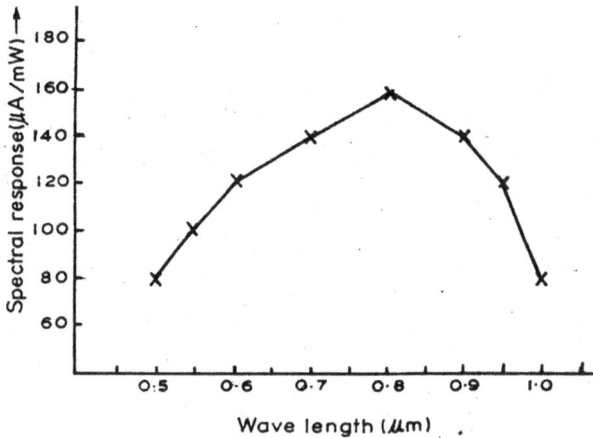

Figure 9: *Typical Spectral Response of a Solar Cell.*

6. Materials for Solar Cells

The theory of photovoltaic predicts the characteristics of materials which would operate with an optimum efficiency as a photovoltaic solar energy convertor. Solar cells convert radiant energy into electrical energy. Therefore it depends on optical and electrical properties of the material. The dominant solar cells parameters are conversion efficiency and output power. The optical properties of semiconductor material determine what fraction of solar radiations is absorbed by the substance and its electrical properties determine how much of this is converted to electricity.

			IIIA	IVA	VA	VIA	VIIA	VIIIA
								2 He 4.003
			5 B 10.811	6 C 12.011	7 N 14.007	8 O 15.999	9 F 18.998	10 Ne 20.183
IB	IIB		13 Al 26.982	14 Si 28.086	15 P 30.974	16 S 32.064	17 Cl 35.453	18 Ar 39.948
29 Cu 63.54	30 Zn 65.37	31 Ga 69.72	32 Ge 72.59	33 As 74.922	34 Se 78.96	35 Br 79.909	36 Kr 83.80
47 Ag 107.870	48 Cd 112.40	49 In 114.82	50 Sn 118.69	51 Sb 121.75	52 Te 127.60	53 I 126.904	54 Xe 131.30
79 Au 196.967	80 Hg 200.59	81 Tl 204.37	82 Pb 207.19	83 Bi 208.980	84 Po (210)	85 At (210)	86 Rn (222)

The Periodic Table

PV cells are made of semiconductor materials. The major types of materials are crystalline and thin films, which vary from each other in terms of light absorption efficiency, energy conversion efficiency, manufacturing technology and cost of production.

The atoms in a semiconductor are materials from either group IV of the periodic table, or from a combination of group III and group V (called III-V semiconductors), or of combinations from group II and group VI (called II-VI semiconductors). Because different semiconductors are made up of elements from different groups in the periodic table, properties vary between semiconductors. Semiconductor materials come from different groups in the periodic table, yet share certain similarities. The properties of the semiconductor material are related to their atomic characteristics, and change from group to group. Researchers and designers take advantage of these differences to improve design and choose the optimal material for a PV application.

The solar spectrum dictates the range of materials which can be used for generation of photo-electricity. The number of photons absorbed from the solar radiations decreases with increasing forbidden energy gap of the semiconductor since the photons having energy less than the forbidden band gap (Eg) are not able to excite the electrons from valence to conduction band and create electron-hole pairs. Therefore, the smaller the band gap of the semiconductor, the larger the portion of solar spectrum utilized and greater number of electron-hole pairs are produced giving rise to high value of photo-current (I_{sc} or J_{sc}); but the maximum voltage obtainable is correspondingly smaller. On the other hand a large forbidden energy band gap material can give higher photo-voltage or V_{oc} and low leakage across the junction; but at the same time smaller portion of the solar radiation is absorbed and hence low current (I_{sc} or J_{sc}) can be obtained. These two offsetting effects show that high efficiency can be obtained at specific forbidden band gap value. Figure 10 shows the ideal (theoretical) solar cell efficiency as a function of Eg of the material for various temperatures. It can be seen that the efficiency curve has a broad maximum. Therefore semiconductor materials with band gap between 1 to 2 eV can be considered as solar cell materials. Many other factors like type of absorption, recombination levels, impurity levels, etc. affect and degrade the actually achievable efficiency. It is also seen that efficiency is reduced at higher temperatures.

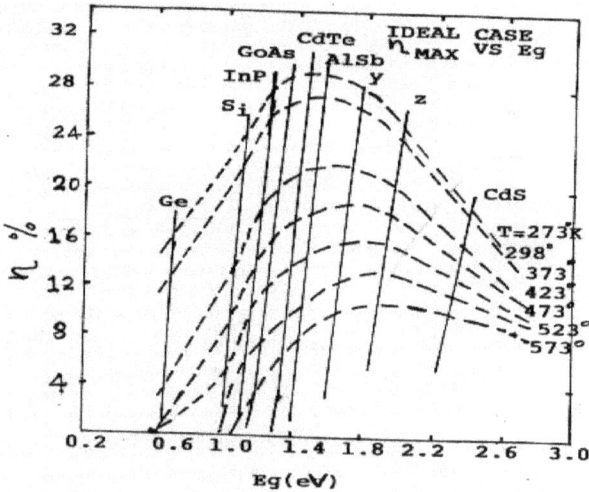

Figure 10: *Ideal Solar Cell Efficiency vs Band Gap of Semiconductor.*

At present there are various photo-voltaic cells available utilising different materials. Photo-voltaic conversion efficiency more than 10% have been achieved mainly with silicon (Si, Eg =1.1eV), gallium arsenide (GaAs, Eg= 1.43eV), indium phosphide (InP, Eg= 1.34eV), cuprous sulphide (Cu_2S, Eg= 1.2eV), cadmium telluride (CdTe, Eg=1.5eV) and copper indium selenide ($CuInSe_2$, Eg=1.01eV).

Various types of photovoltaic systems on the basis of type of junction are shown in Table 1.

(a) Homo-junction

(b) Hetero-junction

(c) Metal-semiconductor junction

(d) Semiconductor-electrolyte junction

Different materials used in solar cells are also listed in Table 1 with respective band gap and obtainable conversion efficiency along with corresponding photovoltaic system.

Table 1: *Materials Used in Various Types of Solar Cells*

S.No.	Material	Band Gap (in eV)	Efficiency
Homojunctions			
1	Ge	0.66	....
2	Si	1.11	....
3	CdTe	1.40	25
4	GaAs	1.43	25
Heterojunctions			
5	n-CdS/p-Cu$_2$S	2.42	15-20
6	n-CdS/p-InP	2.42	20
7	p-ZnTe/n-CdTe	2.25	....
8	p-ZnTe/n-CdSe	2.26	....
9	p-CuInSe$_2$/n-CdS	1.01	20
10	p-Si/n-In$_2$O$_3$	1.11	11
Schottky Diodes			
11	n-GaAs/Au	1.43	20
12	n-Si/Cr((Cu)	1.11	....
13	p-Cu$_2$O/Cu	1.93	10
Photelectrochemical Cells			
14	n-WSe$_2$/2M KI+0.02M I$_2$	1.35	19.8
15	n-GaAs/Ferrocene-Ferrocium in LiClO$_4$ methanol solution	1.42	15.5
16	p-InP/VCl$_3$-VCl$_2$-HCl	1.35	13.0
17	CdSe/Ferri/Ferro cynide	1.7	12.4
18	CuInSe$_2$/2M KI+0.02M In$_2$+0.02M CuI+2M HI	1.01	12.0

The main problem in large scale terrestrial use of solar energy is the cost of photovoltaic modules. Attempts are being made to –

(i)   reduce the cost and increase the efficiency so as to become competitive with the presently available energy technologies,

(ii) develop new technologies based on different architectural design, and

(iii) develop new materials as to serve as light absorbers and charge carriers.

This has given rise to use of-

(a) Amorphous Silicon Solar Cells

(b) Thin Films Solar Cells

(c) Photoelectrochemical Solar Cells

(d) Multi-junction Solar Cells

(e) Nanocrystal Solar Cells

(f) Polymer Solar Cells

(g) Transparent Electrodes

The future development of photovoltaic solar cells for wide-scale use in energy sector lies in the materials and processing techniques. Recent technology indicates bright prospects.

7.   Conclusion

The study of basics of photovoltaic effect used for the development in solar cell with given detailed theoretical and practical aspects will help in understanding and future development of advanced solar cell technology. Efforts are being made to increase the efficiency and reduce the cost of solar cells. Solar cells are reliable, static and require little maintenance and research is going on for other suitable materials and also for better technology. Long term potential in this area is enormous since the Sun is practically inexhaustible source of energy and provides ecologically pure energy.

References

[1]   B. P. Chandra, Recent Advances in Solar Cell Material, National Seminar on Renewable Energy Sources 2008.

[2]   B. O'Regan, M. Grätzel, Nature 353, 737, 1991. https://doi.org/10.1038/353737a0

[3]   D. Anderson, M. D. Archer and R. D. Hill, Clean Electricity from Photovoltaic, Eds. (London Imperial College Press), 2001.

[4]   Green M A, Emery K, Hishikawa Y, Warta W. Solar Cell Efficiency Tables (version 33), Progress in Photovoltaics: Research and Applications 2009; 17: 85-94. https://doi.org/10.1002/pip.880

[5]   Harray A. Sorensen, Energy Conversion Systems.

[6]   J. W. Christain, P. Haasen, T. B. Massalski, Progress in Material Science, Vol. 35 pp-205-418, Pregamon Press 1991.

[7]   K. Yamamoto: Sol. Energy Mater. Sol. Cells 66, 117, 2001. https://doi.org/10.1016/S0927-0248(00)00164-1

[8]   M. A. Green, "Photovolatics: coming of age", Conf. Record 21st IEEE Photovoltaic Specialists Conf., 1-7 (1990).

[9]   M Bazilian, I Onyeji, M Liebreich, I MacGill, J Chase, J Shah, D Gielen (2013). "Re-considering the economics of photovoltaic power". Renewable Energy. https://doi.org/10.1016/j.renene.2012.11.029

[10]   Meera Ramrakhiani, Photovoltaic Effect and Solar Cells, Everymen's Science; 28 (6) pp-209-215; 1994.

[11]   Nelson, Jenny. The Physics of Solar Cells. Imperial College Press, 2003. https://doi.org/10.1142/p276

[12]   P. Maycock, PV market update. Renewable Energy World, July-Aug., 86, 2004.

[13]   Pallab Bhattacharya, Semiconductor Optoelectronic Devices; Prentice-Hall of India Pvt. Ltd. New Delhi 1999.

[14]   Pulfrey, L. D. Photovoltaic Power Generation. New York: Van Nostrand Reinhold Co. ISBN 9780442266400, 1978.

[15]   R. K. Kotnala and N.P. Singh, Essentials of Solar Cell; Allied Publishers Pvt. Ltd. 1986.

[16]   R. Trykozko, Principles of Photovoltaic Conversion of Solar Energy, Opto-Electr. Rev., 5, No. 4, 1997.

[17]   S. Chandra, R. K. Pandey; Phy. Stat. Sol. (a) 59, 787, 1980. https://doi.org/10.1002/pssa.2210590246

Chapter 3

Nanostructured solar cells

Surendra K Pandey

Government Science College, Jabalpur (Madhya Pradesh) - 482 001 India

skpandey.7008@gmail.com

Abstract

Photovoltaics presently represent the fastest growing sector of the electricity generation industry, although growing from a small base. Energy conversion efficiency is a key parameter with this technology since it directly impacts both material and deployment costs. The performance of the traditional bulk semiconductor solar cell is limited to about 33% while thermodynamic limits on the conversion of sunlight to electricity are much higher at 93%. Low dimensional structures appear capable of allowing much of this gap to be bridged. Nanostructure based solar energy is attracting significant attention as possible candidate for achieving drastic improvement in photovoltaic energy conversion efficiency. The use of nanostructures in photovoltaics offers the potential for high efficiency by either using new physical mechanisms or by allowing solar cells which have efficiencies closer to their theoretical maximum, for example by tailoring material properties. At the same time, nanostructures have potentially low fabrication costs, moving to structures or materials which can be fabricated using chemically or biologically formed materials. Despite this potential, there are multiple and significant challenges in achieving viable nanostructured solar cells, ranging from the demonstration of the fundamental mechanisms, device-level issues such as transport mechanisms and device structures and materials to implement nanostructured solar cells, and low cost fabrication techniques to implement high performance designs. The cell designs and enhancements are categorized by the type of nanostructure utilized. These include bulk nanostructured materials (3D), quantum wells (2D), nanowires (1D), and quantum dots/nanoparticles (0D). During the last decade, the development of the photovoltaic device theory and nanofabrication technology enables studies of more complex nanostructured solar cells with higher conversion efficiency and lower production cost. The fundamental principles and important features of these advanced solar cell designs are systematically presented and summarized, with a focus on the function and role of nanostructures and the key factors affecting device performance. Among various nanostructures, special attention has been given to those relying on quantum effect.

Key Words

Photovoltaics, Nanostructures, Low-cost Self-assembled Nanostructures, Nanostructured Solar Cell Systems, Solar Power

Contents

1.	**Introduction**	**60**
2.	**Key Issues**	**67**
3.	**Potential Nanostructures**	**69**
3.1	**Nanocomposites and Nanostructured Polycrystalline Materials**	**70**
3.2	**Quantum Wells**	**72**
3.3	**Nanowires and Tubes**	**73**
3.4	**Quantum Dots and Nanoparticles**	**75**
3.5	**Nanocones**	**78**
3.6	**III-V Semiconductors**	**79**
4.	**Common Nanostructured Solar Cell Systems**	**80**
4.1	**Nanocrystalline Silicon Solar Cell**	**82**
4.2	**Nanostructured Metal Oxides in Dye-Sensitized Solar Cell**	**83**
4.3	**Organic Solar Cells**	**84**
4.4	**Organic/Inorganic Hybrid Solar Cell**	**85**
4.5	**The Third Generation Solar Cells**	**87**
4.6	**Polycrystalline Thin-film Solar Cells**	**87**
4.7	**Quantum Well Solar Cell**	**88**
4.8	**Nanowire Solar Cell**	**91**

4.9   **Quantum Dot Solar Cells** 93

4.10 **All Quantum Dot Tandem Solar Cell** 94

4.11 **Intermediate Band Solar Cell** 96

4.12 **Multiple Exciton Solar Cell** 97

4.13 **Virtual Band Gap Solar Cells** 98

4.14 **Hot Carrier Solar Cell** 100

5.   **Discussion and Conclusion** 101

References 104

1. Introduction

Solar power is poised to enter the mainstream energy market with novel materials that boost energy conversion efficiency and bring down manufacturing costs. The ability of sunlight to generate electricity was first discovered by French physicist Andre-Edmond Becquerel in 1839, when he observed that shining light on certain materials produced an electric current. But it took just over a hundred years to 1941 before Russell Ohl in the United States invented a silicon solar cell. The silicon solar cell is still the predominant model in use today, representing some 94 percent of the global market. But even with energy conversion efficiencies as high as 33 percent, silicon-based solar cells are still too expensive for general use. In recent years, fuelled by the growing global energy demands and to some extent, by the need to reduce carbon emissions to mitigate global warming, solar power is gaining in popularity as improvements in design boost energy conversion efficiency and lower manufacturing costs. There is a trade-off between the cost of manufacture and the efficiency, which is expressed in the unit price of electricity generated. The current cost of about $4/W is still considered too high for the market. A major advantage of solar power is that it has minimum impacts on the environment, which are mostly associated with the manufacturing processes, and do not require major changes in land use. Solar panels can be conveniently integrated into existing building structures and rooftops; and large arrays can be sited in deserts. But to really capture the mainstream energy market, major increases in energy conversion efficiency and/or reductions in manufacturing cost are needed; and the prospects look bright for both [1].

Over the past decade, 'second generation' thin-film technologies have been developed that do not require costly crystalline silicon wafers and can be manufactured much cheaper. These include devices based on a range of new inorganic semi-conducting materials, as well as multi-junction amorphous (non-crystalline) silicon. Thin-film cells are fabricated using techniques such as sputtering, physical vapour deposition and plasma-enhanced chemical vapour deposition. Multi-junction cells based on amorphous silicon have been the most successful second-generation technology to date [2]. Amorphous silicon can be made from waste silicon from the computer chips industry, and devices can be manufactured at relatively low cost and at high speed with roll-to-roll processing on flexible stainless steel and other substrates, which can be easily integrated into roofing materials. These advantages have helped them capture the 5.6-6 percent of the market not dominated by crystalline or polycrystalline silicon. One such product on the market is a triple-junction flexible solar panel made of three separate amorphous silicon layers, each with a different band gap, so as to harvest light from the entire solar spectrum, and works even in cloudy conditions. In May 2005, Sharp Corporation, the world's top manufacturer of solar panels, announced the introduction of a new polycrystalline solar module in Japan with the industry's highest conversion efficiency of 15.8 percent. This sets a benchmark for all second and third generation solar cells.

Third generation technologies are based on new materials, new mechanisms and concepts in light energy harvesting and conversion. They come in two kinds - those aimed at achieving very high efficiencies and the rest aimed at the lowest cost with moderate efficiencies of 15-20 percent. In the first category are approaches based on quantum dots and new mechanisms, such as 'hot carriers', thermovoltaics and multiple electron-hole pair creation. These are at the early research stage and yielding exciting results, but are not yet ready for the market. The second category includes a wide range of applications based on organic material, some of which are near to market, or already in the market [3].

Nanostructured solar cells - a type of third or next generation solar cell [4] include those that are based on nanostructures and/or nanostructured interfaces such as nanowire, mesoscopic and quantum dot solar cells. They hold great promise towards new approaches for converting solar energy into either electricity (in photovoltaic devices) or chemical fuels. Nanostructured materials are being investigated and developed as versatile components of optoelectronic devices with the ability to manipulate light (via plasmonic enhancement, photonic crystals, and so on) and control energy flow at nearly the atomic level.

The efficiency of solar cells is the electrical power it puts out as percentage of the power in incident sunlight. In physics, the **Shockley-Queisser limit** or **Detailed Balance Limit** refers to the maximum theoretical efficiency of a solar cell using a p-n junction to collect power from the cell. It was first calculated by William Shockley and Hans Queisser at Shockley Semiconductor in 1961. The limit is one of the most fundamental to solar energy production, and is considered to be one of the most important contributions in the field. The limit places maximum solar conversion efficiency around 30% assuming a p-n junction band gap of 1.1 eV [5]. One of the most fundamental limitations on the efficiency of a solar cell is the 'band gap' of the semi-conducting material used in conventional solar cells. In a solar cell, negatively doped (n-type) material with extra electrons in its otherwise empty conduction band forms a junction with positively doped (p-type) material, with extra holes in the band otherwise filled with valence electrons. When a photon with energy matching the band gap strikes the semiconductor, it is absorbed by an electron, which jumps to the conduction band, leaving a hole. Both electron and hole migrate in the junction's electric field, but in opposite directions. If the solar cell is connected to an external circuit, an electric current is generated. If the circuit is open, then an electrical potential or voltage is built up across the electrodes. Photons with less energy than the band gap slip right through without being absorbed, while photons with energy higher than the band gap are absorbed, but their excess energy is wasted, and dissipated as heat. It is possible to improve on the efficiency by stacking materials with different band gaps together in multi-junction cells. Stacking dozens of different layers together can increase efficiency theoretically to greater than 70 percent. But this results in technical problems such as strain damages to the crystal layers. The most efficient multi-junction solar cell is one that has three layers - gallium indium phosphide/gallium arsenide/germanium (GaInP/GaAs/Ge) made by the National Center for Photovoltaics in the US, which achieved an efficiency of 34 percent in 2001. Recently, entirely new possibilities for improving the efficiency of photovoltaics have opened up.

Nanoscale systems exhibit different properties than bulk or thin films of the same compound, and have allowed new ways of approaching solar energy conversion for electricity generation or fuels. The large surface-to-volume ratio of nanomaterials can provide various benefits, and, furthermore, objects with a size of approximately 1-20 nm can also exhibit quantization effects, which become more pronounced with decreasing size. Two broad approaches based on nanostructures are being explored for photovoltaics: (i) significant reduction in material usage and/or associated final costs; (ii) photovoltaic devices with a higher limiting efficiency than that determined by the Shockley-Queisser analysis.

Front surface reflectivity in a solar cell leads to losses. If light is not absorbed by the active component, it cannot be converted into electricity. Crystalline Si reflects ~30% of the incoming light, and conventional cells employ anti-reflection layers to reduce these losses to about 3-4%, but at an added manufacturing cost. Nanostructured surfaces can reduce reflection losses when the size of the nanostructures is smaller than the wavelength of the incident light. If those nanostructures are also active components, then the need to deposit costly anti-reflection coatings and texturing can be avoided [6]. Furthermore, patterned nanostructures can be designed to capture even more light via light trapping so that less material is needed to absorb the solar flux [7-8]. The low absorptivity of silicon requires the use of thick films in order to absorb all incidents light. This poses a constraint on material quality, because high-purity materials, with long minority-carrier lifetimes, are needed to enable carrier collection over relatively large distances. Moreover, thicker films result in increased material costs. One approach to relaxing such requirements has been explored by using radial p-n junction Si nanowires [7]. A core-annular p-n junction [Figure 1b] is fabricated along the length of the wire. When the incident light generates charge carriers at the junction, minority carriers only need to traverse the nanowire diameter in order to be collected. With less stringent requirements on the minority-carrier lifetimes, lower grade silicon can be used. By combining light-trapping effects and reduced carrier lifetimes, in nanowire solar cells both the amount and quality of raw materials used can be greatly reduced [8].

In traditional solar cells fabricated from bulk semiconductors, electrons and holes are generated when light is absorbed in the semiconductor. They then separate and migrate to different contacts to produce both a voltage V and current I, and thus power ($P = I \times V$). Because of the large dielectric constants of these materials, electrons and holes are quickly screened from one another and do not interact. Nanostructuring is a way to bypass the requirement of high-dielectric-constant semiconductors and allows use of new classes of materials and device designs. For example, two dissimilar materials, where one is n-type (conducts electrons) and the other is p-type (conducts holes), can be intermixed with nanoscale morphology [Figure 1c]. Absorption of light produces an excited state (exciton) that undergoes rapid charge transfer producing electrons and holes in separate phases, making their interaction less probable. Two types of solar cell are based on this design - organic photovoltaic devices and dye-sensitized solar cells.

Another next-generation approach for photovoltaics is based on semiconductor nanocrystals. Their most important properties for photovoltaic applications are the strong size-dependence of the bandgap, and the large modification of the relaxation dynamics of photoexcited charge carriers that are created by the absorption of photons within energies larger than the bandgap. The bandgap of the absorber layer controls which photons can be absorbed and limits the output voltage of a solar cell [9]. Because only the radiation with higher energy than the bandgap is absorbed, narrower-bandgap materials absorb more solar photons, resulting in higher photocurrents. However, the output voltage is linearly proportional to the bandgap, and thus wider-bandgap materials allow higher voltages. In the Shockley-Queisser analysis there is an optimal bandgap that achieves the highest efficiencies, and ranges between 1.2 and 1.4 eV. Semiconductors with bandgaps lower than 1 eV are generally not employed in single-layer solar cells. Quantum confinement effects in quantum dots (QDs) can increase the bandgap by more than 1 eV compared with the bulk value, expanding the range of semiconductor materials viable for photovoltaics. A prototypical example is PbS. Bulk PbS has a bandgap of 0.4 eV but PbS QDs can have bandgaps from ~0.6 to ~2 eV depending on their size.

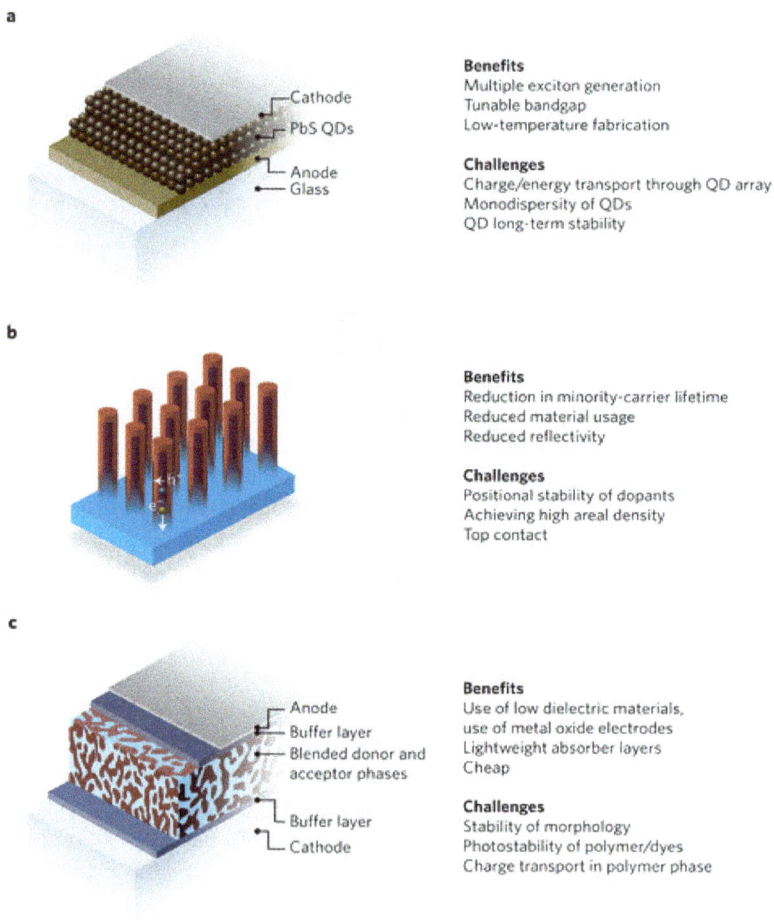

a

Cathode
PbS QDs
Anode
Glass

Benefits
Multiple exciton generation
Tunable bandgap
Low-temperature fabrication

Challenges
Charge/energy transport through QD array
Monodispersity of QDs
QD long-term stability

b

Benefits
Reduction in minority-carrier lifetime
Reduced material usage
Reduced reflectivity

Challenges
Positional stability of dopants
Achieving high areal density
Top contact

c

Anode
Buffer layer
Blended donor and
acceptor phases
Buffer layer
Cathode

Benefits
Use of low dielectric materials,
use of metal oxide electrodes
Lightweight absorber layers
Cheap

Challenges
Stability of morphology
Photostability of polymer/dyes
Charge transport in polymer phase

Figure 1: *Examples of Nanostructured Solar Cells (a) Quantum Dot Solar Cell (b) Nanowire Solar Cell (c) Mesoscopic Solar Cell.*

Quantum dot solar cells [Figure 1a] can be processed from colloidal solutions at ambient temperatures, enabling relatively low manufacturing costs. For example, PbS QDs can be synthesized and processed into films near ambient temperature without vacuum processing [10]. Efforts have focused on the specific deposition steps that can impact both the optical and electrical properties of the QD films [11]. Solar cell efficiencies are now approaching 9% using PbS QDs with a bandgap of 1.25-1.4 eV [12]. Other QDs of interest for photovoltaic applications are crystalline precursor inks, which after annealing produce continuous larger-grain thin films with bulk-like properties [13]. Examples include QDs made of CdTe, copper zinc tin sulfide (CZTS), copper indium gallium selenide (CIGS) and Si. Recent newcomers are solution-processed perovskite solar cells [14], which evolved from nanostructured device concepts. However, like most other single-junction devices, their efficiency is limited by the Shockley-Queisser analysis, and will need to be incorporated into multijunction architectures in order to achieve higher efficiencies.

New photovoltaic technologies should have the potential not only to reduce module costs but also to achieve PCE beyond ~33%. Two critical assumptions of the Shockley-Queisser analysis are - (i) photons with energy less than the semiconductor bandgap are not absorbed and thus cannot contribute to PCE; (ii) energetic electrons created by high-energy photons immediately relax to the band edge, that is, the fraction of energy of photons with energy greater than the bandgap is immediately lost as heat. Approaches to achieve higher limiting efficiencies attempt to either use the high-energy photons more efficiently, or recover the low-energy photons normally not converted. Nanostructures are being explored to eliminate these losses. Multijunction solar cells commonly employ a stack of p-n junctions where the bandgap of the light-absorbing semiconductor decreases in successive layers. In the limit of an infinite number of junctions, the theoretical PCE reaches 68% at 1-sun intensity [15]. A 1-sun PCE of 34.1% has been achieved in a triple-junction solar cell based on III-V semiconductors. Under optical concentration, the efficiency increases, and the record PCE for any photovoltaic cell is 44.4% when measured with light equivalent to 302 suns [16]. However, these high-efficiency cells are far too expensive. Nanomaterials are being explored as part of multiple-junction solar cells, because the same material can be used to produce layers with different bandgaps, and the multiple junctions can be solution-processed at significantly lower cost [17]. Other examples of circumventing assumptions of the Shockley-Queisser analysis using nanostructures include intermediate-band solar cells [18] and solar cells based on multiple exciton generation (MEG) [4]. In addition, nanocrystals can be components of up and/or down conversion layers and fluorescent concentrators [19], all of which attempt to reach higher efficiencies at lower cost. Multiple exciton generation, or carrier multiplication, allows a single photon to produce multiple excitons (electron-hole pairs). This increases the photocurrent and can permit the PCE to exceed the limit imposed by the Shockley-Queisser analysis [20], which assumes that each photon only produces one electron. Multiple exciton generation is greatly enhanced in QDs compared with bulk semiconductors because the competition between the relaxation pathways of the MEG and of phonon emission, as well as the hot-carrier relaxation, can be modified to favour MEG. Quantum confinement also enhances the Coulombic interaction that drives MEG. In addition to single-component nanostructures, multicomponent nanocrystals with selectively arranged individual domains that allow complex functionalities offer new opportunities for photovoltaic research. Such nanoheterostructures allow for further tailoring of the excited-state relaxation dynamics, resulting in additional enhancements of the MEG efficiency [21].

While nanostructures offer significant potential advantages for both increasing the PCE and reducing manufacturing costs, the technology is not yet mature, and the efficiency of proof-of-principle devices lies well below those of established technologies. Furthermore, it is not always possible to independently verify the reported efficiency of such devices. The PCE encloses information on the performance of a solar cell in a single number, and thus allows a direct comparison of devices across many technologies. One issue with emerging photovoltaic technologies made at the lab scale involves the stability of the materials or devices. Devices with poor stability provide a complication for certification purposes, due to different measurement conditions (atmosphere, temperature, humidity, and light source), contact probes, transportation, and so on. Device stability is obviously an important consideration for photovoltaic technologies as the longer the photovoltaic system operates, the lower is the total cost. Many of the same features that make nanostructured solar cells attractive also introduce additional challenges. Nanostructured materials have larger surface area; the surface properties of photovoltaic films and electronic devices in general require attention. For nanostructured devices, improper passivation of internal surfaces can hinder long-term stability. Stability issues are being addressed by chemistry [22], or device configuration [10]. No consensus exists among the research community on how to properly characterize the stability. Low-maintenance, long-lifetime solar simulators are not readily available, and lifetime tests inherently take a long time to perform. Therefore, some approaches involve accelerated lifetime testing in harsh environments, such as 85°C at 85% humidity, but that translation to real-world stability may not be the same for all technologies. Next-generation solar cells must meet stringent requirements in terms of PCE, cost and long-term stability. It is believed that the nanostructured solar cells have the potential to achieve these objectives.

The ability to see nano-sized materials has opened up a world of possibilities in a variety of industries and scientific endeavours. Solar power is poised to enter the mainstream energy market with novel materials that boost energy conversion efficiency and bring down manufacturing costs. The third generation technologies are based on new materials, new mechanisms and concepts in light energy harvesting and conversion. One of the most fundamental limitations on the efficiency of a solar cell is the 'band gap' of the semi-conducting material used in conventional solar cells. It is possible to improve on the efficiency by stacking materials with different band gaps together in multi-junction cells. But this results in technical problems such as strain damages to the crystal layers. Nanostructured systems have quantum optical properties that are absent in the bulk material due to the confinement of electron-hole pairs (called excitons) on the particle, in a region of a few nanometres. The first advantage of such systems is their tunable bandgap. There is an optimum bandgap that corresponds to the highest possible solar-electric energy conversion. Another advantage is that in contrast to traditional semiconductor materials that are crystalline or rigid, they can be molded into a variety of different form, in sheets or three-dimensional arrays. In an ordered three-dimensional array, there will be strong electronic coupling between them so that excitons will have a longer life, facilitating the collection and transport of 'hot carriers' to generate electricity at high voltage. In addition, such an array makes it possible to generate multiple excitons from the absorption of a single photon. Thus nanostructured systems are offering the possibilities for improving the efficiency of solar cells in at least two respects, by extending the band gap of solar cells for harvesting more of the light in the solar spectrum, and by generating more charges from a single photon.

2. Key Issues

The ultimate goal in photovoltaics is to achieve a high efficiency technology which can be manufactured at low cost and large scale. Low-cost self-assembled nanostructures may be more efficient in material usage and may have much lower fabrication costs. Existing low cost approaches typically use colloidal formation of quantum dots, which are then inserted or used in a device structure. Since the colloidal nanocrystals are solution based, they are readily incorporated in organic and dyesensitized approaches, in which case they act as an absorber material with the transport and junction supplied by the remaining device [23]. Other low-cost approaches focus on the development of new self-assembly techniques.

Despite the many potential uses and ways to include nanostructures in photovoltaic devices, these solar cells share several issues and challenges. The most basic issue is that the device design rules for nanostructured solar cells do not exist, and thus many choices or design parameters do not have sufficient theoretical or experimental guidance. Figure 2 shows several of the design issues. The design rules for achieving a high efficiency nanostructured solar cell are substantially different than a conventional device. For example, current collection in conventional devices is typically achieved by increasing the diffusion length (or increasing the electric field). In nanostructured photovoltaics, the collection depends on different factors, such as the time taken to remove the carrier from the nanostructure compared to the recombination, and the transport through the device. Transport is an issue which affects nearly all uses of nanostructures, whether for high efficiency or primarily for low cost. The inherent confining potentials in nanostructures allow tailoring of material properties, but also introduce a barrier to transport of carriers at the low energy levels in the nanostructure. LEDs and lasers avoid this problem since they require carrier injection into, not collection from, the nanostructures. Similarly, photo detectors circumvent this issue by using large electric fields, low confining potentials and intersubband optical absorption. There are two fundamental solutions to the transport problem - (i) use of closely space nanostructured arrays which promote the formation of mini-bands; or (ii) excitation of the carriers in the confining potential to the conduction/valence band of the barrier or matrix material (either thermally, via an electric field, or via photon-induced transitions), which then acts to transport carriers. For high efficiency approaches, the closely spaces array must be a quantum dot (QD) array, since only QDs have a zero density of states between the bands. In other nanostructured arrays, carriers quickly thermalize to the lowest energy level, representing a loss. In addition, the closely spaced array should ideally be periodic, since variations in the QD spacing change the absorption edge. However, closely spaced arrays of QDs with long range order are difficult to fabricate.

The other approach to transport in nanostructured materials is to use photons to excite carries to the upper energy band. This process is used in quantum well and quantum dot intra-red photodetectors (QWIP and QDIPs). Once at this energy, carriers must be prevented from being captured back into the nanostructure. Transport in the barrier allows high performance provided that the barrier or matrix material surrounding the nanostructure has good transport properties, that there is a strong electric field, and that carriers are not transported in the nanostructure where recombination rate can be high. To avoid transporting carriers in the nanostructure, the direction of transport of carriers should be perpendicular to the confinement of the nanostructure, which allows QD and QW structures, but not nanorods aligned parallel to the direction of light absorption.

```
                    ┌─────────────────┐
                    │  Optimum band   │
                    │  structure and  │
                    │   materials     │
                    └─────────────────┘
```

┌──────────────────┐   ┌──────────────────┐
│ 1. Absorption   │   │ 4. FF > 90% of   │
│ of >98% of all   │   │ theoretical   │
│ photons above   │   │ maximum   │
│ band gap   │   │   │
└──────────────────┘   └──────────────────┘

        ┌──────────────────┐   ┌──────────────────┐
        │ 2. Collection   │   │ 3. V_{OC} > 93% of │
        │ of >98% of all   │   │ theoretical   │
        │ generated   │   │ maximum   │
        │ carriers   │   │   │
        └──────────────────┘   └──────────────────┘

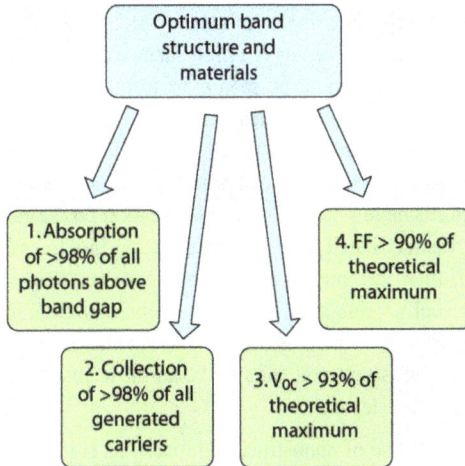

Figure 2: *Design issues and challenges in nanostructured solar cells.*

An additional requirement in nanostructured solar cells is the need to include structures which increase absorption. Due to the low volume of nanostructure material and the need to keep devices thin for transport reasons, a necessity of these approaches is features which promote effective absorption. New approaches to increasing the optical absorption include nanostructured features for light trapping and plasmon absorption.

3. Potential Nanostructures

Nanoscale materials are defined as a set of substances where at least one dimension is less than approximately 100 nm. A nanometer is one millionth of a millimetre that is approximately 100,000 times smaller than the diameter of a human hair. Nanomaterials have extremely small size with at least one dimension of the order of 100 nm or less. Nanostructured materials can be nanoscale in zero dimension (e.g. Quantum dots), one dimension (eg. surface films), two dimensions (eg. strands or fibres), or three dimensions (eg. particles). They can exist in single, fused, aggregated or agglomerated forms with spherical, tubular, and irregular shapes. Common types of nanomaterials include nanotubes, dendrimers, quantum dots and fullerenes. Nanomaterials have the structural features in between those of atoms and the bulk materials. While most of microstructured materials have similar properties to the corresponding bulk materials, the properties of materials with nanometer dimensions are significantly different from those of atoms and bulk materials. This is mainly due to the nanometer size of the materials, which render them - (i) large fraction of surface atoms (ii) high surface energy (iii) spatial confinement and (iv) reduced imperfections cannot be seen exist in the corresponding bulk materials. The use of nanostructures in photovoltaics offers the potential for high efficiency by either using new physical mechanisms or by allowing solar cells which have efficiencies closer to their theoretical maximum, for example by tailoring material properties. At the same time, nanostructures have potentially low fabrication costs, moving to structures or materials which can be fabricated using chemically or biologically formed materials. Despite this potential, there are multiple and significant challenges in achieving viable nanostructured solar cells, ranging from the demonstration of the fundamental mechanisms, device-level issues such as transport mechanisms and device structures and materials to implement nanostructured solar cells, and low cost fabrication techniques to implement high performance designs.

The likely most immediate use of nanostructured materials is their use to realize practical advantages in conventional device structures rather than utilizing new physical mechanisms for efficiency increases. Given the ability of nanostructures to modify material parameters, there are many avenues by which nanostructures can increase solar cell efficiency closer to the theoretical limit for a particular device. Suggested approaches include changing optical/material interactions by approaches such as nano-texturing or plasmon absorption; by using nanostructured material to modify the band gap of existing materials; or by using nanostructures to modify the strain between two materials or the impact of dislocations. While these offer the greatest short-term potential for using nanostructures, they also have the greatest variety, with the benefits and gains dependent on the specific device configuration in which the nanostructures are included. Many types of nanostructures have been applied to solar cells. The potential nanostructures may be classified into four types - (a) nanocomposites [3D], (b) quantum wells [2D], (c) nanowire and nanotubes [(quasi) 1D], and (d) nanoparticles and quantum dots [(quasi) 0D]. These structures have been employed in various functions and for various performance/energy conversion enhancement strategies.

3.1   Nanocomposites and Nanostructured Polycrystalline Materials

Nano-sized phosphor particles absorb part of the spectrum and convert it to the more suitable energy for the solar cell as Nanocomposite systems called as up [24] or down [25] energy convertors. One of the first examples of using 3D nanostructured materials for photovoltaic applications is the dye-sensitized solar cell (DSC), originally developed by Grätzel and O'Regan [26]. This cell utilizes organic dyes to absorb photons and then via electrochemical charge transfer processes separate electrons to a titanium dioxide conducting layer and positive charge within an electrolyte solution typically containing an iodine complex. Such structures, in combination with new absorbing molecules, have helped to improve the performance of DSCs as shown in Figure 3. Furthermore, the presence of a liquid electrolyte in DSCs has stimulated efforts to develop polymer-based solar cells [27].

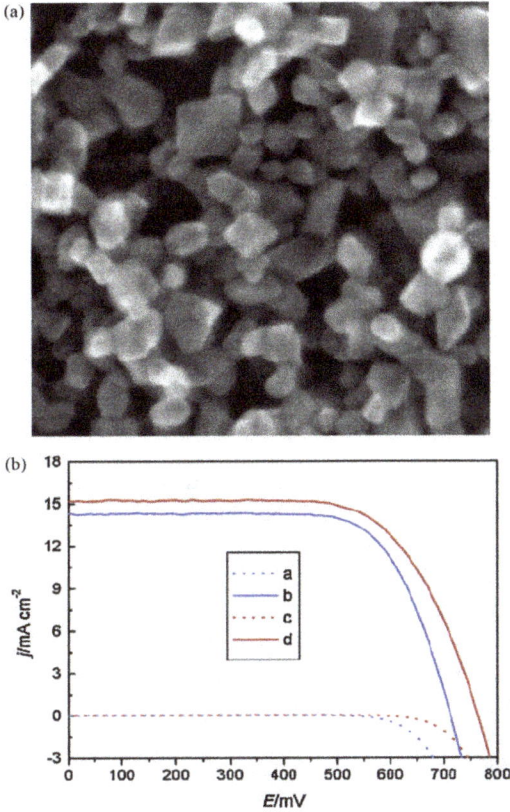

Figure 3: Grätzel (dye-sensitized) solar cell with nano titania electron conductor - (a) typical microstructure (image width is 210 nm) of a nanocrystalline titania layer, and (b) typical current density-voltage characteristics in dark and under AM1.5 conditions showing an efficiency as high as 8% depending on whether there is only a sensitizer molecule present (device A) or if an additional hexadecylmalonic acid absorber molecule is co-grafted to the nano-titania particles [28].

Biologically inspired processes have also been shown to produce inorganic nanostructured materials for hybrid solar cell applications [29]. Nanu et al. [30] spray coated a composite of $CuInS_2$ (CIS) and TiO_2 nanoparticles onto a graphite/nanocrystalline titania electrode structure and completed the structure with a dense titania film followed by a transparent conducting oxide (TCO) with power conversion efficiency of about 5%. Another form of bulk nanostructured solar cells is the use of nanocrystalline silicon (nc-Si) to replace amorphous Si or combine it with a-Si for tandem structures [31]. The major technical challenge with such nanocrystalline solar cells is the fact that the surface area of grain boundaries is very high, significantly increasing the density of recombination centres as well as the probability of recombination due to charge carriers having to traverse so many boundaries.

3.2 Quantum Wells

The key feature of quantum wells is confinement of charge carriers in 1D, creating a sheet of carriers with well-defined energy levels and high mobility owing to modification of the band structure. These structures also have potential for high absorption due to higher density of states at the band edges which lead to high short circuit current. In the past decade there has been a concerted effort to fabricate solar cells that incorporate (multi) quantum wells (MQW) in the active region of the device. These are primarily based on III-V materials, mainly GaAs and related alloys such as AlGaAs and InGaAs [32]. Quantum wells are typically fabricated using metal organic chemical vapor deposition (MOCVD) or molecular beam epitaxy (MBE). These deposition processes allow precise, atomic level control of the thickness of specific alloy layer/composition. For example, it is possible to directly grow alternating layers of InAlAs/InGaAs with Ångstrom level precision [Figure 4], such that the potential barrier between each layer creates quantum-confined states [33]. Owing to the relatively high cost nature of the described processes, it is expected that application of quantum well-based solar cells will be limited to space or concentrator PV systems. Nevertheless, such cells constitute a prototypical quantum confined nanostructured system, and as such are of great fundamental interest.

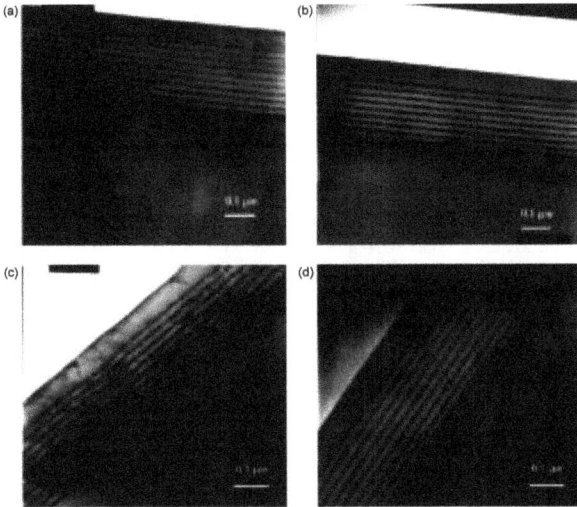

Figure 4: *Typical structure of InGaAsNSb/GaAs multi-quantum wells (MQW) as viewed by transmission electron microscopy - (a) InGaAs/GaAs, (b) InGaAsSb/GaAs, (c) InGaAsN/GaAs, and (d) InGaAsNSb/GaAs, MQWs. The data shows that the introduction of Sb suppresses 3D growth, leading to fewer defects in the MQW stack [32].*

Barnham and co-workers were among the first to study quantum well solar cells (QWSC) [34]. The reason for a potential increase in efficiency using quantum wells is that absorption in quantum wells is very high due to the carrier density obtained by quantum confinement in the plane of the well. The work by Barnham and co-workers has shown that while it is possible to obtain a short-circuit current density enhancement in such cells, there is often a corresponding reduction in the open-circuit voltage. This is associated with recombination of quantum confined charge carriers as they attempt to traverse the multiple wells, as well as defects (dislocations, etc.) that are typically present within the MQW structure.

3.3 Nanowires and Tubes

Nanowires (NWs) are elongated solid nanostructures and carbon nanotubes (CNTs) are perhaps the prototypical nanomaterial - essentially graphene sheets that have been rolled into the form of a tube. The application of nanowires (and nanorods, defined as elongated nanostructures with an aspect ratio of less than or equal to 5:1) to solar cells has been attempted in several device configurations and materials systems. The earliest demonstration was from Alivisatos and co-workers in 2002, in which CdSe nanorods were utilized as the electronic conducting layer of a hole conducting polymer-matrix solar cell [Figure 5]. These cells produced an efficiency of 1.7% for AM1.5 irradiation. The nanowires act as a direct path for transport of charge to the anode without the presence of grain boundaries, thus leading to an enhanced performance compared to solar cells employing nanostructures with aspect ratios approaching 1:1. In addition to the benefit of more efficient charge transport compared to other nanostructured materials configurations, nanowires also offer the potential for enhanced optical absorption characteristics.

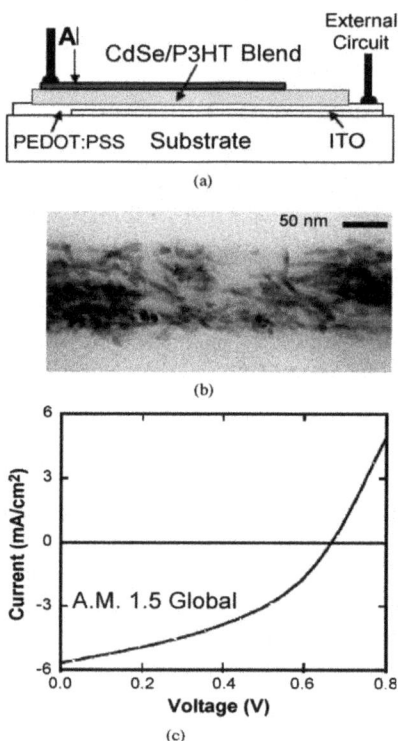

Figure 5: Hybrid organic-inorganic nanorod solar cells demonstrated by Alivisatos and co-workers showing (a) the structure of these devices, (b) the typical microstructure of such layers as viewed in a TEM, and (c) a typical I-V characteristic under AM1.5 illumination with an efficiency of the order of 1.7% [38].

Carbon nanotubes have also been shown to yield a photovoltaic effect. Lee et al. [35] showed that an individual carbon nanotube diode electrostatically doped in a split-gate field effect transistor configuration is an ideal p-n junction, with an ideality factor of 1. It was argued this is due to the fact that there are no surface states on a nanotube since the carbon bonds are well satisfied in the graphene structure of the CNT. Such a device was then shown to possess a small photovoltaic effect and the role of exciton transitions in the photoconductivity spectra of CNT devices, as well as the impact of bandgap renormalization and minority carrier diffusion (n and p), was further elucidated [36]. The problem of separating CNT by their electronic structure remains a challenge to further application of these nanostructures in PV application. Furthermore, Carbon nanotubes are also being explored as electrodes for solar cells [37]. Thus, nanowires and nanorods show great promise for future solar cell devices. The remaining technical challenges include proper surface passivation, shunting, and high quality contacts.

3.4 Quantum Dots and Nanoparticles

The so-called 0D structures have been implemented in various PV applications and enhancement schemes. It is possible to synthesize quantum dots (QD) and nanoparticles (NP) in many compositions, including semiconductors and metals, as well as coat them with dielectrics or additional semiconductors to create core-shell nanoparticles. One of the early efforts in the utilization of quantum dots was in the down-conversion of high-energy photons [39]. In a typical solar cell, high-energy photons are excited well beyond the conduction band edge, where they are mostly lost due to interaction with phonons (thermalization). Down-conversion is aimed at absorbing these high-energy photons and shifting them to lower energies that are matched to the particular absorber material in the solar cell. The opposite of down-conversion is the up-conversion of photons of energy below the bandgap that are typically not absorbed by the semiconductor [40]. The typical mechanism of up-conversion involves absorption of sub-bandgap light into an intermediate state, followed by further absorption of a second photon to the conduction band edge. The excited charge carrier then relaxes back to the valence band edge, emitting a single higher energy photon. Another novel band structure that can be obtained with quantum dots is the intermediate band (IB), which is indeed a form of up-conversion. This is possible because the states of closely spaced QDs can overlap to form an effective band structure (min-bands) that when finely tuned yields an intermediate band. The key challenge with such cells is the relatively low absorption crosssection of the QDs, which to date has limited performance and hindered demonstration of the full potential of intermediate bands in a practical manner.

Multi-exciton generation (MEG) based solar cells is another promising application of quantum dots. MEG, the generation of more than one electron-hole pair per photon, is a process that occurs in all semiconductors, though in the bulk the efficiency of the process is very low [41]. Several quantum dot-based solar cell designs have been proposed in the literature for harnessing the above mentioned high efficiency mechanisms [42]. One of the promising approaches is the structure in which the quantum dots are embedded in a p-i-n structure, similar to the approach of Luque and co-workers [43]. Others include quantum dots as a replacement for organic dyes in a Grätzel-type (dye-sensitized) cell, and in an exciton recycling configuration. In addition to MEG and IB band structures, Green and co-workers have proposed to use multi-layered Si quantum dot arrays to create an all Si-based multi-junction solar cell [44]. Multi-junction solar cells involve stacking of multiple bandgap cells, separated by tunnel junctions, to absorb a broader portion of the solar spectrum [45]. The larger bandgap cell is grown on the top of increasingly lower bandgap cells. This is typically implemented in III-arsenide and phosphide systems with 2-4 bandgaps.

Another mechanism associated with nanoparticles is that of plasmonics [46]. Surface plasmonpolaritons (SPPs) are collective oscillations of the Fermi Sea in metallic materials [47]. In metallic nanostructures/particles these SPPs are resonant at frequencies that are size- and shape-dependent [48-50]. Furthermore, the resonant SPPs created within the nanoparticles by absorption of light can produce electromagnetic fields with a very high local intensity, thus acting to provide a local concentrator effect. At slightly larger particle sizes (20-100 nm), the particles effectively scatter light. Yu and co-workers [51] showed that when Au NPs are placed on top of a standard Si p-n junction, the short-circuit current can be improved by a few percent, and peaks at wavelengths associated with SPP resonant frequency [Figure 6]. The effect is associated with the scattering from larger particles rather than from re-radiation of the electric field, as the length scale over which the field concentration occurs is on the order of 10's of nanometers. This is too short to be of relevance to standard Si p-n junction, in which the junction typically lies 0.2-1 mm below the surface.

Yu and co-workers further showed enhanced photo-current effects in a-Si-based thin film cells and that the concentration of nanoparticles plays a strong role in the observed photocurrent (and power conversion) enhancement [52]. The polarizability of the NPs further contributes strongly, and shaping the particles into ellipsoid or other shapes may further allow enhancement of certain optical modes for improved coupling into the underlying solar cell. The optimum size distribution for absorption of the solar spectrum by plasmonic nanoparticles has been calculated by Cole and Halas [50]. Silver nanoparticles have also been applied to silicon solar cells and showed an absorption enhancement of between 7 and 16 times in the wavelength range of 1050-1200 nm [53].

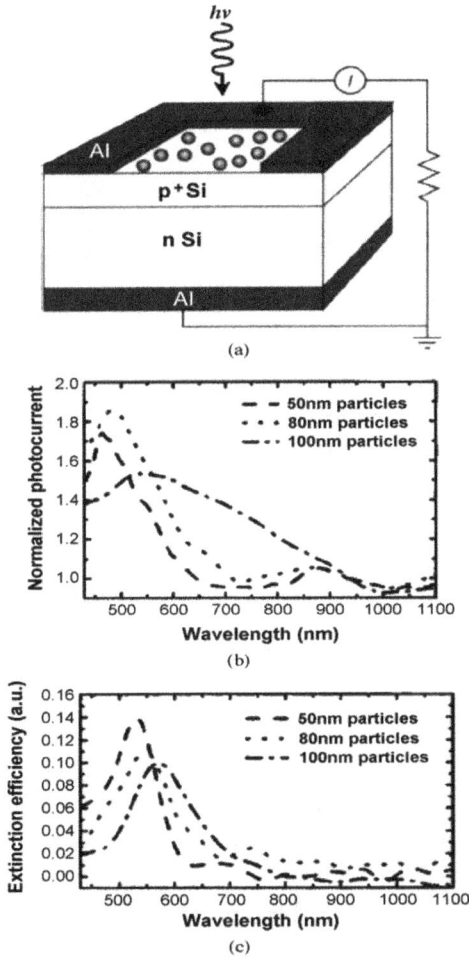

Figure 6: *Demonstration of the use of plasmonic nanoparticles to enhance the photocurrent in PV devices; (a) schematic of the structures used to demonstrate the concept, (b) normalized photocurrent as a function of wavelength and (c) extinction efficiency of nanoparticles of the same size used to show that the peaks directly correlate with plasmon resonance [51].*

Regarding the issue of charge transport in quantum dots, in recent years there has been some work towards fabrication and testing of quantum dots films for transistor application that shows promise for PV applications. Murray and Talapin fabricated PbSe QD films and measured their transistor characteristics [54]. They obtained electron mobilities of up to 0.9 cm^2/(Vs) by linking the QDs with molecules that allow for efficient electron transfer between dots. Similarly, Yu et al. also fabricated CdSe QD films and obtained mobilities on the order of 0.01 cm^2/(Vs) [55]. These values are on the order of typical amorphous silicon thin films and therefore show the potential for using QDs in advanced solar cell application. Efficient charge separation from quantum dots is also an area of fundamental interest. There has also been interesting work in energy transfer of excitons from small to large quantum dots in multi-layer quantum dot composite layers that may also be of interest to PV applications [56].

3.5 Nanocones

The hybrid solar cells make use of nanoscale texturing which has two advantages; it improves light absorption and reduces the amount of silicon material needed. Previous nanoscale texturing of solar cells has involved nanowires, nanodomes, and other structures. Recently, it has been found that a nanocone structure with an aspect ratio (height/diameter of a nanocone) of approximately one provides an optimal shape for light absorption enhancement because it enables both good anti-reflection (for short wavelengths of light) and light scattering (for long wavelengths). In nanoscale texturing, the space between structures has normally been too small to be filled with polymer so that a full second layer is required. The tapered nanocone structure allows the polymer to be coated in the open spaces eliminating the need for other materials. In the formation of the nanocone/polymer hybrid structure with a simple and low-temperature method as shown in Figure 7, processing costs can also be reduced.

After testing the solar cell and making some improvements, a device with an efficiency of 11.1% has been produced, which is the highest among hybrid silicon/organic solar cells to date. In addition, the short-circuit current density which indicates the largest current that the solar cell can generate is slightly lower than the world record for a monocrystalline silicon solar cell and very close to the theoretical limit. Due to the good performance and inexpensive processing of hybrid silicon nanocone-polymer solar cells, it has been predicted that they could be used as economically viable PV devices in the near future.

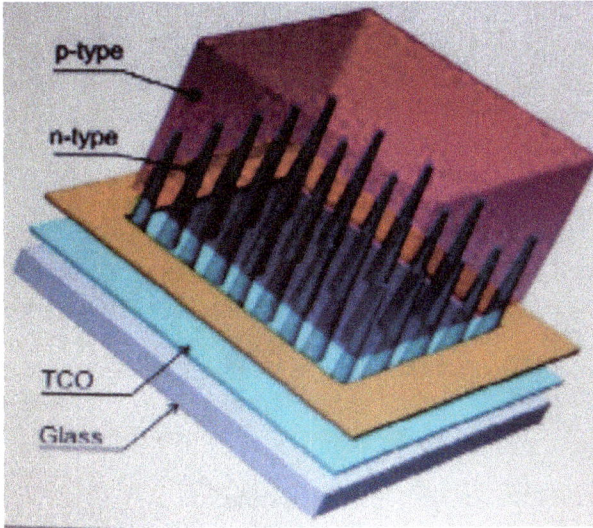

Figure 7: *Nanocones for solar cell applications.*

3.6 III-V Semiconductors

GaAs and InGaP are realized for high efficiency solar cells due to their high reliability and direct band gap and hence used as power sources for space satellites having an overall efficiency of 30%. III-V semiconductors based on GaAs grown on GaAs substrates finds application in photovoltaics because of their improved efficiency rate with respect to Si and enhanced physical properties [57]. In recent days, Silicon is almost replaced by solar devices based on III-V semiconductors due to their less weight, enhanced radiation resistance and better efficiency to be used in flat PV modules for space applications. GaAs employed in high efficiency solar cells is a III-V compound having the direct bandgap of 1.42 eV with high electron mobility and high electron saturation velocity operating at frequencies above 250 GHz. As compared to Si devices, high frequency GaAs make less noise perhaps work at high power due to high breakdown voltage than the equivalent Si device and its electronic properties are comparatively superior to Si [58].

Compared to single-junction solar cells, multijunction solar cells are more adequate due to the fact that there exists tuning of each junction to the wavelength of the light collected. The complex heterostructures with phosphides and arsenides multi-junction solar cells on germanium substrates have been realized for satellite power sources with improved efficiency of 20%. With the help of InGaP/GaAs/Ge triple junction device, efficiency rate of 30% was attained by the year 2000. GaInP/GaAs/Ge with high-efficiency currently finds application in terrestrial concentrators and also in space applications [59]. The three- and four-junction solar cells designed from appropriate bandgap materials with higher efficiencies are needed. In order to reduce the strain-induced defects that causes degradation in the performance of solar cells, III-V semiconductors having bandgap comparably lower than that of GaAs are preferred that could be lattice matched with GaAs. Materials such as GaNAs with anomalous large bandgap bowing [60] and GaInNAs [61-65] having preferable lattice matching to GaAs were used for high efficiency solar cell applications and were found to exhibit short minority carrier diffusion lengths [66-67]. As the solar cells need long diffusion lengths in the effective collection of the photogenerated carriers, short diffusion length of III-N-V semiconductors has to be taken into account for further improvement.

4. Common Nanostructured Solar Cell Systems

Conventional solar cell approaches suffer from two intrinsic major energy loss channels [Figure 8] - (i) their inability to absorb photons with energy less than the device absorption threshold and (ii) the waste of photon energy when photons with energies above the device absorption threshold are absorbed (cooling). These aspects set an upper efficiency limit for solar cell photo-conversion of approx 30% (*the Shockley-Queisser limit*).

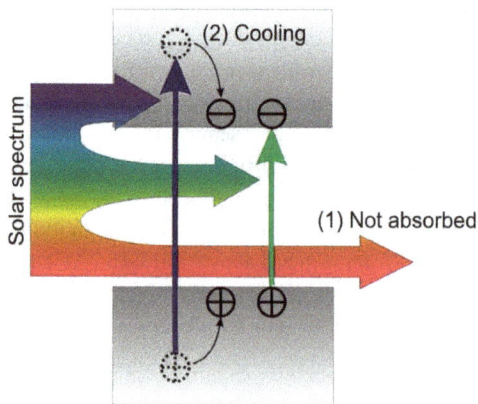

Figure 8: The typical solar spectrum for PV application.

The emergence of high quality nano-structures in recent years has led to development of so called 'third generation photovoltaic concepts' that aim efficiencies exceeding the Shockley-Queisser limit. The routes to surpass 30% efficiencies can be grouped into three generic categories, namely - (i) multiple energy threshold devices (e.g. multi-junction and multi-transition solar cells); (ii) the use of excess thermal energy to enhance voltages or carrier collection (e.g. hot carrier solar cells and carrier multiplication); and (iii) the modification of the incident spectrum (e.g. up-down conversion).

'There is plenty of room at the bottom' - a famous talk given by physicist Richard Feynman, who mentioned the concepts of nanotechnology for the first time about 50 years ago. Since then, people have made continuous efforts to attain structural manipulation of matters on an atomic and molecular scale in order to make use of their unique properties. The emergence of nanotechnology opens a lot of opportunities for new materials and devices with a vast range of applications. Recently, it has been extensively implemented in the development of novel solar cell structure in order to boost the energy conversion efficiency and reduce the production cost.

The nanostructure could refer to any objects in which at least one of the three dimensions has an intermediate size between molecular and micrometer size structure (usually between 0.1~100 nm). In this scale, some interesting physical phenomena that are not present in bulk system become pronounced and the material properties are artificially changed. Although quite a few different kinds of nanostructures have been developed so far, they can be classified into two major categories according to their function and role in the solar cells, those serving as supporting structures and those serving as light absorbers. In either case, different advantages of nanostructures, such as large surface area, good carrier transportation, high absorption coefficient, long-term stability, and tunable band-gap, have been used independently or jointly to enhance the solar cell performance. Among these features, the band-gap modulation by quantum confinement effect (QCE) is a very unique characteristic of nanostructures in which the size of dimension is comparable or smaller than the Bohr radius of the element. This size effect makes nanostructures a perfect candidate for the high efficiency tandem solar cells, because in principle it is possible to use nanostructures to build absorbers with any desired electronic band-gap that can match different parts of the sunlight spectrum. Furthermore, we should realize that besides electronic band structure the QCE is also able to change the phonon band structure of the material.

Nanostructures indeed have been used in the fabrication of conventional solar cells for a long time. However, in those circumstances, only very limited benefits of nanostructures have been used to improve cell performance. During the last decade, the development of the photovoltaic device theory and nanofabrication technology enables studies of more complex nanostructured solar cells with higher conversion efficiency and lower production cost. Outlined below are the fundamental principles and important features of some of these advanced solar cell designs.

An important advantage for nanostructured solar cells is that they can be used to incorporate new physical mechanisms which allow an efficiency greater than that of a one-junction solar cell. While the ultimate thermodynamic efficiency limit is the same for nanostructured solar cells as for a tandem solar cell, nanostructured approaches have several advantages. The most important of these for increasing solar cell efficiency is the possible increase in efficiency for a given number of materials. For example, the quantum dot intermediate band solar cell (IBSC), two materials give a similar efficiency to a three-junction tandem, while hot-carrier approaches allow the use of one absorber material to yield efficiencies over 50% under concentration.

Numerous suggested approaches for increasing the efficiency above the Shockley-Queisser limit have been proposed. Of these approaches, the ones receiving the majority of the attention and focus for experimental implementation are thermophotonics, multiple absorption path solar cells, and multiple energy level solar cells. For these, theoretical calculations which include realistic physical limits show that they have efficiencies similar to a three junction tandem at a given concentration level. In order to reach ultra-high efficiency with these approaches, the most straight forward approach is to use two devices, each receiving half of the solar spectrum. With two solar cells, an efficiency similar to a six junction tandem can be reached. Multiple absorption path (via exciton generation or Auer absorption) or multiple energy level solar cells (via quantum dots or quantum wells) can readily be configured such that two such devices are connected optically in series similar to the way a two junction tandem is connected. Thermophotonic approaches would be more difficult to use in such a configuration, but the goal of thermophotonics relates to increasing the efficiency of existing solar cells rather than reaching the thermodynamic limits. Consequently, the goal of reaching ultra-high efficiency (capable of approaching the thermodynamic limits and efficiency > 50%) in a midterm time frame is most directly addressed by multiple absorption path (MAP) and multiple energy level (MEL) approaches.

4.1 Nanocrystalline Silicon Solar Cell

Hydrogenated amorphous silicon (a-Si:H) has been widely used as the absorbing material in thin film solar cells. In order to boost the conversion efficiency, multijunction structures have been studied and a record stable active area efficiency of 13% has been demonstrated [68]. Further improvement of cell performance requires the discovery of new materials with a low defect density and good transport properties. Hydrogenated nanocrystalline silicon (nc-Si:H) has also been extensively studied as a potential candidate for the low band gap absorbers in the multijunction cells [69].

In fact, nc-Si:H material is a mixture of nanometer-sized crystallites and amorphous tissues [70]. Small crystallites with size of a few nanometers are embedded in amorphous silicon matrix. In nanocrystalline thin film, the presence of small crystallites provides crystalline paths for carrier transport, where the carrier mobility is much higher than pure amorphous system. Hence, the nc-Si:H absorber layer can be thick producing high photocurrent density. Another advantage of nc-Si:H over a-Si:H or hydrogenated amorphous silicon germanium (a-SiGe:H) is the reduced light-induced degradation in the solar cell efficiency. It is shown that the performance improvement in nc-Si:H solar cells is attributed to the presence of small crystallite network and the crystallites are actually not absorbers and do not contribute to the light absorption process. Therefore, it is not that essential to optimize the optical and electrical properties of the crystallites.

4.2   Nanostructured Metal Oxides in Dye-Sensitized Solar Cell

Compared to macroscopic systems, nanostructures have much larger surface area which can alter mechanical, thermal, and catalytic properties of materials. This feature has been widely investigated in order to increase the absorption of dye-sensitized solar cells (DSCs) and hence improve their conversion efficiency. In DSCs, energy conversion is realized by charge separation in sensitizer dyes adsorbed on a wide band gap semiconductor electrode, such as transparent TiO_2 [71]. The energy conversion efficiency of early DSCs was not high though the sensitizer itself exhibited a very efficient charge separation process. This was because of the very limited light absorption by the monolayer of the dyes, resulting in low absorption coefficient of the solar cells. This problem was solved later by the utilization of nanoporous TiO_2 electrode instead of relatively flat polycrystalline electrode as shown in Figure 9 [72]. The huge surface area of nanoporous TiO_2 significantly increases the effective light absorption area of the dyes and thereby improves the cell conversion efficiency to higher than 10% [73].

Unlike in nc-Si:H film, the particle size of TiO_2 nanocrystals plays an important role in the design of DSCs. This is mainly correlated with the electron transport process in the nanoporous TiO_2 electrode, which is evaluated by the diffusion length of electrons $L = (D * \tau)^{1/2}$, where D is the electron diffusion coefficient and τ is the electron lifetime. Experimental results have shown that D increases with nanocrystal size, and this is probably due to the lower number of boundaries between particles. Nevertheless, the increase of particle size also results in a decrease of total cross-sectional area of the boundaries (required for reasonable absorption) which increases the resistance for electron transport. In addition, it is found that τ is inversely proportional to D. These competing effects indicate that there should be an optimal particle size that gives maximum diffusion length and best cell performance. In high efficient DSCs, the typical diameter of the TiO_2 nanoparticles is between 10 nm and 20 nm. Other nanostructured metal oxides have also been investigated as alternative n-type electrodes including ZnO, Nb_2O_5, and SnO_2. Particularly, ZnO has attracted much attention since the 1960s, because of its good carrier mobility and flexibility in synthesis and morphology (spherical particles, rods, wires, and hollow tubes) [74]. However, DSCs using these electrodes have not shown conversion efficiency comparable to TiO_2-based DSCs and a lot of optimization is required to be done.

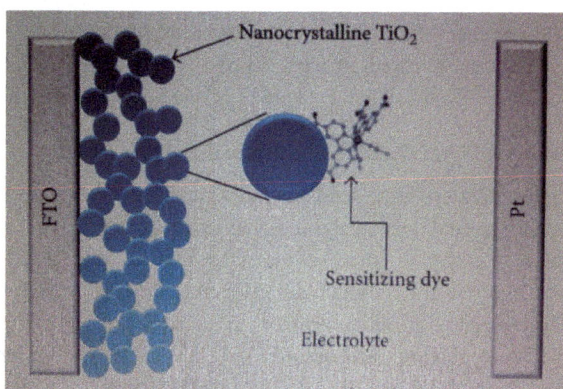

Figure 9: Schematic diagram of the dye-sensitized solar cell with nanoporous TiO_2 electrodes.

4.3 Organic Solar Cells

Besides dye-sensitized solar cells, which may be considered as organic/inorganic hybrid cells, other types of organic solar cells have currently become of broader interest. These cells can be divided roughly into molecular and polymer organic solar cells or into flat-layer systems and bulk heterojunctions. Extremely high optical absorption coefficients are possible with these materials, which offer the possibility for the production of very thin solar cells (far below 1 mm) and therefore only very small amounts of needed materials. Considering the fact that light-emitting films of plastic materials have been realized there is a chance to achieve photovoltaic conversion also in such materials. Only modest solar conversion efficiencies of up to 1% were reached until 1999 but efficiencies then increased rapidly. With molecular flat-layer systems based on molecular organic single crystals made of iodine- or bromine-doped pentacene, efficiencies of up to 3.3% under AM1.5 illumination have been reported at Lucent Technologies. Nearly the same value was reported with improved bulk heterojunctions (interpenetrating network) of conjugated polymers and fullerene derivatives [75]. Before these cells become practical, which at the moment looks still far away, the efficiency will have to be increased further. Also, long-term stability and protection against environmental influences are significant challenges.

4.4 Organic/Inorganic Hybrid Solar Cell

As a low cost alternative to silicon solar cells, hybrid solar cells consisting of organic thin films and inorganic nanostructures are being actively studied due to their promising performance. In this type of cell, inexpensive and easily processable organic materials are used and inorganic semiconductor nanostructures are embedded in the film to improve the absorption capability and tailor the absorption spectra through QCE [76]. One of these hybrid approaches is the nanoparticle sensitized solar cell, which is conceptually similar to conventional DSCs but replaces the organic dyes with inorganic nanoparticles [77]. The band gap of incorporated nanoparticles can be properly tuned by controlling the particle size, and thus desired absorption range and characteristics are able to be obtained. Nanocrystals as absorption material also have the advantage of stability due to the large extinction coefficients and intrinsic dipole moments [78]. Another strategy for hybrid solar cells is to use inorganic nanocrystals/conductive polymer blends which combine the flexibility of polymers with the stability and high mobility of inorganic nanoparticles. This approach is based on the concept of bulk heterojunction that initially originated from organic solar cells but replaces the organic/organic heterojunction with inorganic/organic heterojunction. Various inorganic semiconductor nanocrystals including CdSe, $CuInS_2$, CdS, or PbS have been implemented in such a structure [79-80]. This strategy indeed provides several promising potential advantages - (i) inorganic materials have high absorption coefficients, photoconductivity, and stability; (ii) the doping characteristics of nanostructures can be easily tuned; (iii) the size dependent QCE can change the electronic and optical properties of nanostructures, so that hybrid solar cells containing nanostructures with different sizes can be used for tandem solar cells. The challenge in hybrid solar cells using inorganic nanostructures mainly lies on the accessibility and reproducibility of the nanostructure synthesis routes.

Recently, the emergence of hybrid organic-inorganic perovskite solar cell attracts much interest due to its significant performance advantage over conventional hybrid solar cells [81]. This kind of solar cell utilizes the metal halide based material most commonly

$CH_3NH_3PbI_3$ nanocrystal or a closed variant (such as Cl, Br, I) as the light absorbers, and the perovskite pigment is a mixture of CH_3NH_3X (X = Cl, Br, I) and PbX_2 prepared by sequential vapour deposition or solution process [82]. A structural diagram of the hybrid perovskite solar cell is schematically shown in Figure 10. Since the first report in 2009, the conversion efficiency of hybrid perovskite solar cell has been boosted at an unanticipated rate by exploring novel nanomaterials and device architectures [83-84]. Particularly, the progress in the development of solid state transport layer with high carrier mobility has substantially improved the device performance, and a breaking efficiency of 15.7% has been achieved by employing ZnO nanoparticles as the electron-transport layer of the solar cell [85].

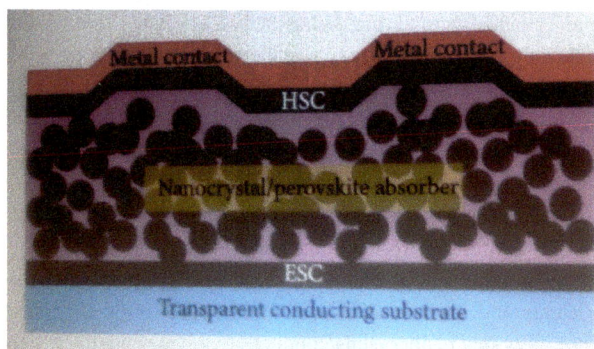

Figure 10: *Schematic diagram of a hybrid perovskite solar cell. The 'HSC' and 'ESC' represent the hole selective contact and electron selective contact, respectively.*

Although efficiencies up to 20% are predicted to be realistically achievable [86], there are still several issues which need to be resolved before hybrid perovskite solar cells can be commercialized. The first one is the development of large-area fabrication method with good uniformity and reproducibility, both of which are important for mass production. Nevertheless, almost all the best-performing cells reported to date only have areas smaller than 0.1 cm^2 [87]. Another more challenging issue involves the fast degradation of the solar cell performance. The structure and function of organic components are easily influenced or even destroyed once the ambient gas or moisture penetrates the packaging materials and reaches the solar cell. This problem in fact has held back the commercialization process of most organic solar cells.

4.5   The Third Generation Solar Cells

Semiconductor nanocrystals are regarded as useful called third-generation solar cells. This utility is due to the fact that their optical band can be tuned by both material selection and quantum confinement and because advances in synthesis allow control over nanocrystal size and shape to optimize performance. Solar cells may be formed using a p-n junction, a Schottky barrier, or a metal insulator semiconductor structure based on various semiconductor materials, such as crystalline silicon, amorphous silicon, germanium, III-V compounds, quantum wells and quantum dots structure. III-V compound semiconductor such as gallium arsenide and indium phosphide has near optimum direct energy band gaps, high optical absorption coefficients and good values of minority carrier lifetimes and mobilities making them better materials than silicon for making high efficiency solar cells. Despite the low optical absorption coefficient resulting from the indirect bandgap of silicon, the mature of crystal growth and fabrication process of silicon semiconductors ensures well control on minimizing defect density, thus minority carriers generated by photons can diffuse into depletion region without excessive losses due to non-radiative recombination. Solar cells can also be made of III-V compound semiconductors, e.g., gallium arsenide (GaAs) and indium phosphide (InP). The materials have high optical absorption coefficient due to their direct bandgaps and near optimal bandgap ~1.4 eV for solar energy conversion. The third generation solar cells, novel solar concepts are proposed to further increase the power conversion efficiency using the low-dimensional structures including hot carriers cell, tandem cell, multiple quantum wells (MQW) cell and intermediate band solar cell. III-V quantum dot superlattice based solar cells are proposed because of their promising potentials in high power conversion efficiency application. The intermediate band solar cell (IBSC) pursues the enhancement of efficiency through the absorption of below bandgap energy photons and production of additional corresponding photocurrent without degrading its output voltage.

4.6   Polycrystalline Thin-film Solar Cells

Polycrystalline thin-film solar cells such as $CuInSe_2$ (CIS), Cu (In, Ga) Se_2 (CIGS), and CdTe compound semiconductors are important for terrestrial solar applications because of their high efficiency, long term stable performance and potential for low-cost production. Because of the high absorption coefficient ($\sim10^5$ cm^{-1}), a thin layer of ~ 2 mm is sufficient to absorb the useful part of the spectrum. Highest record efficiencies of 19.2% for CIGS [88] and 16.5% for CdTe [89] have been achieved. Many groups across the world have developed CIGS solar cells with efficiencies in the range of 15-19%, depending on different growth procedures. Glass is the most commonly used substrate, but recently some effort has been made to develop flexible solar cells on polyimide and metal foils. Highest efficiencies of 12.8% and 17.6% have been reported for CIGS cells on polyimide [90] and metal foil [91] respectively. Similarly, CdTe solar cells having the efficiency in the range of 10-16%, depending on the deposition process, have been developed on glass substrates, while flexible cells with efficiency of 7.8% on metal [92] and 11% on polyimide have been achieved. Currently, these polycrystalline compound semiconductors solar cells are attracting considerable interest for space applications because proton and electron irradiation tests of CIGS and CdTe solar cells have proven that their stability against particle irradiation is superior to Si over III-V solar cells [93]. Moreover, lightweight and flexible solar cells can yield a high specific power (W/kg) and open numerous possibilities for a variety of applications. The super state configuration facilitates low-cost encapsulation of solar modules. This configuration is also important for the development of high-efficiency tandem solar cells effectively utilizing the complete solar spectrum for photovoltaic power conversion [94]. The emphasis is placed on various aspects of solar cell development and most of the efficiencies reported are related to small-area cells (≤ 1 cm^2).

4.7 Quantum Well Solar Cell

The quantum well (QW), a 1D confinement nanostructure, has been applied widely in semiconductor devices since its emergence, including photodetectors, light emitting diodes, lasers, optical modulators, and high mobility transistors. The optical and electrical properties of QWs are engineered to meet the particular requirements of different applications. It is thus reasonable to speculate that there may be some advantages of using QWs in solar cells. The ability to modulate the band gap of QW structures through QCE provides a means of achieving high energy conversion efficiency due to the better band gap match to the solar spectrum. The multiple quantum well (MQW) structure was proposed in the early 1990s, making a solid step in the utility of QW solar cells (QWSCs) [95]. The band diagram of a MQW solar cell is shown schematically in Figure 11.

The QWs are located in the intrinsic region of a p-i-n structure, and carrier photogeneration and recombination occur in both the barrier and QW layers. For the carriers generated in QWs to be effectively collected at the external electrodes and thereby contribute to the photocurrent, the carriers must be able to escape from the QWs. Besides quantum tunnelling process, thermal escape is also crucial for achieving efficient carrier escape from QWs, which indicates a requirement of the thermal energy and transverse electric field in the operation of QWSCs [96-98]. Fortunately, the thermal energy at room temperature is usually sufficient and the probability for carrier escape can be close to unity in the presence of a strong electrical field.

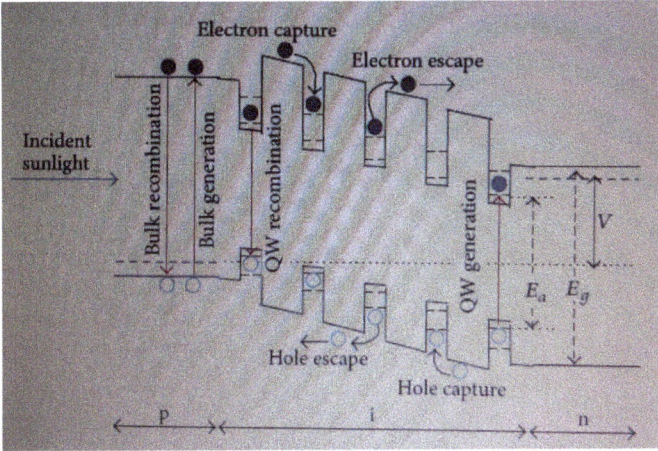

Figure 11: *Band diagram for the QWSC with the MQW structure in the intrinsic region. The photogeneration and recombination processes are shown for both bulk and QW regions. The carrier capture and escape routes are also shown in the figure.*

High quality growth techniques with monolayer accuracy, such as molecular beam epitaxy (MBE) and metal organic vapour phase epitaxy (MOVPE), have been used to fabricate QWSCs. According to lattice constant difference between barrier and well layers, the material systems used in QWSCs can be categorized into lattice matched systems, such as AlGaAs/GaAs, InGaP/GaAs, InP/InGaAs, and InGaAsP/InP, and lattice mismatched systems including GaAs/InGaAs, InP/InAsP, GaAsP/InGaAs, GaAs/InGaAs, and InGaAs/InGaAsP. The initial studies of QWSCs were mainly based on the lattice matched material systems using GaAs or InP as the well material. Therefore, the absorption threshold is adjusted to higher energies over a conventional, homojunction cell formed from the GaAs or InP. Such material systems show good solar cell performance and it is a good way to avoid the complications caused by the strain so that it is easier to reveal the effects from QWs and to understand the underlying operation mechanisms. For most solar cell applications, however, lower absorption threshold is indeed desirable. A good example is to enhance the long wavelength absorption of high efficiency GaAs cells. A major problem here is that there are no lattice-matched, lower band-gap systems in nature to act as the wells. For example, InGaAs is a particular good choice as a well material for devices using GaAs as barriers, but it has larger lattice constant so that compressive strain is required to grow the InGaAs QWs. The strain builds up in each well layer and finally results in numerous threading dislocations through strain relaxation at the top and bottom of the MQW stack, causing larger leakage current, worse short wavelength quantum efficiency, and loss in open circuit voltage [99].

To solve this problem, strain-balanced QWSCs have been proposed to minimize the build-up of strain. The idea is to compensate the strain from QWs by using barrier layers with opposite strain [100]. A typical example of such cell is the strain-balanced GaAsP/InGaAs QWSCs as shown in Figure 12. Over each period, the compressive strain from the InGaAs well layer with larger lattice constant is compensated by the tensile strain from the GaAsP barrier layer with smaller lattice constant. The average lattice constant equals that of GaAs and the average strain in the cell should approach zero, leading to significant reduction of dislocations in this structure.

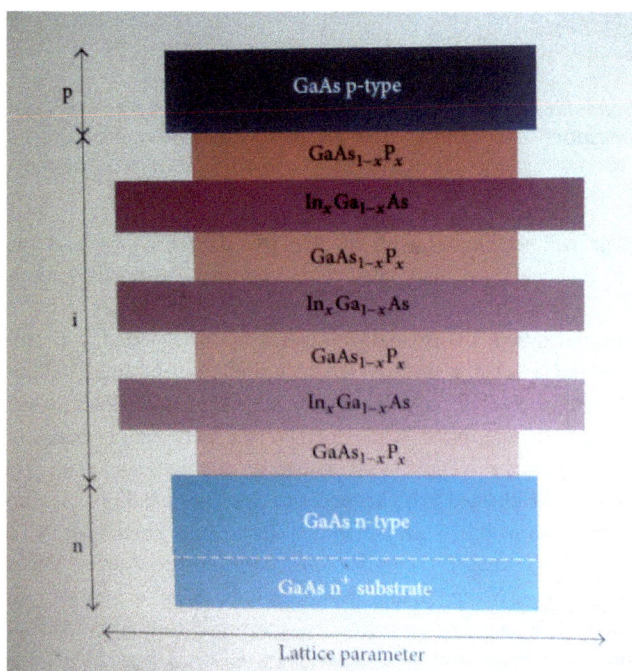

Figure 12: *Schematic diagram of a strain-balanced GaAsP/InGaAs 3-period MQW solar cell.*

As of now, QWSCs have shown advantages in engineering the absorption spectrum and improving the photocurrent. However, there are still a few challenges remaining. The most important one is the increase of absorption of the MQW stack. This may be solved by growing more QWs, but this strongly depends on the availability of high quality and yield growth technology. Another widely studied approach is the use of light trapping schemes [101].

4.8 Nanowire Solar Cell

In general, a nanowire is defined as a wire-like nanostructure with a diameter of the order of several nanometers. Due to the lack of grain boundaries along its length, the use of nanowire as a direct conduction path for charge carriers offers various merits in device performance, such as large charge transport rate and small carrier recombination probability. In addition, the nanowire structure also exhibits superior light trapping characteristics, resulting in much less surface reflection compared with conventional planar structure. These advantages indicate that it is possible to achieve high conversion efficiency in nanowire solar cells, and, more importantly, the employment of nanowire structure reduces the consumption of material and releases the requirement of material quality, both of which give rise to a substantial cost reduction. So far, there are no rigorous classification and criterion on nanowire solar cells. Nevertheless, they can be roughly divided into two categories, the nanowire dye sensitized solar cell and the nanowire solar cell based on p-n junction structure.

The earliest report of nanowire dye sensitized solar cell was from Huynh et al. in 2002, in which CdSe nanowires were utilized as the hole conducting layer and an efficiency of 1.7% was achieved under AM1.5 irradiation [102]. Later Law et al. [103] demonstrated a similar structure using ZnO nanowires instead of CdSe nanowires. This kind of solar cell exhibited a maximum power conversion efficiency of 1.5% under AM1.5 condition. Compared with nanowire dye sensitized solar cells, the nanowire solar cells based on p-n junction structure attract more interest and are being extensively studied. Generally, this kind of nanowire solar cell can be divided into three types according to the p-n junction location as shown in Figure 13.

Direct deposition of nanowire array as an antireflection layer on the surface of traditional cells is the simplest application of nanowires to solar cells. Theoretical analysis suggests that semiconductor nanowires have a very low luminous reflectance, especially in the long wavelength range. For instance, the silicon nanowire has been demonstrated to have much lower reflectance in the infrared range than that of single crystalline and multicrystalline silicon materials [104]. This inherent optical characteristic can be attributed to the multi-scattering of light in the nanowire structure and thereby it is advised that the effect of incident light trapping can be optimized by adjusting the diameter and refractive index of nanowires. So far, the conversion efficiency of nanowire array solar cells is still relatively low compared to conventional silicon solar cells [105-106]. The efficiency loss can be explained by the presence of a large number of surface states which results from the high density of nanowires. These surface states are difficult to be fully passivated, so that the surface recombination velocity is significantly increased. This finding means that a trade-off between the nanowire density and surface passivation effect needs to be resolved to maximize the performance of nanowire array solar cells.

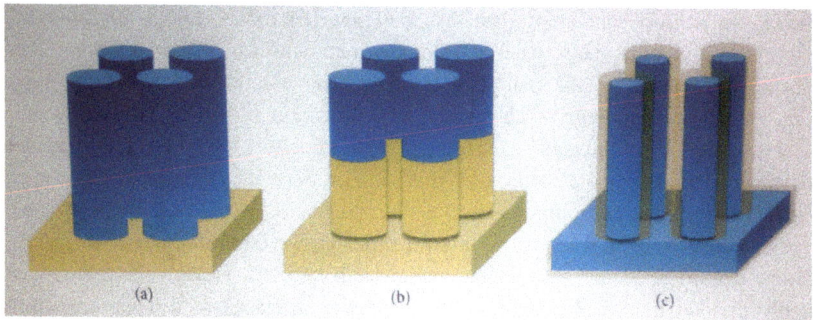

Figure 13: Schematic diagrams of nanowire solar cells based on (a) substrate junction structure, (b) axial junction structure, and (c) radial junction structure.

Besides using nanowire array as antireflection layer, novel device structures with pn junction located on the nanowire surface or inside the nanowire have also attracted increasing interests. Researchers from Harvard University have developed an axial junction nanowire solar cell with a conversion efficiency of 3.4% [107]. Later, Garnett and Yang proposed a novel orthogonal network of photon absorption path and charge carrier transport path in a radial junction nanowire solar cell, in which the long nanowire axial enhances photon absorption and the radial junction improves the charge carrier collection. Theoretical calculation predicts that a conversion efficiency of up to 11% can be obtained in a radial junction nanowire solar cell with an electron diffusion length of 100 nm, which is much higher than that of nanowire solar cells with planar structure [108].

4.9   Quantum Dot Solar Cells

Semiconductor quantum dot (QD) nanocrystals can be exploited as light absorbing entities in nanostructured solar cells. The ability to systematically change the QD absorption threshold with nano-crystal size [Figure 14a] allows fine tuning of its optoelectronic properties. Quantum dot sensitized solar cells (QD-SSCs) represents one of the potential routes for developing thin film-low cost photovoltaic devices (2nd generation photovoltaics). In these solar cells, sunlight is absorbed by a semiconductor nanocrystal sensitizer (QD donor) which is anchored to the surface of a nanostructured mesoporous oxide [Figure 14b].

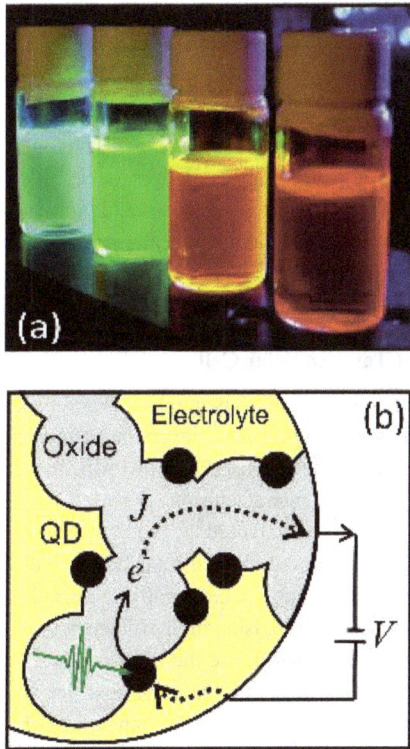

Figure 14: The operation of a QD solar cell: (a) The QD absorption threshold with nano-crystal size, and (b) The QD donor and oxide acceptor.

The photocurrent (J) in QD-SSCs is primarily determined by the efficient transfer of photogenerated electrons from the QD to the oxide. The nanostructured nature of the oxide, which is defined by a high surface to volume ratio, allows high photocurrents in solar cell devices. The output voltage (V) of the solar cell will be ultimately determined by the absorption threshold of the employed QD sensitizer, which can be tuned to match the energetics of electron and hole acceptor contacts (oxide's conduction band and electrolyte's redox potential) for maximum solar cell efficiency. Apart from exciton generation-dissociation, other kinetic processes such as carrier transport and recombination will determine the performance of devices. In nanostructured systems, where attaching contacts to the sample is tricky, it is convenient to interrogate those processes by optical means. A unique tool to do so is THz time domain spectroscopy (THz-TDS), where freely propagating THz pulses are employed as a probe for determining the complex conductivity of the samples. Furthermore, the high time resolution of THz pulses (with durations of ~1 picosecond) allows the study of dynamic processes out of equilibrium using a optical pump-THz probe scheme. Other relevant topics that influence solar cell performance in approaches like the QD-SSCs are long term stability issues (e.g. sensitizers prone to photo-oxidation), surface defect engineering (e.g. the passivation of surface physico-chemical impurities acting as recombination centers), assembling of nanocrystals (e.g. homogenous sensitization of oxides) and device engineering (e.g. encapsulation issues and sample to sample reliability).

4.10 All Quantum Dot Tandem Solar Cell

Among various high efficiency approaches, tandem structure is the only approach that has shown the capability to achieve high conversion efficiencies (44% at ~500 suns) beyond the Shockley-Queisser limitation [109]. However, these cells require the use of expensive III-V materials and manufacturing process, which are only viable for concentrating systems. Therefore, it would be very attractive if alternative low cost materials and process can be adopted in such a tandem cell design. In recent years, all quantum dot tandem solar cells which use band-gap engineered nanocrystals as absorbing material have been proposed as a promising high efficiency approach [110-112]. Most of these tandem solar cells are designed to use the most common elements such as Si, Ge, PbSe, and PbS. This not only reduces the material cost but also secures the long term availability of raw materials. Besides, this kind of tandem solar cell has a good potential for large scale manufacturing, which benefits from the integrated circuit industries. The all quantum dot tandem solar cell with three sub-cells is schematically shown in Figure 15. In the tandem stack, the top and intermediate sub-cells are composed of nanocrystals with engineered band gaps and a conventional single junction solar cell is used as the bottom cell. The effective band gaps of the two upper cells can be tuned by the size of nanocrystals and their optimized values depend on the selection of the bottom cell.

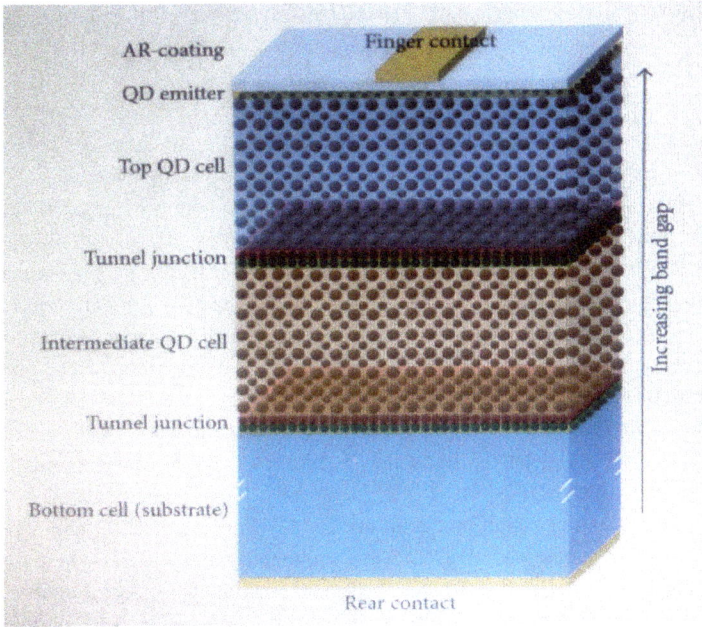

Figure 15: *Schematic diagram of the all quantum dot tandem solar cell with three sub-cells.*

There is a series of steps involved in the realization of a working all quantum dot tandem solar cell, including the fabrication of band gap engineered nanocrystals, formation of a rectifying p-n junction, and the interconnection of cells through a tunneling layer. Of these challenges, the ability to adjust the material's band gap through quantum confinement effect is the basis of the concept. Despite intensive studies, the reliable production of nanocrystals using low cost fabrication techniques is still an area of difficulty. This is because in order to accurately tune the band gap, certain aspects of nanocrystal materials need to be precisely controlled, such as crystallite size and distribution, barrier thickness, and defect passivation. Recently, two interesting techniques have been reported to provide relatively controllable growth of nanocrystals with size-tunable energy gaps [113-115]. The superlattice structure fabricated using tradition thin film techniques such as sputtering allows for an easy integration with current silicon process technology, while the solution-based process which produces colloidal semiconductor nanocrystals shows great advantages for the fabrication of low-cost structures.

4.11 Intermediate Band Solar Cell

The approach of intermediate band solar cells (IBSCs) is to introduce a continuous electronic band within the band gap of another semiconductor material as shown in Figure 16 [116]. In the operation of IBSCs, subband gap photons of specific energy first pump electrons from valence band (VB) to IB (Process (1) in Figure 16). Before occurrence of nonradiative recombination, the electrons in IB are further emitted to CB through absorption of another subband gap photons (Process (2) in Figure 16). The wave function of the electrons in IB should be delocalized, which is similar as the property of the electron wave function in CB and VB, so that radiative absorption and recombination dominate at these energy levels and the photons do not necessarily have to be absorbed by the same electron in the two steps. In order to provide both empty states for capture of electrons from VB and occupied states for injection of electrons to CB, the IB has to be partially filled with electrons (i.e., 'metallic'). Ideally, half-filled IB is needed to ensure that both capture and emission processes are equally likely. The absorption of subband gap photons enhances the photocurrent while the output voltage is preserved. This leads to a limiting efficiency as high as 63.2% (under maximum sun concentration 46,050 suns) for IBSCs. This efficiency is similar as that for a three-cell tandem solar cell and is much higher than the 40.7% conversion efficiency for an ideal single band gap solar cell that is made from the same material.

As an alternative approach, quantum dots (QDs) were also proposed to form IBSCs [117]. In this approach, the energy levels comprising the IB arise from the confined electron states in the QDs. The use of QDs has a few advantages over other nanostructures such as quantum wells or wires. This is because of the following - (i) Only QDs provide a true zero density of states between the IB and CB. This prevents nonradiative recombination between the CB and IB [118-119]. (ii) The optical transition from IB to CB is strictly forbidden by selection rules in quantum wells for light incidence perpendicular to the plane of growth. This is definitely detrimental for the solar cell applications [120]. Although several material systems have been proposed for the implementation of QD IBSCs, the practical experimental work still mainly focuses on the InAs/GaAs QD superlattice structure grown by molecular beam epitaxy using the Stranski-Krastanov growth method [121].

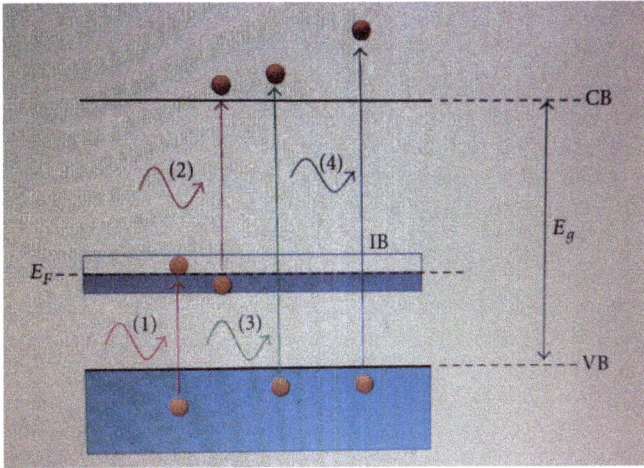

Figure 16: *Band structure of an intermediate band material in equilibrium. EF is the Fermi level of the intermediate band material. The photon absorption processes involving photons with different energies (marked as 1, 2, 3, and 4) are also shown.*

4.12 Multiple Exciton Solar Cell

The most advanced experimental approach to multiple absorption path (MAP) solar cells is multiple exciton generation (MEG), similar to Auger or impact ionization. Ideally, such a device would have a threshold voltage of $n{\times}EG$ and a quantum efficiency of $n{\times}100\%$, where n is the number of electron hole pairs generated per photon. The schematic diagram of a multiple exciton solar cell is shown in Figure 17.

Experimental measurements in colloidal PbS and PbSe quantum dots make exciton generation approaches the only of the new concepts to demonstrate the physical mechanisms on a scale required for high efficiency [122-123]. Despite the experimental results, MAP devices have several challenges. One issue in MAP solar cells, in common with all new concepts approaches, is that the fundamental mechanisms are incompletely understood. For example, demonstrating Auger absorption in materials which have optimum band gaps, and improved understanding of which materials and configurations may demonstrate high Auger absorption are both areas which require additional analysis and experiments. The second issue relates to transport of carriers. Since the metastable energy for the carriers is at the lower energy levels for the quantum dot, either carriers must be excited from this lower energy to the continuum of states (similar to a multiple

97

energy level device), or the band structure must be designed to allow transport from this lower energy level, either via the formation of mini-bands or by designing the band structure such that the lowest energy level is within the thermal energy of the continuum of states. Final issue relates to how to incorporate MAP devices into a shorter-term path in realization. Multiple energy level (MEL) devices presently are most suited towards high energy portion of the spectrum, since processes involving 3 to 4 electrons require band gaps of approximately 1.0 eV or higher. Such a high energy converter fit well with the optimum region of the solar spectrum and moreover the colloidal approaches are also a low cost technology. However, it meshes poorly with existing high efficiency designs, since the portion of the spectrum which it would convert already has well understood components which are relatively close to their optimum. Consequently, MEG devices are most suited to stand-alone low, cost high efficiency concepts.

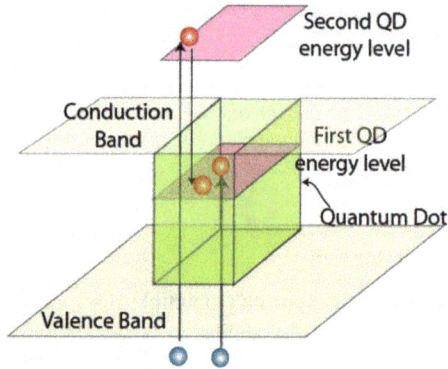

Figure 17: Multiple exciton solar cell.

4.13   Virtual Band Gap Solar Cells

Multiple energy level (MEL) solar cells, in which multiple energy levels or bands are simultaneously radiatively coupled via both generation and recombination, are another approach to exceeding the Shockley-Queisser limit. The multiple quasi-Fermi levels can arise from a band or through localized energy levels, as shown in Figure 18. Further, the band or energy levels may be introduced either by designing a material or defect level which contains multiple bands or by using nanostructures, such as quantum dots, wires or wells. Although materials with inherent multiple band structures offer an elegant and conceptually simper approach, the numerous other requirements for a photovoltaic material, such as the ability to dope it, good mobilities, requirements of the density of states, make such an approach viable primarily for longer term. Similarly, while impurity photovoltaic (IPV) devices could potentially be used if a defect with optimum properties is found, the ability to design or find such a material and defect reduce the utility of IPV solar cells.

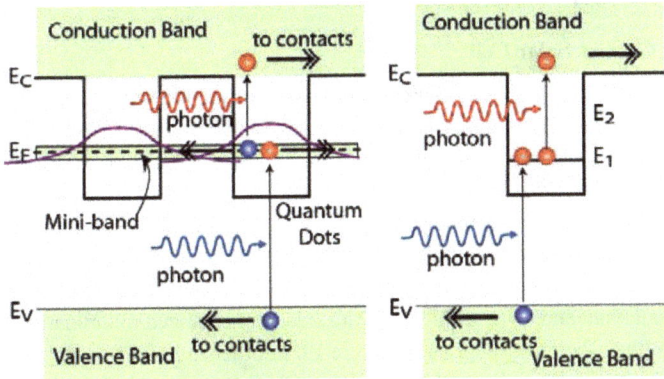

Figure 18: *(a) Mini-band approach and (b) localized approach for virtual band gap solar cells.*

Approaches which utilize nanostructures to implement the energy levels have an 'effective' band gap introduced by the difference between the conduction (or valence) band offsets of other materials. These approaches, here called virtual band gap solar cells, have to-date demonstrated the highest efficiency nanostructures, substantially because the barrier material uses existing solar cell materials and provides a substantial fraction of the total energy conversion. Key challenges remain in the development of virtual band gap solar cells, including the demonstration of the simultaneous optimum radiative coupling between all the bands [124], the identification of optimum material systems and nanostructure configuration, and issues related to implementation in a device, particularly transport and optical absorption. However, a central advantage of virtual band gap solar cells is that they allow the understanding and technology developed for other types of nanostructured devices to be applied to solar cells.

Because of the more established physics and technology surrounding virtual band gap solar cells, they provide the shortest path for realizing a high efficiency nanostructured solar cell. In particular, they have an advantage for low energy photons for solar cells with a large number of band gaps, since the inter-subband transitions required at the low energy photon range are well-documented and commercially using in quantum-well infrared photo detectors. Furthermore, efficient conversion of low energy photons faces challenges using p-n junction approaches, as demonstrated by the difficulty in thermophotovoltaics (TPV) for lower temperature sources. By not requiring a low physical band gap, virtual band gap devices can achieve higher efficiency in this energy range.

4.14 Hot Carrier Solar Cell

In traditional solar cells, photo-carriers relax from their initial energetic position to the band edge by thermal emission process before they can be extracted out from the devices. As a consequence, a huge amount of absorbed energy is wasted especially for high energy photons. In 1982, a kind of photovoltaic device called hot carrier solar cells (HCSCs) was proposed to timely extract the energetic photo-carriers, so that the thermalization loss can be minimized and the solar cell efficiency can be remarkably improved [125]. In principle the efficiency of HCSCs can be as high as 67% under AM1.5 irradiation and even achieve 86% under concentrated illumination [126]. As shown from the schematic diagram in Figure 19, in order to build a complete HCSC, we need to figure out two things: the absorber with extremely slow cooling rate of hot carriers and the extraction contact with a narrow energy range [127-128]. During the last decade, most efforts in this area are devoted to the discovery of appropriate hot carrier absorber material. In fact, since the emergence of HCSC, low dimensional structures are considered to be most promising for reducing the hot carrier cooling rate [129-130].

Theoretical and experimental work have demonstrated that the unique properties of nanostructures are responsible for the reduced cooling rate, such as more stringent conservation rules in the carrier-phonon interaction which couple carriers to fewer vibration modes and the carrier localization effect that prevents carrier cooling by out-diffusion from the hot phonon regions [131-133]. In the meantime, nanostructures have also shown some potential in the fabrication of selective energy contact [134-136]. Unfortunately, however, until now no working device has yet been made for HCSCs, which more or less reflects that the development of HCSCs is now facing serious challenge. This might be related to the fact that it is quite difficult to use current manufacturing techniques to produce the nanostructures required by HCSCs. Moreover, the lack of customized characterization facilities and methods also hinders the development of new materials and device structures.

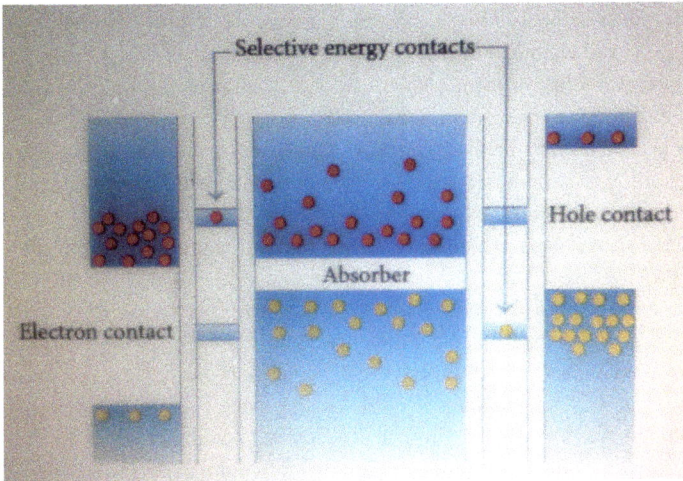

Figure 19: The energy band diagram of a hot carrier solar cell. Photogenerated electron-hole pairs in the absorber are kept hot at an elevated temperature before they are extracted out from the energy-selective contacts.

5. Discussion and Conclusion

Nanostructured photovoltaics aim at developing solar cell architectures that exploit the potential of nanotechnology. The ambition is to unravel and tailor the optoelectronic properties of nanometer sized semiconductors for obtaining low-cost high-efficiency solar cell devices. The approach is multidisciplinary, combining fundamental research on sample's physico-chemistry with solar cell device engineering. The motivation for using nanostructure materials emerges from the specific physical and chemical properties of nanostructures. Special physical effects, related to the nanometer-sized scale, increase interesting macroscopic properties. While there is substantial research required to realize a fully a nanostructured, low cost solar cell, certain aspects are more immediately applicable to high efficiency solar cells and serve as a route towards a high performance low-cost nanostructured solar cell. In order to have the largest impact, the demonstration of nanostructures using low cost approaches with improved transport properties is essential. A closely-spaced ordered array of quantum dots allows implementation of any of the potential uses of nanostructures. Nanostructures without long range order have a challenge in demonstrating efficient transport.

The growing interest in applying nanoscale materials for solving the problems in solar energy conversion technology can be enhanced by the introduction of new materials such as quantum dots, multilayer of ultrathin nanocrystalline materials and the availability of sufficient quantities of raw materials. The inexpensive purification or synthesis of nanomaterials, deposition methods for the fabrication of thin film structures and easy process control in order to achieve a large-area production within acceptable performance tolerances and high life time expectancy are still the main challenges for the realization (fabrication) of solar cells. Therefore in attaining the main objectives of photovoltaics, the efficiency of solar cells should be improved without any compromise on the processing cost of these devices. Nanotechnology incorporation into the films shows special promise in enhancing the efficiency of solar energy conservation and also reducing the manufacturing cost. Its efficiency can be improved by increasing the absorption efficiency of the light as well as the overall radiation-to-electricity. This would help to preserve the environment, decrease soldiers carrying loads, provide electricity for rural areas and have a wide array of commercial applications due to its wireless capabilities. The solar energy, a boon to the mankind has to be properly channelized to meet the energy demand in the developing countries and solar cell industry can reach greater heights by the incorporation of third generation solar cell devices and panels based on nanostructures.

There has been a long history of using nanostructures in photovoltaic devices. The first application of such strategy was made to improve the performance of conventional thin film solar cells. Although some positive effects of nanostructures are clearly observed in these devices, they are still less competitive due to the inherent limitation of the absorber performance. The development of nanostructured absorbers which combines advantages of quantum confined energy levels and low cost processes is considered to be rather important for the realization of next-generation photovoltaic concepts. Many diverse designs have been proposed to utilize the unique flexibility of nanostructures to optimize photon absorption, carrier generation, and charge separation. Theoretical calculation illustrates promising high efficiencies for these devices despite of their different approaches. In practice, however, a competitive technology needs to meet other requirements besides efficiency such as spectral robustness, ease of manufacture, and long-term stability.

Hybrid solar cells combining inorganic nanoparticles and conductive polymer blends are the best developed so far and further improvement is likely to reduce the overall dollar per watt cost. Recently, the discovery of perovskite material significantly improves the absorption capacity of hybrid solar cells thus leading to an efficiency boost. This makes hybrid perovskite solar cell an ideal candidate for the portable or flexible applications. In such kind of solar cells, long-term stability seems to be a more challenging issue than conversion efficiency according to existing research and experiment results. The development of more stable hybrid structure and appropriate packaging technique is important for the future development of hybrid perovskite solar cell. Tandem solar cells with multiple energy thresholds in theory have very high efficiency. More importantly, their embedded QW and QD absorbers could be made from low cost, abundant, nontoxic, and stable materials. Therefore, tandem solar cells are considered to be a competitive technology that can significantly increase the implementation of photovoltaics. Nevertheless, despite the well-developed methods for the preparation of nanostructures, these devices still have difficulties in carrier transportation and extraction. Further research should focus on the suppression of carrier recombination and the formation of high quality junctions and metal contacts.

The nanowire solar cells suffer from relatively low efficiency limit, and intermediate level devices and hot carrier cells still have serious theoretical questions to answer. In addition, some more advanced concepts can be realized by the implementation of nanostructures, such as multiple carrier generation, up/down conversion, thermophotonics, and thermophotovoltaics [137-139]. Besides being used as absorber materials, nanostructures are attracting increasing attention as alternative light trapping structures for thin film solar cells. During the last few years, several different approaches have been developed for nanophotonic light trapping. The periodic structure is the best developed approach with good understanding of fundamental mechanisms and mature processing technologies. According to the number of dimensions, there are three kinds of periodic structures, Bragg stacks, gratings, and photonic crystals [140-143]. Plasmonic structure is another promising approach for nanophotonic light trapping. The discrete metal nanoparticles on the surface randomly scatter incident light into the active layer of solar cell, thus resulting in enhancement of light absorption [144-146]. In contrast to the traditional method such as surface texturing, the use of nanophotonic structures is a better choice for thin film solar cells, since they will not affect the surface topography of the thin films, which has a notable impact on the device performance.

Quantum well solar cells have proven to be an excellent structure for understanding the fundamentals of applying quantum confinement to photovoltaics as well as the effect of photon recycling. Various elongated nanostructures such as nanorods and nanowires have been applied to novel cell designs. Nanoparticles and quantum dots have been shown to be useful in PV devices in various modes, including as plasmonic structures, multi-exciton generating structures, demonstration of intermediate (mini) bands, spectral converters to be applied to conventional solar cells, and as the active layers in novel solar cell designs. There are a number of high-efficiency concepts that have been discussed in the literature for achieving the so-called third generation solar cell technologies, i.e. structures that can yield both high efficiency and low cost. These include intermediate bands, carrier multiplication, up/down-conversion, and multi-junctions. The use of low-dimensional nanostructures, especially quantum wells, nanowires/tubes, and nanoparticles/quantum dots may ultimately play a key role in implementing these novel band structures and conversion mechanisms in future high efficiency solar cells.

Finally, the field of nano-photovoltaics is relatively new, and there are many exciting prospects for applying nanostructures and nanotechnology to one of the most promising renewable energy technologies, i.e. photovoltaics and related applications such as photo-electrolysis. Although, many technical challenges remain and much fundamental and technological work must be done to properly apply nanostructures to photovoltaics, the future of nanostructures for photovoltaics is bright.

References

[1]   Travis Bradford, Solar Revolution: The Economic Transformation of the Global Energy Industry, MIT Press (2006).

[2]   R. M. Swanson, 'The Promise of Concentrators', Progress in Photovoltaics: Res. Appl., 8, 93-111 (2000). https://doi.org/10.1002/(SICI)1099-159X(200001/02)8:1<93::AID-PIP303>3.0.CO;2-S

[3]   http://optics.org/cws/article/research/23961

[4]   M. C. Beard, J. M. Luther, O. E. Semonin, and A. J. Nozik, J. Acc. Chem. Res., 46, 1252-1260 (2013). https://doi.org/10.1021/ar3001958

[5]   http://www.specmat.com/Overview%20of%20Solar%20Cells.htm

[6]   J. Oh, H.-C. Yuan, and H. M. Branz, Nature Nanotech., 7, 743-748 (2012). https://doi.org/10.1038/nnano.2012.166

[7]   M. D. Kelzenberg et al., Nature Mater., 9, 239-244 (2010). https://doi.org/10.1038/nmat2727

[8]   E. Garnett, and P. Yang, Nano Lett., 10, 1082-1087 (2010). https://doi.org/10.1021/nl100161z

[9]   P. Wurfel, Physics of Solar Cells: From Basic Principles to Advanced Concepts, Wiley (2009).

[10]   C.-H. M. Chuang, P. R. Brown, V. Bulović, and M. G. Bawendi, Nature Mater., 13, 796-801 (2014). https://doi.org/10.1038/nmat3984

[11] M. C. Beard, A. H. Ip, J. M. Luther, E. H. Sargent, and A. J. Nozik, Advanced Concepts in Photovoltaics (eds A. J. Nozik, G. Conibeer, and M. C. Beard) Ch. 11, Royal Society of Chemistry (2014).

[12] Z. Ning et al., Nature Mater., 13, 822-828 (2014). https://doi.org/10.1038/nmat4007

[13] H. W. Hillhouse, and M. C. Beard, Curr. Opin. Colloid Interface Sci., 14, 245-259 (2009). https://doi.org/10.1016/j.cocis.2009.05.002

[14] N.-G. Park, Advanced Concepts in Photovoltaics (eds A. J. Nozik, G. Conibeer, and M. C. Beard) Ch. 7, Royal Society of Chemistry (2014).

[15] A. J. Devos, Phys. D, 13, 839-46 (1980).

[16] M. A. Green, K. Emery, Y. Hishikawa, W. Warta, and E. D. Dunlop, Prog. Photovolt., 22, 701-710 (2014). https://doi.org/10.1002/pip.2525

[17] X. Wang et al., Nature Photon., 5, 480-484 (2011). https://doi.org/10.1038/nphoton.2011.123

[18] Y. Okada, T. Sogabe, and Y. Shoji, Advanced Concepts in Photovoltaics (eds A. J. Nozik, G. Gonibeer, and M. C. Beard) Ch. 13, Royal Society of Chemistry (2014).

[19] F. Meinardi et al., Nature Photon., 8, 392-399 (2014). https://doi.org/10.1038/nphoton.2014.54

[20] M. C. Beard et al., Nano Lett., 9, 836-845 (2009). https://doi.org/10.1021/nl803600v

[21] C. M. Cirloganu et al., Nature Commun., 5, 4148 (2014). https://doi.org/10.1038/ncomms5148

[22] J. Zhang, J. Gao, E. A. Miller, J. M. Luther, and M. C. Beard, ACS Nano, 8, 614-622 (2013). https://doi.org/10.1021/nn405236k

[23] R. Plass, S. Pelet, J. Krueger, M. Gratzel, and U. Bach, 'Quantum dot sensitization of organic-inorganic hybrid solar cells', Journal of Physical Chemistry B, 106, No. 31, 7578-7580 (2002). https://doi.org/10.1021/jp0204531

[24] T. Trupke, M. Green, and P. Wurfel, 'Improving solar cell efficiencies by up-conversion of sub-band-gap light', J. Appl. Phys., 92, 4117-4122 (2002). https://doi.org/10.1063/1.1505677

[25] T. Trupke, M. Green, and P. Wurfel, 'Improving solar cell efficiencies by down-conversion of high-energy photons', J. Appl. Phys., 92, 1668-1674 (2002). https://doi.org/10.1063/1.1492021

[26] B. O'Regan, and M. Grätzel, Nature, 353, 737 (1991). https://doi.org/10.1038/353737a0

[27] J. Y. Kim, K. Lee, N. E. Coates, D. Moses, T.-Q. Nguyen, M. Dante, and A. J. Heeger, Science, 317, 222 (2007). https://doi.org/10.1126/science.1141711

[28] P. Wang, S. M. Zakeeruddin, P. Comte, R. Charvet, R. Humphry-Baker, and M. Grätzel, J. Phys. Chem. B, 107, 14336 (2003). https://doi.org/10.1021/jp0365965

[29] B. Schwenzer, K. M. Roth, J. R. Gomm, M. Murr, and D. E. Morse, J. Mater. Chem., 16, 401 (2006). https://doi.org/10.1039/B512900A

[30] M. Nanu, J. Schoonman, and A. Goossens, Nano Lett., 5, 1716 (2005). https://doi.org/10.1021/nl0509632

[31] S. Guha, and J. Yang, J. Non-cryst. Solids, 352, 1917 (2006). https://doi.org/10.1016/j.jnoncrysol.2006.01.048

[32] X. Yang, J. B. He'roux, L. F. Mei, and W. I. Wang, Appl. Phys. Lett., 78, 4068 (2001). https://doi.org/10.1063/1.1379787

[33] L. S. Yu, S. S. Li, and P. Ho, Proc. SPIE, 1675, 255 (1992). https://doi.org/10.1117/12.137615

[34] N. J. Ekins-Daukes, K. W. J. Barnham, J. P. Connolly, J. S. Roberts, J. C. Clark, G. Hill, and M. Mazzer, Appl. Phys. Lett., 75, 4195 (1999). https://doi.org/10.1063/1.125580

[35] J. U. Lee, P. P. Gipp, and C. M. Heller, Appl. Phys. Lett., 85, 145 (2004). https://doi.org/10.1063/1.1769595

[36] J. U. Lee, Phys. Rev. B, 75, 75409 (2007). https://doi.org/10.1103/PhysRevB.75.075409

[37] A. Du Pasquier, H. E. Unalan, A. Kanwal, S. Miller, and M. Chhowalla, Appl. Phys. Lett., 87, 203511 (2005). https://doi.org/10.1063/1.2132065

[38] W. U. Huynh, J. J. Dittmer, and A. P. Alivisatos, Science, 295, 2425 (2002). https://doi.org/10.1126/science.1069156

[39] T. Trupke, M. A. Green, and P. Wurfel, J. Appl. Phys., 92, 1668 (2002). https://doi.org/10.1063/1.1492021

[40] T. Trupke, M. A. Green, and P. Wurfel, J. Appl. Phys., 92, 4117 (2002). https://doi.org/10.1063/1.1505677

[41] M. Wolf, R. Brendel, J. H. Werner, and H. J. Queisser, J. Appl. Phys., 83, 4213 (1998). https://doi.org/10.1063/1.367177

[42] I. Gur, N. A. Fromer, M. L. Geier, and A. P. Alivisatos, 'Air-stable all-inorganic nanocrystal solar cells processed from solution', Science, 310, 462-465 (2005). https://doi.org/10.1126/science.1117908

[43] A. Martı', E. Antolın, C. R. Stanley, C. D. Farmer, N. Lopez, P. Dıaz, E. Canovas, P. G. Linares, and A. Luque, Phys. Rev. Lett., 97, 247701 (2006). https://doi.org/10.1103/PhysRevLett.97.247701

[44] G. Conibeer, M. Green, R. Corkish, Y. Cho, E.-C. Cho, C.-W. Jiang, T. Fangsuwannarak, E. Pink, Y. Huang, T. Puzzer, T. Trupke, B. Richards, A. Shalav, and K.-I. Lin, Thin Solid Films, 511-512, 654 (2006). https://doi.org/10.1016/j.tsf.2005.12.119

[45] J. E. Granata, J. H. Ermer, P. Hebert, M. Haddad, R. R. King, D. D. Krut, M. S. Gillanders, N. H. Karam, and B. T. Cavicchi, Proceedings of the 29th IEEE Photovoltaic Specialist Conference, IEEE, New York, 824 (2002).

[46] S. A. Maier, and H. A. Atwater, Plasmonics: J. Appl. Phys., 98, 011101 (2005). https://doi.org/10.1063/1.1951057

[47] C. Kittel, Introduction to Solid State Physics, 8th edition, Wiley, New York (2004).

[48] B.-H. Choi, H.-H. Lee, S. Jin, S. Chun, and S.-H. Kim, Nanotechnology, 18, 075706 (2007). https://doi.org/10.1088/0957-4484/18/7/075706

[49] S. Berciaud, L. Cognet, P. Tamarat, and B. Lounis, Nano Lett., 5, 515 (2005). https://doi.org/10.1021/nl050062t

[50] J. R. Cole, and N. J. Halas, Appl. Phys. Lett., 89, 153120 (2006). https://doi.org/10.1063/1.2360918

[51] D. M. Schaadt, B. Feng, and E. T. Yu, Appl. Phys. Lett., 86, 063106 (2005). https://doi.org/10.1063/1.1855423

[52] D. Derkacs, S. H. Lim, P. Matheu, W. Mar, and E. T. Yu, Appl. Phys. Lett., 89, 093103 (2006). https://doi.org/10.1063/1.2336629

[53] S. Pillai, K. R. Catchpole, T. Trupke, and M. A. Green, J. Appl. Phys., 101, 093105 (2007). https://doi.org/10.1063/1.2734885

[54] D. V. Talapin, and C. B. Murray, Science, 310, 86 (2005). https://doi.org/10.1126/science.1116703

[55]   D. Yu, B. L. Wehrenberg, P. Jha, J. Ma, and P. Guyot-Sionnest, J. Appl. Phys., 99, 104315 (2006). https://doi.org/10.1063/1.2192288

[56]   T. A. Klar, T. Franzl, A. L. Rogach, and J. Feldmann, Adv. Mater., 17, 769 (2005). https://doi.org/10.1002/adma.200401675

[57]   Kazuo Nakajima, Keisuke Ohdaira, Kozo Fujiwara, and Wugen Pan, Solar Energy Materials & Solar Cells, 88, 323 (2005). https://doi.org/10.1016/j.solmat.2005.03.012

[58]   D. J. Friedman, J. F. Geisz, S.R. Kurtz, and J. M. Olson, J. Cryst. Growth, 195, 409 (1998). https://doi.org/10.1016/S0022-0248(98)00561-2

[59]   H. Q. Hou, K. C. Reinhardt, S. R. Kurtz, J. M. Gee, A. A. Allerman, B. E. Hammons, P. C. Chang, and E. D. Jones, Proc. 2nd World Conf. on Photovoltaic Energy Conversion, IEEE, 3600 (1998).

[60]   D. J. Friedman, J. F. Geisz, S. R. Kurtz, and J. M. Olson, Proc. 2nd World Conf. on Photovoltaic Energy Conversion, IEEE, 3 (1998).

[61]   S. R. Kurtz, A. A. Allerman, E. D. Jones, J. M. Gee, J. J. Banas, and B. E. Hammons, Appl. Phys. Lett., 74, 729 (1999). https://doi.org/10.1063/1.123105

[62]   J. F. Geisz, D. J. Friedman, J. M. Olson, S. R. Kurtz, and B. M. Keyes, J. Cryst. Growth, 195, 401 (1998). https://doi.org/10.1016/S0022-0248(98)00563-6

[63]   S. R. Kurtz, A. A. Allerman, C. H. Seager, R. M. Sieg, and E. D. Jones, Appl. Phys. Lett., 77, 400 (2000). https://doi.org/10.1063/1.126989

[64]   V. Nadenau, D. Braunger, D. Hariskos, M. Kaiser, C. Koble, M. Ruckh, R. Schaffer, D. Schmid, T. Walter, S. Zwergart, and H. W. Schock, Prog. Photogr. Res. Appl., 3, 363 (1995). https://doi.org/10.1002/pip.4670030602

[65]   S. Niki, P. J. Fons, A. Yamada, O. Hellman, T. Kurafuji, S. Chichibu, and H. Nakanishi, Appl. Phys. Lett., 69, 647 (1996). https://doi.org/10.1063/1.117793

[66]   M. A. Contreras, Brian Egaas, K. Ramanathan, J. Hiltner, A. Swartzlander, F. Hasoon, and Rommel Noufi, Prog. Photovoltaics, 7, 311 (1999). https://doi.org/10.1002/(SICI)1099-159X(199907/08)7:4<311::AID-PIP274>3.0.CO;2-G

[67]   Elif Arici, Dieter Meissner, F. Schäffler, and N. Serdar Sariciftci, International Journal of Photoenergy, 5, 199 (2003). https://doi.org/10.1155/S1110662X03000333

[68] J. Yang, A. Banerjee, and S. Guha, 'Triple-junction amorphous silicon alloy solar cell with 14.6% initial and 13.0% stable conversion efficiencies', Applied Physics Letters, 70, No. 22, 2975-2977 (1997). https://doi.org/10.1063/1.118761

[69] J. Meier, R. Flückiger, H. Keppner, and A. Shah, 'Complete microcrystalline p-i-n solar cell-crystalline or amorphous cell behavior', Applied Physics Letters, 65, No. 7, 860-862 (1994). https://doi.org/10.1063/1.112183

[70] C. S. Jiang, Y. F. Yan, H. R. Moutinho et al., 'Phosphorus and boron doping effects on nanocrystalline formation in hydrogenated amorphous and nanocrystalline mixed-phase silicon thin films', MRS Proceedings, 1153, Article ID A17-07 (2009). https://doi.org/10.1557/proc-1153-a17-07

[71] N. Vlachopoulos, P. Liska, J. Augustynski, and M. Grätzel, 'Very efficient visible light energy harvesting and conversion by spectral sensitization of high surface area polycrystalline titanium dioxide films', Journal of the American Chemical Society, 110, No. 4, 1216-1220 (1988). https://doi.org/10.1021/ja00212a033

[72] B. O'Regan, and M. Grätzel, 'A low-cost, high-efficiency solar cell based on dye-sensitized colloidal TiO_2 films', Nature, 353, No. 6346, 737-740 (1991). https://doi.org/10.1038/353737a0

[73] M. Grätzel, 'Conversion of sunlight to electric power by nanocrystalline dye-sensitized solar cells', Journal of Photochemistry and Photobiology A, 164, No. 1-3, 3-14 (2004). https://doi.org/10.1016/j.jphotochem.2004.02.023

[74] A. Tsukazaki, A. Ohtomo, T. Onuma et al., 'Repeated temperature modulation epitaxy for p-type doping and light-emitting diode based on ZnO', Nature Materials, 4, No. 1, 42-45 (2005). https://doi.org/10.1038/nmat1284

[75] N. S. Sariciftci, L. Smilowitz, A. J. Heeger, and F. Wudl, Science, 258, 1474 (1992). https://doi.org/10.1126/science.258.5087.1474

[76] W. U. Huynh, J. J. Dittmer, and A. P. Alivisatos, 'Hybrid nanorod-polymer solar cells', Science, 295, No. 5564, 2425-2427 (2002). https://doi.org/10.1126/science.1069156

[77] A. Zaban, O. I. Mićić, B. A. Gregg, and A. J. Nozik, 'Photosensitization of nanoporous TiO_2 electrodes with InP quantum dots', Langmuir, 14, No. 12, 3153-3156 (1998). https://doi.org/10.1021/la9713863

[78] A. P. Alivisatos, 'Semiconductor clusters, nanocrystals, and quantum dots', Science, 271, No. 5251, 933-937 (1996). https://doi.org/10.1126/science.271.5251.933

[79] E. Arici, N. S. Sariciftci, and D. Meissner, 'Hybrid solar cells based on nanoparticles of CuInS$_2$ in organic matrices', Advanced Functional Materials, 13, No. 2, 165-170 (2003). https://doi.org/10.1002/adfm.200390024

[80] N. C. Greenham, X. Peng, and A. P. Alivisatos, 'Charge separation and transport in conjugated-polymer/semiconductor-nanocrystal composites studied by photoluminescence quenching and photoconductivity', Physical Review B: Condensed Matter and Materials Physics, 54, No. 24, 17628-17637 (1996). https://doi.org/10.1103/PhysRevB.54.17628

[81] V. Gonzalez-Pedro, E. J. Juárez-Pérez, W.-S. Arsyad, et al., 'General working principles of CH$_3$NH$_3$PbX$_3$ perovskite solar cells', Nano Letters, 14, No. 2, 888-893 (2014). https://doi.org/10.1021/nl404252e

[82] M. Liu, M. B. Johnston, and H. J. Snaith, 'Efficient planar heterojunction perovskite solar cells by vapour deposition', Nature, 501, No. 7467, 395-398 (2013). https://doi.org/10.1038/nature12509

[83] A. Kojima, K. Teshima, Y. Shirai, and T. Miyasaka, 'Organometal halide perovskites as visible-light sensitizers for photovoltaic cells', Journal of the American Chemical Society, 131, No. 17, 6050-6051 (2009). https://doi.org/10.1021/ja809598r

[84] J.-H. Im, C.-R. Lee, J.-W. Lee, S.-W. Park, and N.-G. Park, '6.5% efficient perovskite quantum-dot-sensitized solar cell', Nanoscale, 3, No. 10, 4088-4093 (2011). https://doi.org/10.1039/c1nr10867k

[85] H.-S. Kim, C.-R. Lee, J.-H. Im et al., 'Lead iodide perovskite sensitized all-solid-state submicron thin film mesoscopic solar cell with efficiency exceeding 9%', Scientific Reports, 2, Article 591 (2012). https://doi.org/10.1038/srep00591

[86] N.-G. Park, 'Organometal perovskite light absorbers toward a 20% efficiency low-cost solid-state mesoscopic solar cell', The Journal of Physical Chemistry Letters, 4, No. 15, 2423-2429 (2013). https://doi.org/10.1021/jz400892a

[87] O. Malinkiewicz, A. Yella, Y. H. Lee et al., 'Perovskite solar cells employing organic charge-transport layers', Nature Photonics, 8, No. 2, 128-132 (2014). https://doi.org/10.1038/nphoton.2013.341

[88] K. Ramanathan, M. A. Contreras, C. L. Perkins, S. Asher, F. S. Hasoon, J. Keane, D. Young, M. Romero, W. Metzger, R. Noufi, J. Ward, and A. Duda, Research and Applications, 11, 225 (1999).

[89]   X. Wu, J. C. Kane, R. G. Dhere, C. DeHart, D. S. Albin, A. Duda, T. A. Gessert, S. Asher, D. H. Levi, and P. Sheldon, Proceedings of the 17th European Photovoltaic Solar Energy Conference and Exhibition, Munich, 995 (2002).

[90]   A. N. Tiwari, M. Krejci, F. J. Haug, and H. Zogg, Research and Applications, 7, 393 (1999).

[91]   J. R. Tuttle, A. Szalaj, and J. A. Keane, Proceedings of the 28th IEEE Photovoltaic Specialists Conference, Anchorage, 1042 (2000).

[92]   I. Matulionis, S. Han, J. A. Drayton, K. J. Price, and A.D. Compaan, Proceedings of the 2001 MRS Spring Meeting, San Francisco, 1 (2001).

[93]   A. Romeo, M. Arnold, D. L. Batzner, and H. Zogg, Proceedings of the PV in Europe-From PV Technology to Energy Solutions Conference and Exhibition, 377 (2002).

[94]   D. L. Batzner, A. Romeo, M. Terheggen, M. Dobeli,H. Zogg, and A.N. Tiwari, Thin Solid Films, 451-452, 536 (2004). https://doi.org/10.1016/j.tsf.2003.10.141

[95]   K. W. J. Barnham, and G. Duggan, 'A new approach to high-efficiency multi-band-gap solar cells', Journal of Applied Physics, 67, No. 7, 3490-3493 (1990). https://doi.org/10.1063/1.345339

[96]   J. Nelson, M. Paxman, K. W. J. Barnham, J. S. Roberts, and C. Button, 'Steady-state carrier escape from single quantum well', IEEE Journal of Quantum Electronics, 29, No. 6, 1460-1468 (1993). https://doi.org/10.1109/3.234396

[97]   J. Barnes, J. Nelson, K. W. J. Barnham et al., 'Characterization of GaAs/InGaAs quantum wells using photocurrent spectroscopy', Journal of Applied Physics, 79, No. 10, 7775-7779 (1996). https://doi.org/10.1063/1.362383

[98]   I. Serdiukova, C. Monier, M. F. Vilela, and A. Freundlich, 'Critical built-in electric field for an optimum carrier collection in multiquantum well p-i-n diodes', Applied Physics Letters, 74, No. 19, 2812-2814 (1999). https://doi.org/10.1063/1.124022

[99]   P. R. Griffin, J. Barnes, K. W. J. Barnham et al., 'Effect of strain relaxation on forward bias dark currents in GaAs/InGaAs multiquantum well p-i-n diodes', Journal of Applied Physics, 80, No. 10, 5815-5820 (1996). https://doi.org/10.1063/1.363574

[100] N. J. Ekins-Daukes, J. M. Barnes, K. W. J. Barnham et al., 'Strained and strain-balanced quantum well devices for high-efficiency tandem solar cells', Solar Energy Materials and Solar Cells, 68, No. 1, 71-87 (2001). https://doi.org/10.1016/S0927-0248(00)00346-9

[101] D. B. Bushnell, K. W. J. Barnham, J. P. Connolly et al., 'Light-trapping structures for multi-quantum well solar cells', Proceedings of the 29th IEEE Photovoltaic Specialists Conference, 1035-1038 (2002). https://doi.org/10.1109/pvsc.2002.1190782

[102] W. U. Huynh, J. J. Dittmer, and A. P. Alivisatos, 'Hybrid nanorod-polymer solar cells', Science, 295, No. 5564, 2425-2427 (2002). https://doi.org/10.1126/science.1069156

[103] M. Law, L. E. Greene, J. C. Johnson, R. Saykally, and P. Yang, 'Nanowire dye-sensitized solar cells', Nature Materials, 4, No. 6, 455-459 (2005). https://doi.org/10.1038/nmat1387

[104] S. K. Srivastava, D. Kumar, P. K. Singh, M. Kar, V. Kumar, and M. Husain, 'Excellent antireflection properties of vertical silicon nanowire arrays', Solar Energy Materials & Solar Cells, 94, No. 9, 1506-1511 (2010). https://doi.org/10.1016/j.solmat.2010.02.033

[105] K.-Q. Peng, and S.-T. Lee, 'Silicon nanowires for photovoltaic solar energy conversion', Advanced Materials, 23, No. 2, 198-215 (2011). https://doi.org/10.1002/adma.201002410

[106] K. Peng, Y. Xu, Y. Wu, Y. Yan, S.-T. Lee, and J. Zhu, 'Aligned single-crystalline Si nanowire arrays for photovoltaic applications', Small, 1, No. 11, 1062-1067 (2005). https://doi.org/10.1002/smll.200500137

[107] B. Tian, X. Zheng, T. J. Kempa et al., 'Coaxial silicon nanowires as solar cells and nanoelectronic power sources', Nature, 449, No. 7164, 885-889 (2007). https://doi.org/10.1038/nature06181

[108] E. C. Garnett, and P. Yang, 'Silicon nanowire radial p-n junction solar cells', Journal of the American Chemical Society, 130, No. 29, 9224-9225 (2008). https://doi.org/10.1021/ja8032907

[109] D. C. Law, R. R. King, H. Yoon et al., 'Future technology pathways of terrestrial III-V multijunction solar cells for concentrator photovoltaic systems', Solar Energy Materials and Solar Cells, 94, No. 8, 1314-1318 (2010). https://doi.org/10.1016/j.solmat.2008.07.014

[110] G. Conibeer, M. Green, R. Corkish et al., 'Silicon nanostructures for third generation photovoltaic solar cells', Thin Solid Films, 511-512, 654-662 (2006). https://doi.org/10.1016/j.tsf.2005.12.119

[111] B. Zhang, Y. Yao, R. Patterson, S. Shrestha, M. A. Green, and G. Conibeer, 'Electrical properties of conductive Ge nanocrystal thin films fabricated by low temperature in situ growth', Nanotechnology, 22, No. 12, Article ID 125204 (2011). https://doi.org/10.1088/0957-4484/22/12/125204

[112] J. J. Choi, Y.-F. Lim, M. B. Santiago-Berrios et al., 'PbSe nanocrystal excitonic solar cells', Nano Letters, 9, No. 11, 3749-3755 (2009). https://doi.org/10.1021/nl901930g

[113] F. Gao, M. A. Green, G. Conibeer et al., 'Fabrication of multilayered Ge nanocrystals by magnetron sputtering and annealing', Nanotechnology, 19, No. 45, Article ID 455611 (2008). https://doi.org/10.1088/0957-4484/19/45/455611

[114] M. Buljan, U. V. Desnica, M. Ivanda et al., 'Formation of three-dimensional quantum-dot superlattices in amorphous systems: experiments and Monte Carlo simulations', Physical Review B: Condensed Matter and Materials Physics, 79, No. 3, Article ID 035310 (2009). https://doi.org/10.1103/PhysRevB.79.035310

[115] M. Ardyanian, H. Rinnert, and M. Vergnat, 'Structure and photoluminescence properties of evaporated GeOx/SiO$_2$ multilayers', Journal of Applied Physics, 100, No. 11, Article ID 113106 (2006). https://doi.org/10.1063/1.2400090

[116] A. Luque, and A. Martí, 'Increasing the efficiency of ideal solar cells by photon induced transitions at intermediate levels', Physical Review Letters, 78, No. 26, 5014-5017 (1997). https://doi.org/10.1103/PhysRevLett.78.5014

[117] A. Martí, L. Cuadra, and A. Luque, 'Quantum dot intermediate band solar cell', Proceedings of the 28th IEEE Photovoltaic Specialists Conference, Anchorage, Alaska, USA, 940-943 (2000). https://doi.org/10.1109/pvsc.2000.916039

[118] G. W. Bryant, and G. Solomon, Optics of Quantum Dots and Wires, Artech House, Norwood, Mass, USA (2005).

[119] A. J. Nozik, A. Martí, and A. Luque, Next Generation Photovoltaics: High Efficiency through Full Spectrum Utilization, Artech House, Series in Optics and Optoelectronics, Institute of Physics Publishing (2003).

[120] J. P. Loehr, and M. O. Manasreh, Semiconductor Quantum Wells and Superlattices for Long-Wavelength Infrared Detectors, Artech House, Boston, Mass, USA (1993).

[121] A. Martí, N. López, E. Antolín, et al., 'Novel semiconductor solar cell structures: the quantum dot intermediate band solar cell', Thin Solid Films, 511, 638-644 (2006). https://doi.org/10.1016/j.tsf.2005.12.122

[122] R. D. Schaller, and V. I. Klimov, 'High efficiency carrier multiplication in PbSe nanocrystals: implications for solar energy conversion', Physical Review Letters, 92, No. 18, 186601/1-4 (2004). https://doi.org/10.1103/PhysRevLett.92.186601

[123] R. J. Ellingson, M. C. Beard, J. C. Johnson, P. Yu, O. I. Micic, A. J. Nozik, A. Shabaev, and A. L. Efros, 'Highly Efficient Multiple Exciton Generation in Colloidal PbSe and PbS Quantum Dots', Nano Letters, 5, No. 5, 873-878 (2005). https://doi.org/10.1021/nl0502672

[124] N. J. Ekins-Daukes, C. B. Honsberg, and M. Yamaguchi, 'Signature of Intermediate Band Materials from Luminescence Measurements', Proceedings of the 31st Photovoltaic Specialists Conference, Orlando, 49-54 (2005). https://doi.org/10.1109/pvsc.2005.1488066

[125] R. T. Ross, and A. J. Nozik, 'Efficiency of hot-carrier solar energy converters', Journal of Applied Physics, 53, No. 5, 3813-3818 (1982). https://doi.org/10.1063/1.331124

[126] J. F. Guillemoles, G. Conibeer, and M. A. Green, 'Phononic engineering with nanostructures for hot carrier solar cells', Proceedings of the 15th International Photovoltaic Science and Engineering Conference, Shanghai, China (2005).

[127] M. A. Green, Third Generation Solar Cells, Springer, Berlin, Germany (2003).

[128] G. J. Conibeer, C.-W. Jiang, D. König, S. Shrestha, T. Walsh, and M. A. Green, 'Selective energy contacts for hot carrier solar cells', Thin Solid Films, 516, No. 20, 6968-6973 (2008). https://doi.org/10.1016/j.tsf.2007.12.031

[129] D. Knig, K. Casalenuovo, Y. Takeda et al., 'Hot carrier solar cells: principles, materials and design', Physica E: Low-Dimensional Systems and Nanostructures, 42, No. 10, 2862-2866 (2010). https://doi.org/10.1016/j.physe.2009.12.032

[130] A. J. Nozik, C. A. Parsons, D. J. Dunlavy, B. M. Keyes, and R. K. Ahrenkiel, 'Dependence of hot carrier luminescence on barrier thickness in GaAs/AlGaAs superlattices and multiple quantum wells', Solid State Communications, 75, No. 4, 297-301 (1990). https://doi.org/10.1016/0038-1098(90)90900-V

[131] D. J. Westland, J. F. Ryan, M. D. Scott, J. I. Davies, and J. R. Riffat, 'Hot carrier energy loss rates in GaInAs/InP quantum wells', Solid State Electronics, 31, No. 3-4, 431-434 (1988). https://doi.org/10.1016/0038-1101(88)90311-5

[132] Y. Rosenwaks, M. C. Hanna, D. H. Levi, D. M. Szmyd, R. K. Ahrenkiel, and A. J. Nozik, 'Hot-carrier cooling in GaAs: quantum wells versus bulk', Physical Review B, 48, No. 19, 14675-14678 (1993). https://doi.org/10.1103/PhysRevB.48.14675

[133] G. Conibeer, R. Patterson, L. Huang et al., 'Modelling of hot carrier solar cell absorbers', Solar Energy Materials & Solar Cells, 94, No. 9, 1516-1521 (2010). https://doi.org/10.1016/j.solmat.2010.01.018

[134] G. J. Conibeer, D. König, M. A. Green, and J. F. Guillemoles, 'Slowing of carrier cooling in hot carrier solar cells', Thin Solid Films, 516, No. 20, 6948-6953 (2008). https://doi.org/10.1016/j.tsf.2007.12.102

[135] P. Aliberti, Y. Feng, Y. Takeda, S. K. Shrestha, M. A. Green, and G. Conibeer, 'Investigation of theoretical efficiency limit of hot carriers solar cells with a bulk indium nitride absorber', Journal of Applied Physics, 108, No. 9, Article ID 094507 (2010). https://doi.org/10.1063/1.3494047

[136] Y. Takeda, T. Ito, T. Motohiro, D. König, S. Shrestha, and G. Conibeer, 'Hot carrier solar cells operating under practical conditions', Journal of Applied Physics, 105, No. 7, Article ID 074905 (2009). https://doi.org/10.1063/1.3086447

[137] A. Franceschetti, J. M. An, and A. Zunger, 'Impact ionization can explain carrier multiplication in PbSe quantum dots', Nano Letters, 6, No. 10, 2191-2195 (2006). https://doi.org/10.1021/nl0612401

[138] D. Chen, Y. Yu, Y. Wang, P. Huang, and F. Weng, 'Cooperative energy transfer up-conversion and quantum cutting down-conversion in $Yb^{3+}:TbF_3$ nanocrystals embedded glass ceramics', Journal of Physical Chemistry C, 113, No. 16, 6406-6410 (2009). https://doi.org/10.1021/jp809995f

[139] V. Švrček, A. Slaoui, and J.-C. Muller, 'Silicon nanocrystals as light converter for solar cells', Thin Solid Films, 451-452, 384-388 (2004). https://doi.org/10.1016/j.tsf.2003.10.133

[140] Y. Kanamori, K. Hane, H. Sai, and H. Yugami, '100 nm period silicon antireflection structures fabricated using a porous alumina membrane mask', Applied Physics Letters, 78, No. 2, 142-143 (2001). https://doi.org/10.1063/1.1339845

[141] J. Gjessing, E. S. Marstein, and A. Sudbø, '2D back-side diffraction grating for improved light trapping in thin silicon solar cells', Optics Express, 18, No. 6, 5481-5495 (2010). https://doi.org/10.1364/OE.18.005481

[142] Z. Yu, A. Raman, and S. Fan, 'Fundamental limit of light trapping in grating structures', Optics Express, 18, No. 19, A366-A380 (2010). https://doi.org/10.1364/OE.18.00A366

[143] Y. Park, E. Drouard, O. E. Daif et al., 'Absorption enhancement using photonic crystals for silicon thin film solar cells', Optics Express, 17, No. 16, 14312-14321 (2009). https://doi.org/10.1364/OE.17.014312

[144] A. Aubry, D. Y. Lei, A. I. Fernández-Domínguez, Y. Sonnefraud, S. A. Maier, and J. B. Pendry, 'Plasmonic light-harvesting devices over the whole visible spectrum', Nano Letters, 10, No. 7, 2574-2579 (2010). https://doi.org/10.1021/nl101235d

[145] K. R. Catchpole, S. Mokkapati, F. Beck et al., 'Plasmonics and nanophotonics for photovoltaics', MRS Bulletin, 36, No. 6, 461-467 (2011). https://doi.org/10.1557/mrs.2011.132

[146] H. Tan, R. Santbergen, A. H. M. Smets, and M. Zeman, 'Plasmonic light trapping in thin-film silicon solar cells with improved self-assembled silver nanoparticles', Nano Letters, 12, No. 8, 4070-4076 (2012). https://doi.org/10.1021/nl301521z

Chapter 4

Quantum dot as light harvester nanocrystals for solar cell applications

M. Patel[1,2,*], S. Sahu[1], A. K. Verma[1], P. Agnihotri[1] ,Surya Prakash Singh[3], Ramanuj Narayan[4] and Sanjay Tiwari[1,*]

[1] Photonics Research laboratory, S.O.S. in Electronics and Photonics, Pt. Ravishankar Shukla University, Raipur (C.G.)-492010 India

[2] Department of Physics and Electronics, St. Thomas College, Bhilai (C.G.)-490006,India

[3]Network of Institutes for Solar Energy, Inorganic & Physical Chemistry Division, CSIR-Indian Institute of Chemical Technology, Uppal Road, Tarnaka, Hyderabad 500607, India

[4]Polymers & Functional Materials Division, Indian Institute of Chemical Technology, Hyderabad 500007, India

drsanjaytiwari@gmail.com, patel.mohan@gmail.com

Abstract

In this article we are reviewing the application of quantum dot nanocrystals as light harvesters for solar cell applications. Three foremost ways to make use of semiconductor quantum dots in solar cells are metal-semiconductor photovoltaic cell, polymer-semiconductor solar cell and quantum dot sensitized solar cell. Band energies can be controlled by size change in quantum dots which gives new ways to control the response and efficiency of the solar cell. Quantum dot solar cell reduces heat waste by multiple electron generation (MEG) and converts up to three electrons per photon. Therefore, more than 100% quantum efficiency is possible for quantum dot solar cells. Furthermore Quantum dot forms one or more intermediate bands (IBs) in the host semiconductor bandgap, enabling two-step absorption of sub-band gap photons. Since the IBs are electrically isolated from Valance Band and Conduction Band, their introduction increases short circuit current (I_{sc}) and keeps open circuit voltage (V_{oc}) unreduced.

Keywords

Nanocrystals, Solar Cell, Quantum Dot, Multiple Electron Generation, Intermediate Bands

Contents

1.   Introduction..118

2.   Quantum Dot Solar Cells..121

2.1   Metal-Semiconductor Photovoltaic Cells 121

2.2   Polymer-Semiconductor Solar Cells..122

2.3   Quantum Dot Sensitized Solar Cells..122

2.4   Intermediate Band Solar Cells 124

3.   Conclusion 124

References...125

1.   Introduction

Nanomaterials are among the new promising energy harvester for solar cells along with organic-inorganic hybrid structures [1-12]. These structures and materials improved efficiency and exhibit enhanced selectivity. Different sizes and shapes (e.g., spheres, tubes rods, wires, hexagonals and pyramids) of nanomaterials structures are being used for development of solar energy converters [13-27].

Shockley and Queisser in 1961 calculated the maximum conversion efficiency using a single junction solar cell with band gap of 1.1eV is about 30% assuming that each absorbed photon (energy above the band gap) produces just one electron-hole pair [28]. This is relatively low because photons with energy below the band-gap are not able to create electron-hole pair and lost (transmission effect). Again photons having energy much higher than band-gap energy increases the temperature of the cell (thermionic effect). Since conversion efficiency is most important parameter for the solar cell several schemes have been proposed to increase the efficiency beyond Shockley–Queisser (S-Q) limit i.e. tandem cells [29-35], hot carrier solar cells [36-43], intermediate band solar cell [44-46], solar cells producing multiple electron-hole pairs per photon [47-50] etc.

The solar spectrum contains photons with different frequencies and hence different energies. This energy ranges from 0.5eV to 3.5eV. Sub band-gap photons (energy below the band gap) are not able to create any electron-hole pair because of insufficient energy and are not absorbed in the solar cell. On the other hand photons with energies higher than the band gap are absorbed and create electrons and holes with excess kinetic energy.

The kinetic energy of electron (or hole) is equal to the photon energy and band gap energy. Because of these excess kinetic energy electrons and holes are having temperature higher than the lattice temperature. These electrons and holes are called "hot electrons and hot holes". Upon absorption of photon the with lattice temperature of 300 K the "hot electrons and hot holes" temperature may be as high as 3000 K. Distribution of this kinetic energy depends on the effective masses of charge carriers. Lower the effective mass higher the excess kinetic energy.

Another reason for energy loss in solar cells is phonon emission caused by thermal relaxation of photogenerated hot electrons and holes when they relax to the respective band edges. There are two fundamental methods to utilize the energy of hot electrons and holes to improve photoconversion efficiency: by enhanced photovoltage, and enhanced photocurrent. Photovoltage generation requires the carrier extraction before they cool down. For Photovoltage generation the generated carriers must be separated, transported and transferred across the contacts before they cool down. Photocurrent generation requires the hot carriers to produce second electron-hole pair generation through impact ionization (inverse Auger process). This requires the impact ionization rate to be higher than the carrier cooling for hot carriers.

Carrier relaxation of photogenerated carriers is distinctly affected by quantization effect in semiconductor nanostructures [36]. When the carriers in the semiconductor are confined [51] (one dimensional confinement in semiconductor quantum wells, two dimensional confinements in quantum wires, three dimensional confinements in quantum dots) by potential barriers to the smaller space than their deBroglie wavelength or to the Bohr radius of excitons in the semiconductor bulk their relaxation dynamics changes [36].

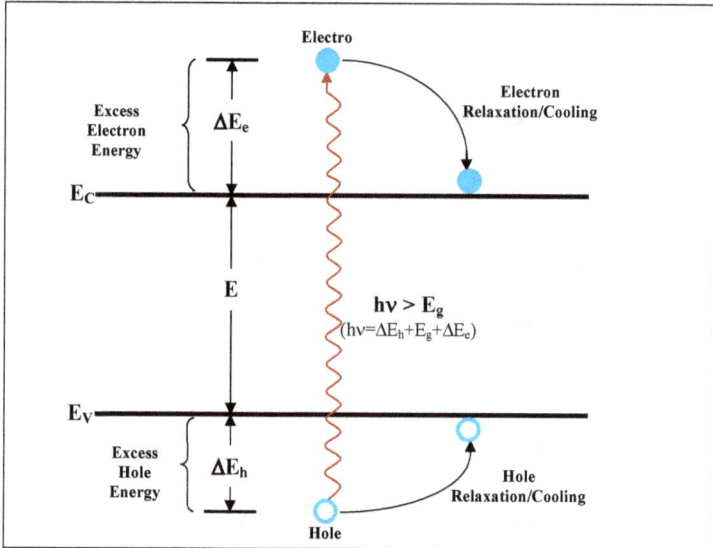

Figure 1: *Hot carrier relaxation/cooling in semiconductors.*

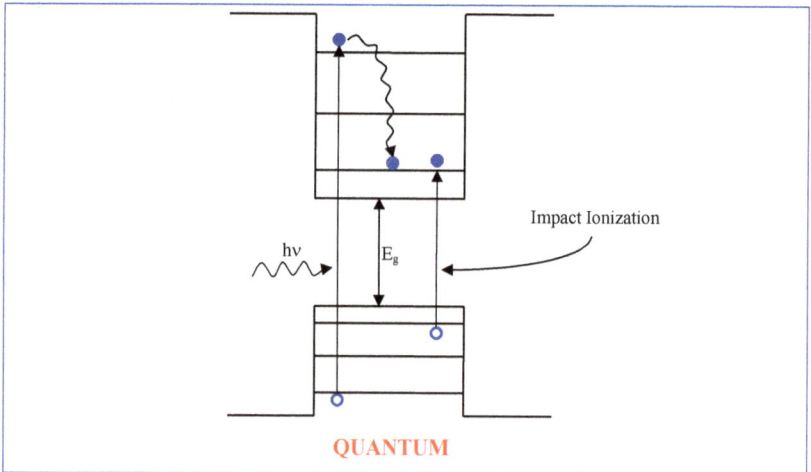

Figure 2: *Enhanced photovoltaic efficiency in QD solar cells by impact ionization (inverse Auger effect).*

2. Quantum Dot Solar Cells

2.1 Metal-Semiconductor Photovoltaic Cells

When semiconductor and metal come in contact with each other the process of Fermi level equilibration undertake. The band bending of valance band and conduction band in space charge region occurs. Width of the space charge region depends on the charge carrier density. When this semiconductor-metal junction is illuminated (photovoltaic cell), due to band bending the generated charge carriers flow in opposite directions to construct the current in the photovoltaic cell. In semiconductor nanocrystals the electrons are confined (one-D, two-D and three-D confinement) and each nanocrystal is isoenergetic. Because of this limitation in size the bands remain flat in semiconductor nanocrystals. The separation of charge is due to Fermi level equilibration [52]. Nanocrystals attaining the Fermi level depends on the degree of electron accumulation. This creates an energy gradient to drive the charge carrier towards the electrode (Figure 3) [53]. Such Schottky photovoltaic device prepared PbS nanocrystal quantum dot films with aluminium and indium tin oxide contacts exhibit infrared power conversion efficiency [54].

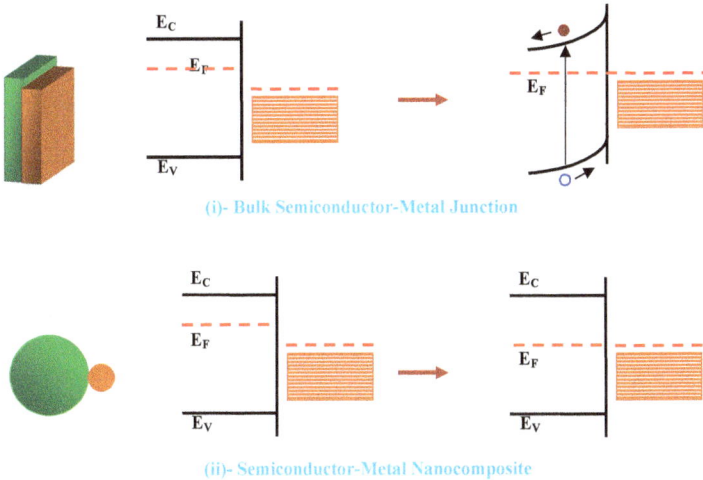

(i)- Bulk Semiconductor-Metal Junction

(ii)- Semiconductor-Metal Nanocomposite

Figure 3: *Fermi level equilibration (i) bulk metal-nanocrystal junction, (ii) metal nanoparticle-semiconductor nanoparticle junction.*

2.2   Polymer-Semiconductor Solar Cells

Inorganic materials are commonly used for photovoltaics, but because of solution processibility and attractive photoconversion efficiency organic materials as gained significant attention. Organic materials are having current carrying property in the ultraviolet-visible region of solar spectrum due to the sp^2-hybridized carbon atoms. The electron in p_z-orbitals of two adjacent sp^2 hybridized carbon atoms in a linear chain form π-bonds, which subsequently results in dimerization (giving rise to a structure having single and double bonds in an alternating fashion i.e., Peierls distortion). As a consequence of the above said isomeric effect, the π-electrons are delocalized and this results in high electronic polarizability [55].

Compared to inorganic semiconducting materials organic materials are having poor charge-carrier mobility but strong absorption coefficients (usually $\geq 10^5$ cm^{-1}). Another difference to inorganic semiconductors is comparatively small diffusion length of excitons. Due to high absorption coefficient and small diffusion length of excitons organic semiconducting materials lead to very small layer thickness (<100nm) and hence reduces the material requirements. Most of the organic semiconductors are having energy band gap around 2 eV which is higher than inorganic silicon and are hole conductors. The high band gap limits the efficiency of cell as photons having lower energy than band gap energy thus will not be absorbed. However the flexibility of modification on organic semiconductors through chemical synthesis is also possible.

Photovoltaic effects of structures consisting of quantum dots (QDs) along with organic semiconductor polymers have been intensively studied [56-64]. In one configuration instead of QDs CdSe nanorods are used along with conjugated polymer poly-3(hexylthiophene) to develop a photovoltaic device [65]. By controlling the nanorod length and radius the electron transport distance and bandgap are controlled respectively. Recently amorphous-Si:H and polymer is used to form a tandem cell with power conversion efficiency of 10.5% [66]. Amorphous-Si:H with 200nm thickness and PDTP-DFBT:fullerene bulk heterojunction with 120nm thickness are used in tandem cell as front and back cells respectively. Where PDTP-DFBT is poly[2,7-(5,5-bis-(3,7-dimethyloctyl)-5H-dithieno[3,2-b:2',3'-d]pyran)-alt-4,7-(5,6-difluoro-2,1,3-benzothia diazole).

2.3   Quantum Dot Sensitized Solar Cells

Photoelectrochemical solar cell includes dye-sensitized solar cells (DSSCs) [67] and quantum dot sensitized solar cells (QDSSCs) [68]. These cells are having high photo conversion efficiency as well as low cost of fabrication. Among these QDSSCs are having more concentration because the QDs are having their properties of size dependent band gap and hence the tuneable absorption spectrum, large absorption coefficient, multiple electron generation [36, 47-49], hot electron injection [36].

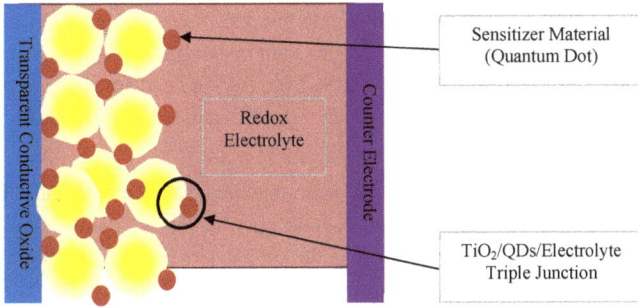

Figure 4: *Representation of the main interfaces/materials constituting the QDSSC.*

The main constituting interfaces/materials of QDSSC are sensitizer materials, TiO2/QDs/electrolyte triple junction, redox electrolyte, and counter electrode and are schematically represented in Figure 4 [69].

In CdSe quantum-dot-sensitized solar cell internal quantum efficiency of ~100% was reported [70]. Again in PbSe-based solar cell over 100% quantum efficiency using the MEG mechanism was reported [71]. Recently power conversion efficiencies beyond 8% were reported for quantum dot sensitized solar cells by recombination control [72].

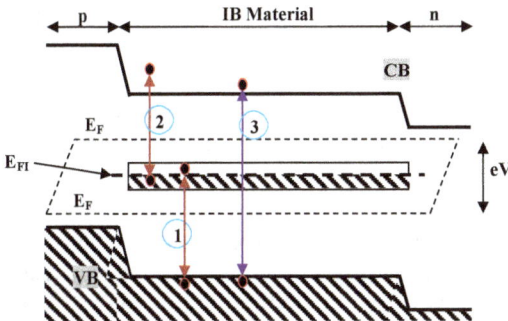

Figure 5: *Band structure of an IBSC.*

2.4   Intermediate Band Solar Cells

The Shockley and Queisser efficiency limit can also be increased by intermediate band solar cells (IBSC) for single junction solar cells by introducing a intermediate level in the semiconductor band gap [73]. Because in single-gap solar cells the major factor limiting the efficiency is that the sub-bandgap photons cannot be absorbed, their energy being wasted. The intermediate band solar cell (IBSC) provides a band (intermediate band-IB) located in the forbidden energy gap between valance band (VB) and conduction band (CB).

This intermediate band provides an energy band (situated between VB and CB) for the electrons excited by sub-bandgap photons to jump into the valance band (jump-1). Again the absorption of anther sub-bandgap photon the electron will jump into the conduction band from the intermediate band (jump-2). Thus consecutive absorption of two sub-bandgap photons creates electron hole pairs. This energy of sub-bandgap photons of solar spectrum were wasted in single-gap solar cells and photons having higher energy than the band gap were absorbed (jump-3) to create electron-hole pairs. The IB should be half-filled with electrons as it supplies the electrons to CB as well as accepts electrons from VB [74]. On the other hand photocurrent enhancement due sub-bandgap photon absorption alone does not improve the efficiency the solar cell if the output voltage is not conserved. So the concept of quasi Fermi level was proposed which is shown in Figure 5 [75]. These intermediate bands are electrically isolated from the valance band and conduction band and their introduction increases I_{sc} and keeps V_{oc} unreduced.

Marti et al. in 2006 demonstrated the IBSC photocurrent generation via two photons excitation. In their experiment they used quantum dot (QD) structures grown by molecular beam epitaxy and included with delta doping layers for half-fill the IB with electrons [76]. Ahsan et al. used an n-GaNAs intermediate band layer which was sandwiched between an emitter of a p-AlGaAs and a barrier layer of an n-AlGaAs. The emitter layer was used to collect holes and the barrier layer with variable doping level to block the electron escape from IB to the bottom of n-GaAs substrate [77].

3.   Conclusion

Clean and renewable energy demand is increasing day by day. As solar energy is an infinite source of energy it can play a major role in our energy requirement. Jawaharlal Nehru National Solar Mission (JNNSM) was launched by Ministry of New and Renewable Energy Government of India in 2010 with an ambitious target of 20,000MW of electricity generation by 2022 and with the objective to establish India as a solar energy global leader. To obtain this ambitious target solar module requirement has increased, inspiring the solar cells demand. On the other hand, delivering solar cells with low cost is a big task. Efficiency is a factor which is mainly influencing cost. Introduction of organic-inorganic nanostructures in hybrid cells and new design and architecture have been made and also new nanomaterials are introduced. Many of these materials are very permissive of supplying clean energy. Quantum dot is one of them and it has unique ability to improve the cell performance. Commercialization of nanostructured solar cells in large areas remains a major challenge. As the new materials and techniques are being introduced day by day we can expect a major breakthrough from these devices.

References

[1] Kamat, P. V., 2006. Harvesting photons with carbon nanotubes. Nano Today, 1(4), pp. 20-27. https://doi.org/10.1016/S1748-0132(06)70113-X

[2] Guchhait, A., Rath, A. K. and Pal, A. J., 2011. To make polymer: quantum dot hybrid solar cells NIR-active by increasing diameter of PbSnanoparticles. Solar energy materials and solar cells, 95(2), pp. 651-656. https://doi.org/10.1016/j.solmat.2010.09.034

[3] Kim, H., Jeong, H., An, T. K., Park, C. E. and Yong, K., 2012. Hybrid-type quantum-dot cosensitized ZnO nanowire solar cell with enhanced visible-light harvesting. ACS applied materials & interfaces, 5(2), pp. 268-275. https://doi.org/10.1021/am301960h

[4] Chen, H. C., Lin, C. C., Han, H. V., Chen, K. J., Tsai, Y. L., Chang, Y. A., Shih, M. H., Kuo, H. C. and Yu, P., 2012. Enhancement of power conversion efficiency in GaAs solar cells with dual-layer quantum dots using flexible PDMS film. Solar Energy Materials and Solar Cells, 104, pp. 92-96. https://doi.org/10.1016/j.solmat.2012.05.003

[5] Wright, M. and Uddin, A., 2012. Organic-inorganic hybrid solar cells: A comparative review. Solar energy materials and solar cells, 107, pp. 87-111. https://doi.org/10.1016/j.solmat.2012.07.006

[6] Ryan, J. W., Marin-Beloqui, J. M., Albero, J. and Palomares, E., 2013. Nongeminate Recombination Dynamics-Device Voltage Relationship in Hybrid PbS Quantum Dot/C60 Solar Cells. The Journal of Physical Chemistry C, 117(34), pp. 17470-17476. https://doi.org/10.1021/jp4059824

[7] Laghumavarapu, R. B., Liang, B. L., Bittner, Z. S., Navruz, T. S., Hubbard, S. M., Norman, A. and Huffaker, D. L., 2013. GaSb/InGaAs quantum dot-well hybrid structure active regions in solar cells. Solar Energy Materials and Solar Cells, 114, pp. 165-171. https://doi.org/10.1016/j.solmat.2013.02.027

[8]   McDaniel, H., Fuke, N., Pietryga, J. M. and Klimov, V. I., 2013. Engineered CuInSe x S2–x Quantum Dots for Sensitized Solar Cells. The journal of physical chemistry letters, 4(3), pp. 355-361. https://doi.org/10.1021/jz302067r

[9]   Li, Z., Yu, L., Liu, Y. and Sun, S., 2014. CdS/CdSe quantum dots co-sensitized TiO2 nanowire/nanotube solar cells with enhanced efficiency. Electrochimica Acta, 129, pp. 379-388. https://doi.org/10.1016/j.electacta.2014.02.145

[10]   Yu, Y., Kamat, P. V. and Kuno, M., 2010. A CdSe nanowire/quantum dot hybrid architecture for improving solar cell performance. Advanced Functional Materials, 20(9), pp. 1464-1472. https://doi.org/10.1002/adfm.200902372

[11]   Ren, S., Chang, L. Y., Lim, S. K., Zhao, J., Smith, M., Zhao, N., Bulovic, V., Bawendi, M. and Gradecak, S., 2011. Inorganic–organic hybrid solar cell: bridging quantum dots to conjugated polymer nanowires. Nano letters, 11(9), pp. 3998-4002. https://doi.org/10.1021/nl202435t

[12]   Robel, I., Subramanian, V., Kuno, M. and Kamat, P. V., 2006. Quantum dot solar cells. Harvesting light energy with CdSe nanocrystals molecularly linked to mesoscopic TiO2 films. Journal of the American Chemical Society, 128(7), pp. 2385-2393. https://doi.org/10.1021/ja056494n

[13]   Ma, W., Swisher, S. L., Ewers, T., Engel, J., Ferry, V. E., Atwater, H. A. and Alivisatos, A. P., 2011. Photovoltaic performance of ultrasmall PbSe quantum dots. ACS nano, 5(10), pp. 8140-8147. https://doi.org/10.1021/nn202786g

[14]   Htoon, H., Malko, A. V., Bussian, D., Vela, J., Chen, Y., Hollingsworth, J. A. and Klimov, V. I., 2010. Highly emissive multiexcitons in steady-state photoluminescence of individual "giant" CdSe/CdS core/shell nanocrystals. Nano letters, 10(7), pp. 2401-2407. https://doi.org/10.1021/nl1004652

[15]   Jiao, S., Shen, Q., Mora-Seró, I., Wang, J., Pan, Z., Zhao, K., Kuga, Y., Zhong, X. and Bisquert, J., 2015. Band engineering in core/shell ZnTe/CdSe for photovoltage and efficiency enhancement in exciplex quantum dot sensitized solar cells. ACS nano, 9(1), pp. 908-915. https://doi.org/10.1021/nn506638n

[16]   Yao, M., Cong, S., Arab, S., Huang, N., Povinelli, M. L., Cronin, S. B., Dapkus, P. D. and Zhou, C., 2015. Tandem solar cells using GaAs nanowires on Si: Design, Fabrication, and Observation of voltage addition. Nano letters, 15(11), pp. 7217-7224. https://doi.org/10.1021/acs.nanolett.5b03890

[17]   Raïssi, M., Vignau, L., Cloutet, E. and Ratier, B., 2015. Soluble carbon nanotubes/phthalocyanines transparent electrode and interconnection layers for

flexible inverted polymer tandem solar cells. Organic Electronics, 21, pp. 86-91. https://doi.org/10.1016/j.orgel.2015.03.003

[18]   Li, Z., Yu, L., Liu, Y. and Sun, S., 2014. CdS/CdSe quantum dots co-sensitized TiO2 nanowire/nanotube solar cells with enhanced efficiency. Electrochimica Acta, 129, pp. 379-388. https://doi.org/10.1016/j.electacta.2014.02.145

[19]   Chen, Y., Tao, Q., Fu, W., Yang, H., Zhou, X., Zhang, Y., Su, S., Wang, P. and Li, M., 2014. Enhanced solar cell efficiency and stability using ZnS passivation layer for CdS quantum-dot sensitized actinomorphic hexagonal columnar ZnO. Electrochimica Acta, 118, pp. 176-181. https://doi.org/10.1016/j.electacta.2013.10.081

[20]   Dimitrov, D. Z. and Du, C. H., 2013. Crystalline silicon solar cells with micro/nano texture. Applied Surface Science, 266, pp. 1-4. https://doi.org/10.1016/j.apsusc.2012.10.081

[21]   Laghumavarapu, R. B., Liang, B. L., Bittner, Z. S., Navruz, T. S., Hubbard, S. M., Norman, A. and Huffaker, D. L., 2013. GaSb/InGaAs quantum dot–well hybrid structure active regions in solar cells. Solar Energy Materials and Solar Cells, 114, pp. 165-171. https://doi.org/10.1016/j.solmat.2013.02.027

[22]   Wang, H., Kubo, T., Nakazaki, J., Kinoshita, T. and Segawa, H., 2013. PbS-quantum-dot-based heterojunction solar cells utilizing ZnO nanowires for high external quantum efficiency in the near-infrared region. The Journal of Physical Chemistry Letters, 4(15), pp. 2455-2460. https://doi.org/10.1021/jz4012299

[23]   Liu, B., Wang, D., Wang, L., Sun, Y., Lin, Y., Zhang, X. and Xie, T., 2013. Glutathione-assisted hydrothermal synthesis of CdS-decorated TiO 2 nanorod arrays for quantum dot-sensitized solar cells. Electrochimica Acta, 113, pp. 661-667. https://doi.org/10.1016/j.electacta.2013.09.143

[24]   Lai, Y., Lin, Z., Zheng, D., Chi, L., Du, R. and Lin, C., 2012. CdSe/CdS quantum dots co-sensitized TiO 2 nanotube array photoelectrode for highly efficient solar cells. Electrochimica Acta, 79, pp. 175-181. https://doi.org/10.1016/j.electacta.2012.06.105

[25]   Yang, J., Pan, L., Zhu, G., Liu, X., Sun, H. and Sun, Z., 2012. Electrospun TiO2 microspheres as a scattering layer for CdS quantum dot-sensitized solar cells. Journal of Electroanalytical Chemistry, 677, pp. 101-104. https://doi.org/10.1016/j.jelechem.2012.05.018

[26] Deng, J., Wang, M., Song, X., Shi, Y. and Zhang, X., 2012. CdS and CdSe quantum dots subsectionally sensitized solar cells using a novel double-layer ZnO nanorod arrays. Journal of colloid and interface science, 388(1), pp. 118-122. https://doi.org/10.1016/j.jcis.2012.08.017

[27] Kim, H., Jeong, H., An, T. K., Park, C. E. and Yong, K., 2012. Hybrid-type quantum-dot cosensitized ZnO nanowire solar cell with enhanced visible-light harvesting. ACS applied materials & interfaces, 5(2), pp. 268-275. https://doi.org/10.1021/am301960h

[28] Shockley, W. and Queisser, H. J., 1961. Detailed balance limit of efficiency of p-n junction solar cells. Journal of applied physics, 32(3), pp. 510-519. https://doi.org/10.1063/1.1736034

[29] Han, H. Y., Yoon, H. and Yoon, C. S., 2015. Parallel polymer tandem solar cells containing comb-shaped common electrodes. Solar Energy Materials and Solar Cells, 132, pp. 56-66. https://doi.org/10.1016/j.solmat.2014.08.018

[30] You, J., Chen, C. C., Hong, Z., Yoshimura, K., Ohya, K., Xu, R., Ye, S., Gao, J., Li, G. and Yang, Y., 2013. 10.2 % Power Conversion Efficiency Polymer Tandem Solar Cells Consisting of Two Identical Sub-Cells. Advanced Materials, 25(29), pp. 3973-3978. https://doi.org/10.1002/adma.201300964

[31] Cheyns, D., Rand, B. P. and Heremans, P., 2010. Organic tandem solar cells with complementary absorbing layers and a high open-circuit voltage. Appl. Phys. Lett, 97(3), p. 033301. https://doi.org/10.1063/1.3464169

[32] Kim, J. Y., Lee, K., Coates, N. E., Moses, D., Nguyen, T. Q., Dante, M. and Heeger, A. J., 2007. Efficient tandem polymer solar cells fabricated by all-solution processing. Science, 317(5835), pp. 222-225. https://doi.org/10.1126/science.1141711

[33] Zhao, D. W., Sun, X. W., Jiang, C. Y., Kyaw, A. K. K., Lo, G. Q. and Kwong, D. L., 2008. Efficient tandem organic solar cells with an Al/MoO3 intermediate layer. Applied Physics Letters, 93(8), p. 83305. https://doi.org/10.1063/1.2976126

[34] You, J., Chen, C. C., Hong, Z., Yoshimura, K., Ohya, K., Xu, R., Ye, S., Gao, J., Li, G. and Yang, Y., 2013. 10.2% Power Conversion Efficiency Polymer Tandem Solar Cells Consisting of Two Identical Sub-Cells. Advanced Materials, 25(29), pp. 3973-3978. https://doi.org/10.1002/adma.201300964

[35] Chou, C. H., Kwan, W. L., Hong, Z., Chen, L. M. and Yang, Y., 2011. A metal-oxide interconnection layer for polymer tandem solar cells with an inverted

architecture. Advanced Materials, 23(10), pp. 1282-1286.
https://doi.org/10.1002/adma.201001033

[36]   Nozik, A. J., 2001. Spectroscopy and hot electron relaxation dynamics in
        semiconductor quantum wells and quantum dots. Annual review of physical
        chemistry, 52(1), pp. 193-231. https://doi.org/10.1146/annurev.physchem.52.1.193

[37]   Shrestha, S. K., Aliberti, P. and Conibeer, G. J., 2010. Energy selective contacts
        for hot carrier solar cells. Solar Energy Materials and Solar Cells, 94(9), pp. 1546-
        1550. https://doi.org/10.1016/j.solmat.2009.11.029

[38]   Luque, A. and Martí, A., 2010. Electron–phonon energy transfer in hot-carrier
        solar cells. Solar Energy Materials and Solar Cells, 94(2), pp. 287-296.
        https://doi.org/10.1016/j.solmat.2009.10.001

[39]   Takeda, Y., Ito, T., Motohiro, T., König, D., Shrestha, S. and Conibeer, G., 2009.
        Hot carrier solar cells operating under practical conditions. Journal of Applied
        Physics, 105(7), p. 074905. https://doi.org/10.1063/1.3086447

[40]   Würfel, P., Brown, A. S., Humphrey, T. E. and Green, M. A., 2005. Particle
        conservation in the hot-carrier solar cell. Progress in Photovoltaics: Research and
        Applications, 13(4), pp. 277-285. https://doi.org/10.1002/pip.584

[41]   König, D., Casalenuovo, K., Takeda, Y., Conibeer, G., Guillemoles, J. F.,
        Patterson, R., Huang, L. M. and Green, M. A., 2010. Hot carrier solar cells:
        Principles, materials and design. Physica E: Low-dimensional Systems and
        Nanostructures, 42(10), pp. 2862-2866.
        https://doi.org/10.1016/j.physe.2009.12.032

[42]   Conibeer, G. J., König, D., Green, M. A. and Guillemoles, J. F., 2008. Slowing of
        carrier cooling in hot carrier solar cells. Thin Solid Films, 516(20), pp. 6948-6953.
        https://doi.org/10.1016/j.tsf.2007.12.102

[43]   Le Bris, A. and Guillemoles, J. F., 2010. Hot carrier solar cells: achievable
        efficiency accounting for heat losses in the absorber and through contacts. Applied
        Physics Letters, 97(11), p. 113506. https://doi.org/10.1063/1.3489405

[44]   Martí, A., López, N., Antolín, E., Cánovas, E., Stanley, C., Farmer, C., Cuadra, L.
        and Luque, A., 2006. Novel semiconductor solar cell structures: The quantum dot
        intermediate band solar cell. Thin Solid Films, 511, pp. 638-644.
        https://doi.org/10.1016/j.tsf.2005.12.122

[45] Wang, W., Lin, A. S. and Phillips, J. D., 2009. Intermediate-band photovoltaic solar cell based on ZnTe: O. Applied Physics Letters, 95(1), p. 011103. https://doi.org/10.1063/1.3166863

[46] Luque, A., Panchak, A., Vlasov, A., Martí, A. and Andreev, V., 2016. Four-band Hamiltonian for fast calculations in intermediate-band solar cells. Physica E: Low-dimensional Systems and Nanostructures, 76, pp. 127-134. https://doi.org/10.1016/j.physe.2015.10.019

[47] Nozik, A. J., 2008. Multiple exciton generation in semiconductor quantum dots. Chemical Physics Letters, 457(1), pp. 3-11. https://doi.org/10.1016/j.cplett.2008.03.094

[48] Nozik, A. J., Beard, M. C., Luther, J. M., Law, M., Ellingson, R. J. and Johnson, J. C., 2010. Semiconductor quantum dots and quantum dot arrays and applications of multiple exciton generation to third-generation photovoltaic solar cells. Chemical reviews, 110(11), pp. 6873-6890. https://doi.org/10.1021/cr900289f

[49] Stolle, C. J., Harvey, T. B., Pernik, D. R., Hibbert, J. I., Du, J., Rhee, D. J., Akhavan, V. A., Schaller, R. D. and Korgel, B. A., 2014. Multiexciton solar cells of CuInSe2 nanocrystals. The journal of physical chemistry letters, 5(2), pp. 304-309. https://doi.org/10.1021/jz402596v

[50] Davis, N. J., Böhm, M. L., Tabachnyk, M., Wisnivesky-Rocca-Rivarola, F., Jellicoe, T. C., Ducati, C., Ehrler, B. and Greenham, N. C., 2015. Multiple-exciton generation in lead selenide nanorod solar cells with external quantum efficiencies exceeding 120 %. Nature communications, 6. https://doi.org/10.1038/ncomms9259

[51] Bera, D., Qian, L., Tseng, T. K. and Holloway, P. H., 2010. Quantum dots and their multimodal applications: a review. Materials, 3(4), pp. 2260-2345. https://doi.org/10.3390/ma3042260

[52] Wood, A., Giersig, M. and Mulvaney, P., 2001. Fermi level equilibration in quantum dot-metal nanojunctions. The Journal of Physical Chemistry B, 105(37), pp. 8810-8815. https://doi.org/10.1021/jp011576t

[53] Kamat, P. V., 2008. Quantum dot solar cells. Semiconductor nanocrystals as light harvesters. The Journal of Physical Chemistry C, 112(48), pp. 18737-18753. https://doi.org/10.1021/jp806791s

[54] Johnston, K. W., Pattantyus-Abraham, A. G., Clifford, J. P., Myrskog, S. H., MacNeil, D. D., Levina, L. and Sargent, E. H., 2008. Schottky-quantum dot

photovoltaics for efficient infrared power conversion. Applied Physics Letters, 92(15), p. 151115. https://doi.org/10.1063/1.2912340

[55] Hoppe, H. and Sariciftci, N. S., 2004. Organic solar cells: An overview. Journal of Materials Research, 19(07), pp. 1924-1945. https://doi.org/10.1557/JMR.2004.0252

[56] Greenham, N. C., Peng, X. and Alivisatos, A. P., 1996. Charge separation and transport in conjugated-polymer/semiconductor-nanocrystal composites studied by photoluminescence quenching and photoconductivity. Physical review B, 54(24), p. 17628. https://doi.org/10.1103/PhysRevB.54.17628

[57] Cui, D., Xu, J., Zhu, T., Paradee, G., Ashok, S. and Gerhold, M., 2006. Harvest of near infrared light in PbSe nanocrystal-polymer hybrid photovoltaic cells. Applied Physics Letters, 88(18), p. 183111. https://doi.org/10.1063/1.2201047

[58] Jiang, X., Schaller, R. D., Lee, S. B., Pietryga, J. M., Klimov, V. I. and Zakhidov, A. A., 2007. PbSe nanocrystal/conducting polymer solar cells with an infrared response to 2 micron. Journal of materials research, 22(08), pp. 2204-2210. https://doi.org/10.1557/jmr.2007.0289

[59] Liu, C. Y., Holman, Z. C. and Kortshagen, U. R., 2008. Hybrid solar cells from P3HT and silicon nanocrystals. Nano letters, 9(1), pp. 449-452. https://doi.org/10.1021/nl8034338

[60] Guchhait, A., Rath, A. K. and Pal, A. J., 2011. To make polymer: quantum dot hybrid solar cells NIR-active by increasing diameter of PbSnanoparticles. Solar energy materials and solar cells, 95(2), pp. 651-656. https://doi.org/10.1016/j.solmat.2010.09.034

[61] Ren, S., Chang, L. Y., Lim, S. K., Zhao, J., Smith, M., Zhao, N., Bulovic, V., Bawendi, M. and Gradecak, S., 2011. Inorganic–organic hybrid solar cell: bridging quantum dots to conjugated polymer nanowires. Nano letters, 11(9), pp. 3998-4002. https://doi.org/10.1021/nl202435t

[62] Chen, H. C., Lin, C. C., Han, H. V., Chen, K. J., Tsai, Y. L., Chang, Y. A., Shih, M. H., Kuo, H. C. and Yu, P., 2012. Enhancement of power conversion efficiency in GaAs solar cells with dual-layer quantum dots using flexible PDMS film. Solar Energy Materials and Solar Cells, 104, pp. 92-96. https://doi.org/10.1016/j.solmat.2012.05.003

[64] Sehgal, P. and Kumar Narula, A., 2015. Quantum dot sensitized solar cell based on poly (3-hexyl thiophene)/CdSe nanocomposites. Optical Materials, 48, pp. 44-50. https://doi.org/10.1016/j.optmat.2015.07.027

[65] Huynh, W. U., Dittmer, J. J. and Alivisatos, A. P., 2002. Hybrid nanorod-polymer solar cells. science, 295(5564), pp. 2425-2427.

[66] Kim, J., Hong, Z., Li, G., Song, T. B., Chey, J., Lee, Y. S., You, J., Chen, C. C., Sadana, D. K. and Yang, Y., 2015. 10.5% efficient polymer and amorphous silicon hybrid tandem photovoltaic cell. Nature communications, 6. https://doi.org/10.1038/ncomms7391

[67] Grätzel, M., 2003. Dye-sensitized solar cells. Journal of Photochemistry and Photobiology C: Photochemistry Reviews, 4(2), pp. 145-153. https://doi.org/10.1016/S1389-5567(03)00026-1

[68] Kamat, P. V., Tvrdy, K., Baker, D. R. and Radich, J. G., 2010. Beyond photovoltaics: semiconductor nanoarchitectures for liquid-junction solar cells. Chemical reviews, 110(11), pp. 6664-6688. https://doi.org/10.1021/cr100243p

[69] Hod, I. and Zaban, A., 2013. Materials and interfaces in quantum dot sensitized solar cells: challenges, advances and prospects. Langmuir, 30(25), pp. 7264-7273. https://doi.org/10.1021/la403768j

[70] Fuke, N., Hoch, L. B., Koposov, A. Y., Manner, V. W., Werder, D. J., Fukui, A., Koide, N., Katayama, H. and Sykora, M., 2010. CdSe quantum-dot-sensitized solar cell with~ 100% internal quantum efficiency. Acs Nano, 4(11), pp. 6377-6386. https://doi.org/10.1021/nn101319x

[71] Semonin, O. E., Luther, J. M., Choi, S., Chen, H. Y., Gao, J., Nozik, A. J. and Beard, M. C., 2011. Peak external photocurrent quantum efficiency exceeding 100% via MEG in a quantum dot solar cell. Science, 334(6062), pp. 1530-1533. https://doi.org/10.1126/science.1209845

[72] Zhao, K., Pan, Z., Mora-Seró, I., Cánovas, E., Wang, H., Song, Y., Gong, X., Wang, J., Bonn, M., Bisquert, J. and Zhong, X., 2015. Boosting power conversion efficiencies of quantum-dot-sensitized solar cells beyond 8% by recombination control. Journal of the American Chemical Society, 137(16), pp. 5602-5609. https://doi.org/10.1021/jacs.5b01946

[73] Luque, A. and Martí, A., 1997. Increasing the efficiency of ideal solar cells by photon induced transitions at intermediate levels. Physical Review Letters, 78(26), p. 5014. https://doi.org/10.1103/PhysRevLett.78.5014

[74] Cuadra, L., Martı, A. and Luque, A., 2004. Present status of intermediate band solar cell research. Thin Solid Films, 451, pp. 593-599. https://doi.org/10.1016/j.tsf.2003.11.047

[75] Luque, A., Martí, A., López, N., Antolín, E., Cánovas, E., Stanley, C., Farmer, C., Caballero, L. J., Cuadra, L. and Balenzategui, J. L., 2005. Experimental analysis of the quasi-Fermi level split in quantum dot intermediate-band solar cells. Applied Physics Letters, 87(8), p. 083505. https://doi.org/10.1063/1.2034090

[76] Martí, A., Antolín, E., Stanley, C. R., Farmer, C. D., López, N., Diaz, P., Cánovas, E., Linares, P. G. and Luque, A., 2006. Production of photocurrent due to intermediate-to-conduction-band transitions: a demonstration of a key operating principle of the intermediate-band solar cell. Physical Review Letters, 97(24), p. 247701. https://doi.org/10.1103/PhysRevLett.97.247701

[77] Ahsan, N., Miyashita, N., Islam, M. M., Yu, K. M., Walukiewicz, W. and Okada, Y., 2012. Two-photon excitation in an intermediate band solar cell structure. Applied Physics Letters, 100(17), p. 172111. https://doi.org/10.1063/1.4709405

Chapter 5

Photoelectrochemical solar cells using nanocrystalline copper selenide photo electrode

Kavita Gour*[1], Preeti Pathak[1], M. Ramrakhiani[2] and P. Mor[2]

[1]Mata Gujri Mahila Mahavidyalaya, Jabalpur (MP), India

[2]Department of Physics and Electronics, Rani Durgawati Vishwavidyalaya, Jabalpur (MP), India

*kavitagour119@rediffmail.com

Abstract

The development of thin film solar cells continues to be an active area of research. Nanocrystalline thin films of copper selenide have been grown on glass and indium tin oxide (ITO) substrates using a chemical method. The golden films have been synthesized at different temperatures, for different deposition times and from different concentration of solutions and annealed at 200°C for 2H. They were then examined by means of X-ray diffraction (XRD) and (AFM) micrographs for their structural and morphological properties. Average spherical grains of the order of 25 nm to 35 nm in size aggregated over about 120 ± 10 nm for different concentration islands are visible in the AFM images. The conductivity in copper selenide thin films makes it a suitable candidate for solar cells. Their photoelectrochemical performance was investigated in a standard two electrodes configuration with redox electrolyte. The investigation may be useful in obtaining efficient, stable and low cost solar cells to compete with the existing technology.

Keywords

Copper Selenide Thin Films, Photoelectrochemical Cells, XRD, Surface Morphology

Contents

1. Introduction 135

1.1 Thin-Film Photovoltaics 137

2. Synthesis of Samples 138

2.1   Reaction Mechanism of Nanocrystalline CuSe Thin
        Film Deposition 139

3.   Thickness Measurement...143

4.   X-Ray Diffraction Studies (XRD) 145

5.   Scanning Electron Microscopy (SEM) 157

6.   Atomic Force Microscopy (AFM) 162

7.   Optical Absorption 167

8.   Photoeletrochemical Cell Studies.....................................173

8.1   CuSe Nanocrystalline Photoelectrode Deposited for
        Different Durations..174

8.2   CuSe Nanocrystalline Photoelectrode deposited at
        Different Temperatures 176

8.3   CuSe Nanocrystalline Photoelectrode Deposited from Various
        Concentration of Precursor Solutions 179

9.   Conclusions..182

References..183

1.   Introduction

Solar photovoltaic energy conversion is a one-step conversion process which generates electrical energy from light energy. The explanation relies on ideas from the quantum theory. Light is made up of packets of energy, called photons, whose energy depends only upon the frequency, or color, of the light. The energy of visible photons is sufficient to excite electrons, bound into solids, up to higher energy levels where they are more free to move. An extreme example of this is the photoelectric effect, the celebrated experiment which was explained by Einstein in 1905, where blue or ultraviolet light provides enough energy for electrons to escape completely from the surface of a metal. Normally, when light is absorbed by matter, photons are given up to excite electrons to higher energy states within the material, but the excited electrons quickly relax back to their ground state.

Solar energy source is a long term natural source of energy. The photo electrochemical cells are considered as potential energy source for the future. The photo electrochemical cell (PEC) is an attractive means of converting solar energy into electricity. It is considered a major candidate for obtaining energy from the sun, since it can convert sunlight directly into electricity. It can provide nearly permanent power and is virtually free of pollution.

Photovoltaic effect is a process in which two dissimilar materials in close contact produce an electrical voltage when struck by light. The research paper describes the photovoltaic effect and characteristics of CuSe thin film solar cells. The photo electrochemical effect is exhibited by the semiconductor (SC) and electrolytic junction in a cell consisting of photo-responsive electrode, an electrolyte and a suitable counter electrode. When SC-electrolyte junction is illuminated by the high energy photon of hv > Eg, photons are absorbed by SC electrode producing electron-hole pairs. This separation of electron hole pairs results in photo-voltage. The working of solar cell is related to the photo voltaic effect. Copper selenide has many phases and structural forms: stoichiometric Cu_2Se, Cu_3Se_2, CuSe and $CuSe_2$, as well as non-stoichiometric, $Cu_{2-x}Se$. The thermal stability of these compounds varies depending on their composition. Among various techniques for preparation of nanometer size materials, chemical methods offer better orientation, which are least expensive, non-polluting and easy to incorporate suitable doping materials for altering the film properties These processes are the low-temperature processes which enables formation of thin films onto plastic as well as glass substrates in addition to the conducting films (substrates) which can potentially lead to a new generation of photovoltaic devices that are light in weight, foldable, flexible and moldable. The inorganic thin films of copper selenide is focused in this work by direct chemical deposition method and annealed at 200°C for 2 h for crystallinity improvement. These nanocrystalline film are used as photo electrode in PEC cell to study the photovoltaic effect. Solar cells are the type of devices which efficiently and directly converts the sunlight into electricity by the use of photovoltaic technology. These devices are specifically designed for the energy storage significantly in the situations where the source of light is not specific. Solar panels are developed in the cells that are responsible for storage and the solar power is generated by these solar panels. Solar panels are developed in the cells that are responsible for storage and the solar power is generated by these solar panels.

The nanoscale is the size scale at which nanotechnology operates. Though we have a lower limit on this scale size (the size of one atom), pinning down an upper limit on this scale is more difficult. A useful and well-accepted convention is that for something to exist on the nanoscale at least one of its dimensions (height, width, or depth) must be less than about 100 nanometers. In fact, it is these limits to the nanoscale that the National Nanotechnology Initiative (NNI) uses for its definition of nanotechnology:

"Nanotechnology is the understanding and control of matter at dimensions of roughly 1 to 100 nm, where unique phenomena enable novel applications." To this, it is useful to add two other statements to form a complete definition. First, nanotechnology includes the forming and use of materials, structures, devices, and systems that have unique properties because of their small size. Also, nanotechnology includes the technologies that enable the control of materials at the nanoscale, because of their small size. Also, nanotechnology includes the technologies that enable the control of materials at the nanoscale. Nanotechnology is based on the manipulation, control, and integration of atoms and molecules to form materials, structures, components, devices, and systems at the nanoscale. Thin film nanotechnology is the application of nanoscience, especially to industrial and commercial objectives.

1.1   Thin-Film Photovoltaics

From the early days of photovoltaics until today, thin-film solar cells have always competed with technologies based on single-crystal materials such as Si and GaAs. Owing to their amorphous or polycrystalline nature, thin-film solar cells always suffered from power conversion efficiencies lower than those of the bulk technologies. This drawback was and still is counterbalanced by several inherent advantages of thin-film technologies. Since in the early years of photovoltaics space applications were the driving force for the development of solar cells, the argument in favor of thin films was their potential lighter weight as compared with bulk materials. An extended interest in solar cells as a source of renewable energy emerged in the mid-seventies as the limitations of fossil energy resources were widely recognized. For terrestrial power applications, the cost arguments and the superior energy balance strongly favored thin films. However, from the various materials under consideration in the fifties and sixties, only four thin-film technologies, namely amorphous hydrogen alloyed (a-)Si:H and the polycrystalline heterojunction systems CdS/CuxS, CdS/CdTe, and CdS/CuInSe2, entered pilot production. Activities in the CdS/CuxS system stopped at the beginning of the eighties because of stability problems. At that time, amorphous silicon became the front runner in thin-film technologies keeping almost constantly a share of about 10% in a constantly growing photovoltaic market, the remaining 90% kept by crystalline Si. Despite their high-efficiency potential, polycrystalline heterojunction solar cells based on CdTe and CuInSe2 did not play an economic role until the turn of the century. During the accelerated growth of the worldwide photovoltaic market in the first decade of the new century, the three inorganic thin-film technologies increased their market share to 14%, where approximately 9% are covered by CdTe modules (numbers from 2008). With annual production figures in the GW range, inorganic thin-film photovoltaics have become a multibillion dollar business. In order to expand this position, further dramatic cost-reduction is required combined with a substantial increase in module efficiency. In this context, material and device characterization becomes an important task not only for quality control in an expanding industry but also remains at the very heart of further technological progress. The thin film technologies have in common that they consist of layer sequences from disordered semiconductor materials that are deposited onto a supporting substrate or superstrate with the help of vacuum technologies. This layer structure and the use of disordered materials defines a fundamental difference to devices based on crystalline c-Si where a self-supporting Si wafer is transformed into a solar cell via a solid-state diffusion of dopant atoms. Thus, there are only the front and the back surface as critical interfaces in the classical wafer solar cell (with the notable exception of the a-Si:H/c-Si heterojunction solar cell). In thin-film solar cells, the number of functional layers can amount to up to eight and more. Some of these layers have thicknesses as low as 10 nm. In large-area modules, these layers homogenously cover areas of up to $6m^2$.

Properties of thin film solar cells:

- Less material use and less energy intensive processing leads to lower cost.

- Easier to fabricate large area devices required for PV applications.

- Performance comparable to single crystal materials.

- Manufacturing Technology is well established.

- Easier Integration of device structure.

- Less stringent requirement on material properties.

2.   Synthesis of Samples

In the present work nanocrystalline thin film of copper-selenide were grown on glass and indium tin oxide (ITO) substrates using chemical method. Preparation of photo- anodes is done with different temperature, different deposition time and different concentration of solution and golden films are synthesized. The sample of CuSe thin films were annealed at 200°C for 2H in a Oven.

Conditions for preparation of photoelectrode:

Concentration- B (Normal) Temperature- 45°C [Fixed]	Concentration- B (Normal) Deposition time- 120 min [Fixed]	Temperature- 45°C Deposition time- 120 min [Fixed]
Deposition time- 60 min	Temperature- 45°C	Concentration- A (Half)
Deposition time- 120 min	Temperature- 60°C	Concentration- B (Normal)
Deposition time- 180 min	Temperature- 90°C	Concentration- C (Double)

2.1 Reaction Mechanism of Nanocrystalline CuSe Thin Film Deposition

It is well known that sodium selenosulphate hydrolyzes in alkaline medium to give Se^{2-} ions as

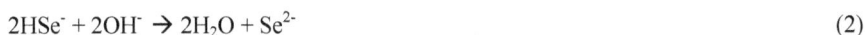

$$Na_2SeSO_3 + OH^- \rightarrow Na_2SO_4 + HSe^- \tag{1}$$

$$2HSe^- + 2OH^- \rightarrow 2H_2O + Se^{2-} \tag{2}$$

In presence of Cu^{2+} ions in the bath, copper selenide will be formed, if the ionic product of Cu^{2+}, Se^{2-} exceeds the solubility product of copper selenide. The formation of a particular phase such as Cu_2Se, Cu_2Se_3, $Cu_{2-x}Se$ will be governed by copper ion concentration, deposition temperature, rate of release of Cu^{2+} from the tartaric acid complex as facilitating the desired ion transport.

A reaction vessel containing glass and indium doped tin oxide (ITO) substrates was used in the experiment. For preparation of sodium salenosulfite (Na_2SeSO_3) first of all 100ml aqueous solution of sodium sulfite (1M) are prepared, then 0.1M of selenium powder are added with constant stirring, the mixture are heated to 60°C till the selenium is dissolved in the solution. Upon filtration, sodium salenosulfite solutions are obtained. The reaction vessel was filled with the composition of solution: 0.1 M 50 ml $CuSO_4$ + 1 M, 2 ml tartaric acid and 0.1M 6 ml Na_2SeSO_3 solution. The glass slides were cleaned with a suitable cleanser, scrubbed with soft cotton, washed thoroughly with de-ionized water followed by rinsing and drying in air. These glass and ITO coated sides were used as substrates for deposition, which were fixed to the circular holder (shown in Figure 1). The temperature of the reaction solution was kept as 45°C during the deposition.

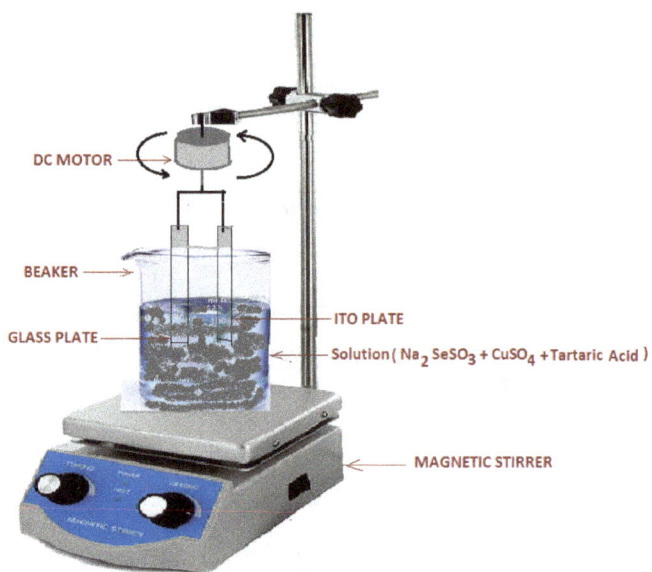

Figure 1(a): *Experimental setup for preparation of photoelectrode.*

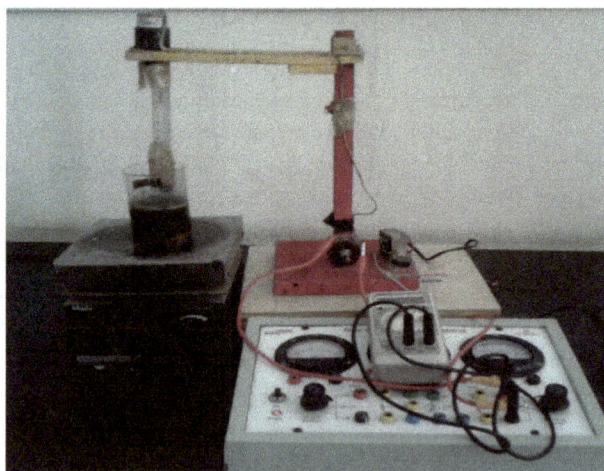

Figure 1(b): *Experimental setup for preparation of photoelectrode (photograph).*

The ITO side was kept facing towards the solution and allowed to rotate with a speed of 25 rpm. The thermostat was set to a temperature of 45°C and the reaction was carried out for 60 min, 90 min, 120 min and 180 min with constant stirring of the solution throughout the experiment.

In the present work nanocrystalline thin films of copper selenide are grown on glass and indium tin oxide (ITO) substrates using a chemical method. The photo anodes are also prepared at different temperatures 20°C, 45°C, 60°C & 90°C for deposition time is 2 hour. The samples of thin films are annealed at 200°C for 2 hour in a oven.

The photo anodes are also prepared for different concentration of precursor solution at temperature of 45°C for deposition time of 2 hour.

Composition of solutions for:

(A) Half Concentration: The reaction vessel was filled with the composition of solution: 0.05 M (50 ml) $CuSO_4$, 0.5 M (2 ml) tartaric acid and 0.05 M (6 ml) Na_2SeSO_3 solution.

(B) Normal Concentration: The reaction vessel was filled with the composition of solution: 0.1 M (50 ml) $CuSO_4$, 1 M (2 ml) tartaric acid and 0.1M (6 ml) Na_2SeSO_3 solution.

(C) Double concentration: The reaction vessel was filled with the composition of solution: 0.2 M (50 ml) $CuSO_4$ 2M (2ml) tartaric acid and 0.2M (6 ml) Na_2SeSO_3 solution.

The samples of thin films are annealed at 200°C for 2 hour in a oven. Good golden adherent films were deposited onto both glass and ITO substrates. The samples of CuSe thin film deposited onto glass substrates for different deposition time, different temperature, and different concentration of solution were examined by means of X-ray diffraction (XRD), SEM and (AFM) micrographs for their structural and morphological properties.

The thin films of CuSe on ITO substrate for different temperature, different deposition time and different concentration of precursor solution were used in PEC cells and photovoltaic effect was investigated. Good golden adherent films were deposited onto both glass and ITO substrates. The configuration of PEC cell (Figure 2) is a single glass vessel surrounded by dark, using CuSe nanocrystalline film on ITO substrate as photo anode, graphite as counter electrode and polysulphide solution (using Sulpher-S, NaOH, Na_2S) as electrolyte.

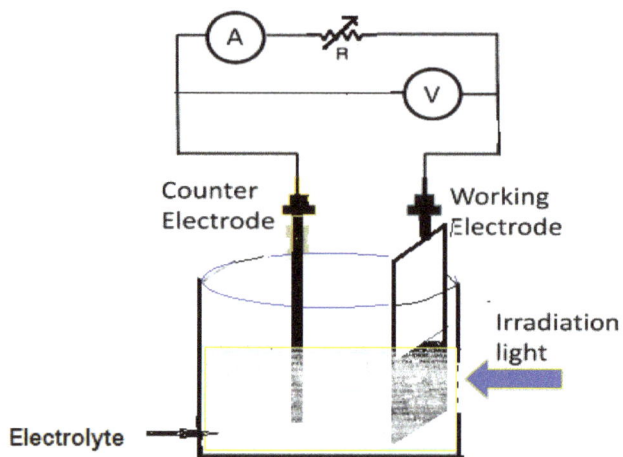

Figure 2(a): *Experimental setup for PEC cell.*

Figure 2(b): *Experimental setup for PEC cell (photograph).*

The photo-electrochemical (PEC) performance of copper selenide film was studied. The distance between working electrode and counter electrode was kept 1 cm. Photocurrent–voltage (I–V) performances of annealed copper selenide photo electrodes were measured under dark and 315 lux light illumination intensity.

3.   Thickness Measurement

The thickness of the film is the most significant parameter that affects the properties of the thin film. It may be measured either by in situ monitoring of the rate of the deposition or after the film is taken out from deposition chamber. Technique of the first type often referred to as monitor methods generally allow both monitoring and controlling of deposition rate and film thickness. Any known physical quantity related to film thickness can be used to measure the thickness. The method chosen should be convenient, reliable and simple.

One of the most convenient and earliest methods for determining film thickness is the gravimetric method. In this method, area and weight of the film are measured. The thickness was obtained using the formula.

$t = M / \rho * A$

$M = m_1 - m_2$

where, t is the film thickness, M is mass of the film material (in gm), A is area of the film (in cm^2), m_1 is mass of the substrate with film, m_2 is mass of the substrate without film and ρ is density of the film material (gm/cm^3). The density of the CuSe film material is 6.0 (gm/cm^3).

The thickness of the film is the most significant parameter that affects the properties of the thin film. The thickness of the deposited copper selenide thin films is measure for different temperature, different deposition time and different concentration of solution.

Table 1A: *Thickness of Film for Different Deposition Time.*

S.N.	Deposition Time	Thickness of the Film [μm]
1	60 min	34.6
2	120 min	146.7
3	180 min	281.1

Figure 3(a): *Variation of thickness of the film with deposition time.*

The thickness of the deposited copper selenide thin film increases with increase in deposition time. The increase in thickness is nearly linear with the deposition time [Figure 3(a)].

Table 1B: *Thickness of Films for Various Temperatures during Deposition.*

S.N.	Temperature	Thickness of the Film [μm]
1	45°C	146.7
2	60°C	85.6
3	90°C	24.4

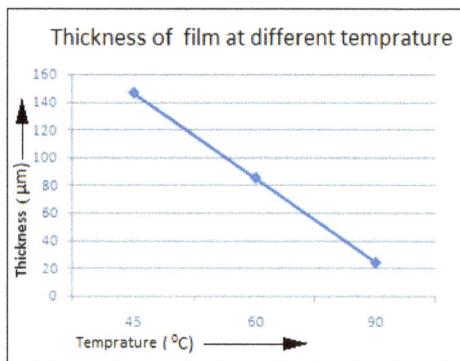

Figure 3(b): *Variation of thickness of the film deposited at different temperatures.*

Figure 3(b) shows graph of thickness of the film for various deposition temperature. The thickness of the deposited copper selenide thin film decreases with deposition temperature linearly.

Table 1C: *Thickness of Film for Different Concentrations of Precursor Solution.*

S.N.	Sample Concentration	Thickness of the Film [μm]
1	Half concentration (A)	22.2
2	Normal concentration (B)	146.7
3	Double concentration (C)	218.9

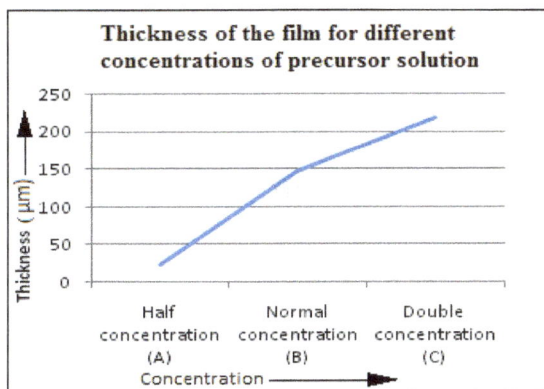

Figure 3(c): *Thickness of the film for different concentrations of precursor solution.*

The thickness of the deposited copper selenide thin film increases with increase in concentrations of precursor solution [Figure 3(c)].

4.   X-Ray Diffraction Studies (XRD)

For finding the results of X-ray diffraction studies of CuSe deposited thin film on glass plate, XRD (Rikagu) Epifluorescence Microscopy (NIKON) is used. PDF card number: 9000063 is used for comparing the standard values with the experimental data. A narrowed peak (horizontal) width in annealed film confirms grain size growth. From the XRD profiles, the inter-planar spacing and miller indices (h k l) are calculated using the Bragg's relation, where 2θ is the angle between the incident and the scattered X-ray. The wavelength of the X-ray is 1.540559 A° (given by Center of Nanoscience & Nanotechecnology, Sathyabama University, Chennai).

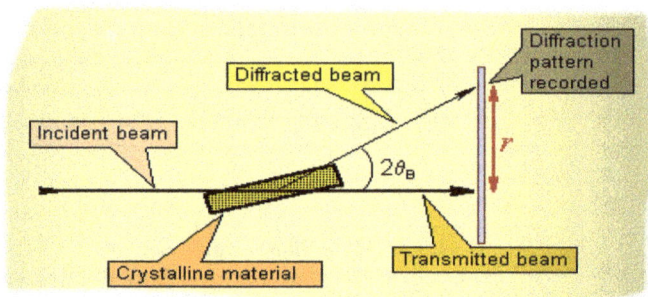

Figure 4: *Principle of X-ray diffraction.*

Figure 4 shows, ray diagram for basic principle of X-ray diffraction. The grain size (D) is calculated using the Scherrer's formula from the full-width at half maximum (FWHM) of corresponding peak of the XRD pattern.

Scherrer's formula:

$$D = C \lambda / (B \cos (\theta))$$

where λ is the wavelength (Å), B is FWHM (radians) corrected for instrument broadening, θ is the Bragg angle, C is a crystal shape factor from 0.9~1.

Table 2: *Measurement Conditions.*

X-Ray	30 kV, 100 mA	Scan speed/Duration time	4.0000 deg/min
Goniometer	Smart Lab	Step width	0.0200 deg
Attachment	Standard	Scan axis	Theta/2-Theta
Filter	Cu_K-beta	Scan range	20.0000-80.0000 deg
CBO selection slit	BB	Incident slit	2/3 deg
Diffracted beam mono	None	Length limiting slit	10.0 mm
Detector	SC-70	Receiving slit #1	2/3 deg
Scan mode	CONTINUOUS	Receiving slit #2	0.150 mm

For XRD characterization, thin films of CuSe were deposited on glass plate and deposition time is taken 60 min, 120 min and 180 min.

(i)   XRD Characterization for 60 min deposition time

Figure 5(a) shows X-ray diffraction pattern of CuSe thin film deposited for 60 min. From the XRD profiles a narrowed peak (horizontal width) in CuSe thin film confirms grain size growth. The inter-planar spacing for (h k l) plane is calculated using the Bragg's relation.

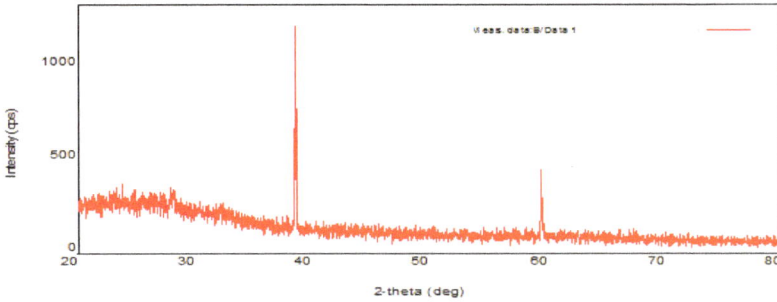

Figure 5(a): *X-ray diffraction pattern of CuSe thin flim (Deposition time - 60 min).*

Table 3A: *Peak List.*

No.	2-theta (deg)	d (ang.)	Height (cps)	FWHM (deg)	Int. I (cps deg)	Int. W (deg)	Asym. factor
1	38.605	2.33034	1173	0.051	80	0.068	1.1
2	59.428	1.55405	389	0.066	33.9	0.087	0.40

The structure of the thin film of CuSe has predominant orientation along the (1, 0, 5) and (1, 0, 10) planes.

(ii)   XRD Characterization for 120 min. deposition time

Figure 5(b) shows the X-ray diffraction pattern of deposited CuSe thin film for deposition time 120 min.

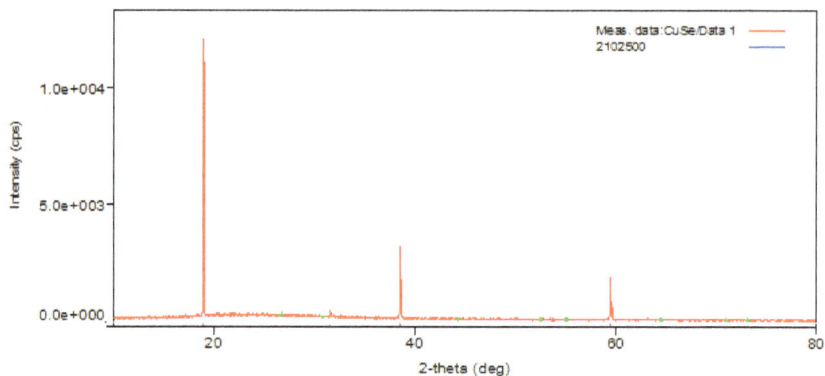

Figure 5(b): *X-ray diffraction pattern of CuSe thin film (Deposition time - 120 min).*

Table 3B: *Peak List.*

No.	2-theta (deg)	d (ang.)	Height (cps)	FWHM (deg)	Int. I (cps deg)	Int. W (deg)	Asym. factor
1	19.0269	4.66059	12537	0.0430	632	0.0504	1.00
2	31.675	2.8226	287	0.029	15.8	0.055	0.8
3	38.6053	2.33029	3233	0.0501	203	0.063	1.27
4	59.433	1.55392	1764	0.061	139	0.079	0.73

The structure of the thin film of CuSe is predominant orientation along the (0, 0, 4), (0, 0, 6), (1, 0, 5) and (1, 0, 10) planes.

(iii) XRD Characterization for 180 min deposition time

Figure 5(c) shows X-ray diffraction pattern of deposited CuSe thin film for deposition time is180 min.

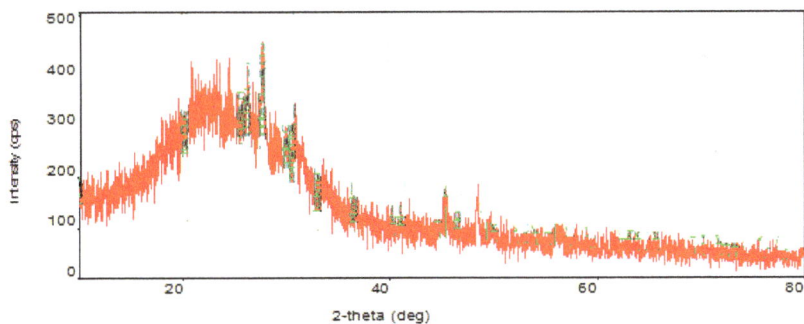

Figure 5(c): *X-ray diffraction pattern of CuSe thin film (Deposition time - 180 min).*

Table 3C: *Peak List.*

No.	2-theta (deg)	d (ang.)	Height (cps)	FWHM (deg)	Int. I (cps deg)	Int. W (deg)	Asym. factor
1	20.8424	4.25853	256.315	0.05	48.7064	0.190025	1
2	27.6902	3.21898	256.025	0.05	51.5558	0.201370	1
3	31.019	2.88071	215.506	0.05	44.3567	0.205826	1
4	45.4755	1.99293	45.3322	0.05	9.92131	0.218858	1
5	48.6141	1.87135	56.065	0.05	12.3738	0.220705	1
6	56.3179	1.63226	21.5311	0.05	4.82917	0.224288	1

The structure of the thin films of CuSe has predominant orientation along the (004), (101) and (006), (107), (108) and (203) planes.

PDF Card No. : 9000063 Quality:C

Sub-File Name:					
Formula:	Cu Se				
Name:	Klockmannite			I/Ic (RIR)= 4.70	
Crystal System:	Space Group: P 63/m m c(194)			Dmeas:	
Cell Parameters:	a= 3.9380	b= 3.9380	c= 17.2500		
	Alpha= 90.000	Beta= 90.000	Gamma= 120.000		
	Volume= 231.671	Z= 1			
Reference:					
Radiation:	Wavelength=				
2Theta range:	10.25 - 99.99				
Database comments:					

Relative Intensity

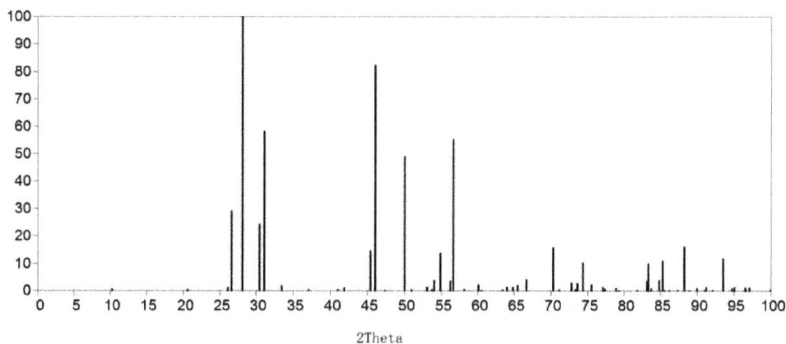

2Theta

No.	2Theta	d-Value	Intensity	h	k	l	No.	2Theta	d-Value	Intensity	h	k	l
1	10.25	8.625	0.4	0	0	2	21	54.91	1.671	0.1	1	0	9
2	20.58	4.313	0.2	0	0	4	22	56.22	1.635	3.3	2	0	3
3	26.11	3.410	1.0	1	0	0	23	56.61	1.625	55.2	1	1	6
4	26.62	3.346	29.0	1	0	1	24	58.13	1.586	0.3	2	0	4
5	28.11	3.171	100.0	1	0	2	25	60.06	1.539	1.9	1	0	10
6	30.45	2.933	24.2	1	0	3	26	60.52	1.529	0.0	2	0	5
7	31.08	2.875	58.1	0	0	6	27	63.37	1.467	0.1	2	0	6
8	33.47	2.675	1.7	1	0	4	28	63.98	1.454	1.1	1	1	8
9	37.04	2.425	0.3	1	0	5	29	64.80	1.438	1.2	0	0	12
10	41.03	2.198	0.2	1	0	6	30	65.46	1.425	1.8	1	0	11
11	41.86	2.156	0.9	0	0	8	31	66.64	1.402	4.0	2	0	7
12	45.37	1.997	14.4	1	0	7	32	70.33	1.338	15.7	2	0	8
13	46.06	1.969	82.2	1	1	0	33	71.11	1.325	0.1	1	0	12
14	47.32	1.920	0.0	1	1	2	34	72.84	1.298	2.7	1	1	10
15	50.00	1.823	48.8	1	0	8	35	73.39	1.289	0.2	2	1	0
16	50.94	1.791	0.1	1	1	4	36	73.63	1.285	2.6	2	1	1
17	53.04	1.725	1.2	0	0	10	37	74.35	1.275	10.1	2	1	2
18	53.71	1.705	0.2	2	0	0	38	74.40	1.274	0.0	2	0	9
19	53.99	1.697	3.6	2	0	1	39	75.53	1.258	2.0	2	1	3
20	54.84	1.673	13.7	2	0	2	40	77.06	1.237	1.2	1	0	13

Note: 2theta are calculated with wavelength = 1.54059

2014-Feb-17 16:05:09 Page-1/2

Crystal structure of CuSe deposited on glass plate is hexagonal, as obtained with the help of PDF Card No.: 9000063. The value of cell parameter is found to be, $a = b = 3.9380$ A°, $c = 17.2500$ A° and $\alpha = 90°$, $\beta = 90°$, $\gamma = 120°$. The diffractrogram of the deposited copper selenide thin films seems to exhibit nanocrystalline in nature.

Table 4: *Particle size for various deposition time.*

S.N.	Deposition Time	Particle Size
1	60 min	26.49 nm
2	120 min	29.31 nm
3	180 min	29.568 nm

The calculated average grain size using observed peaks for deposited copper selenide thin film is shown in Table 4. It is observed that the calculated average grain size for deposited copper selenide thin film increases with increase in deposition time.

(iv) XRD Characterization for photo electrode deposition done at 45°C temperature

Figure 6(a) shows the X-ray diffraction pattern of deposited CuSe thin film for photo electrode deposition done at 45°C temperature.

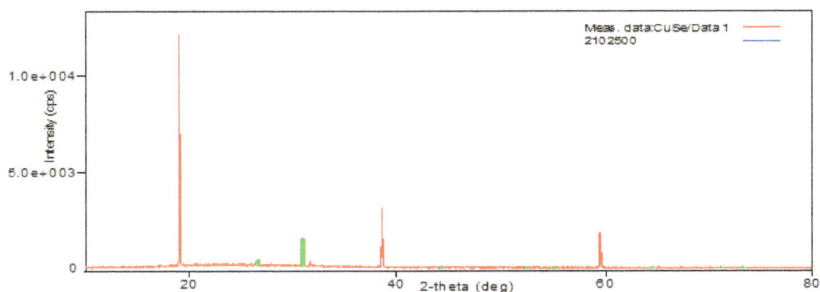

Figure 6(a): *X-ray diffraction pattern of CuSe thin film (temperature - 45°C).*

Table 5A: *Peak List*

No.	2-theta (deg)	D (angle)	Height (cps)	FWHM (deg)	Int. I (cps deg)	Int. W (deg)	Asym. factor
1	19.0269	4.66059	12537	0.0430	632	0.0504	1.00
2	31.675	2.8226	287	0.029	15.8	0.055	0.8
3	38.6053	2.33029	3233	0.0501	203	0.063	1.27
4	59.433	1.55392	1764	0.061	139	0.079	0.73

The structure of the thin film of CuSe is hexagonal with predominant orientation along the (0, 0, 4), (0, 0, 6), (1, 0, 5) and (1, 0, 10) planes.

(v) XRD Characterization for photo electrode deposition done at 60°C temperature

Figure 6(b) shows the X-ray diffraction pattern of deposited CuSe thin film for photo electrode deposition done at 60°C temperature.

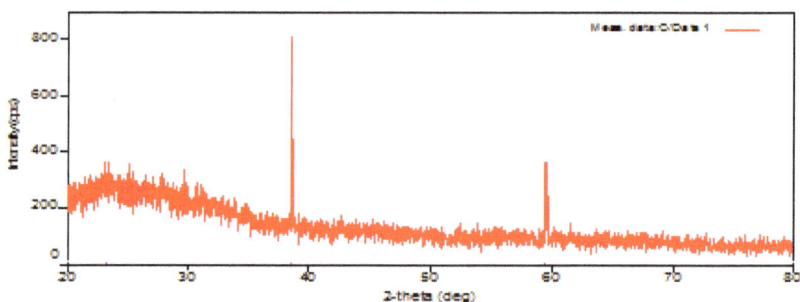

Figure 6(b): *X-ray diffraction pattern of CuSe thin film (temperature - 60°C).*

Table 5B: *Peak List*

No	2-theta (deg)	D (ang.)	Height (cps)	FWHM (deg)	Int.I (cps deg)	Int. W (deg)	Asym. factor
1	23.31	3.81	29	7.8	244	8	0.20
2	38.598	2.3307	614	0.055	37.3	0.061	1.0
3	59.430	1.55401	358	0.034	22.2	0.062	0.7

The structure of the thin film of CuSe is hexagonal with predominant orientation along the (100), (105) and (1, 0, 10) planes.

(vi) XRD Characterization for photo electrode deposition done at 90°C temperature

Figure 6(c) shows the X-ray diffraction pattern of deposited CuSe thin film for photo electrode deposition done at 70°C temperature.

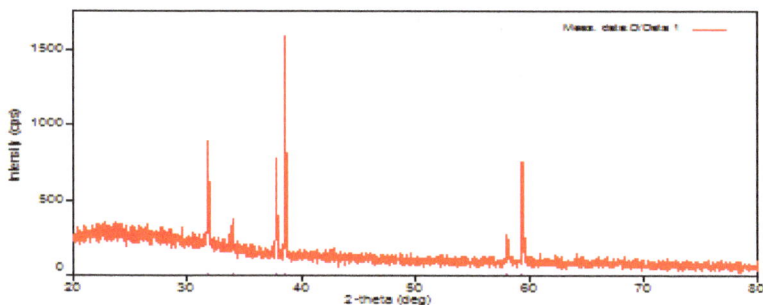

Figure 6(c): *X-ray diffraction pattern of CuSe thin film (temperature - 90°C).*

Table 5C: *Peak List*

No.	2-theta (deg)	d (ang.)	Height (cps)	FWHM (deg)	Int. I (cps deg)	Int. W (deg)	Asym. factor
1	31.851	2.8073	733	0.062	58	0.080	0.9
2	34.001	2.6346	138	0.10	17	0.12	2.1
3	37.803	2.3779	585	0.063	45	0.077	1.6
4	38.591	2.3311	1510	0.043	84	0.056	0.36
5	58.120	1.5859	196	0.077	24.5	0.12	0.6
6	59.427	1.55406	729	0.048	55.0	0.076	0.43

The structure of the thin film of CuSe is hexagonal with predominant orientation along the (006), (104), (105), (106), (204) and (1, 0, 10) planes.

Table 6: *Particle size for different temperature*

S.N.	Temperature	Particle Size
1	45°c	29.31nm
2	60°c	25.031nm
3	90°c	24.825nm

The diffractrogram of the deposited copper selenide thin films seems to exhibit nanocrystalline in nature. Crystal structure of CuSe, deposited on glass plate is found to be hexagonal. The calculated average grain size using observed peaks for deposited copper selenide thin film is shown in Table 6. It is observed that the calculated average grain size for deposited copper selenide thin film decreases with increase in temperature.

(vii) XRD Characterization for A-concentration (half con.) of precursor solution of CuSe photoelectrodes

Figure 7(a) shows X-ray diffraction patterns of CuSe thin film deposited for A-concentration (Half con.) of precursor solution of CuSe photoelectrodes.

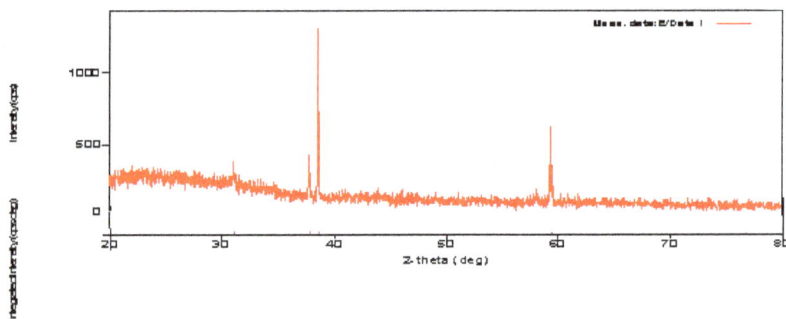

Figure 7(a): *X-ray diffraction of CuSe films from A-concentration (Half con).*

Table 7A: *Peak List*

No.	2-theta (deg)	d (ang.)	Height (cps)	FWHM (deg)	Int. I (cps deg)	Int. W (deg)	Asym. factor
1	31.14	2.870	30	2.5	155	5	4
2	37.786	2.3789	316	0.062	22.8	0.072	0.41
3	38.606	2.3302	1174	0.050	75	0.064	1.6
4	59.431	1.55399	553	0.068	49	0.088	0.8

The structure of the thin film of CuSe has predominant orientation along the (006), (105), (106) and (1, 0, 10) planes.

(viii) XRD Characterization for B-concentration (Normal con.) of precursor solution of CuSe photo electrodes

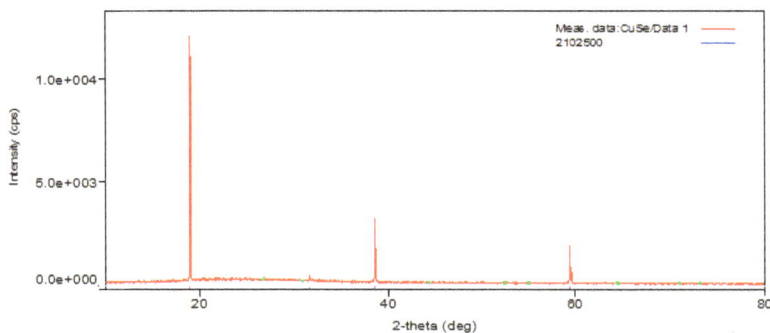

Figure 7(b): *X-ray diffraction of CuSe film from B-concentration (Normal con).*

Table 7B: *Peak List*

No.	2-theta (deg)	d (ang.)	Height (cps)	FWHM (deg)	Int. I (cps deg)	Int. W (deg)	Asym. factor
1	19.0269	4.66059	12537	0.0430	632	0.0504	1.00
2	31.675	2.8226	287	0.029	15.8	0.055	0.8
3	38.6053	2.33029	3233	0.0501	203	0.063	1.27
4	59.433	1.55392	1764	0.061	139	0.079	0.73

The structure of the thin film of CuSe is predominant orientation along the (0, 0, 4), (0, 0, 6), (1, 0, 5) and (1, 0, 10) planes.

(ix) XRD Characterization for C-concentration (Double) of precursor solution of CuSe photo electrodes

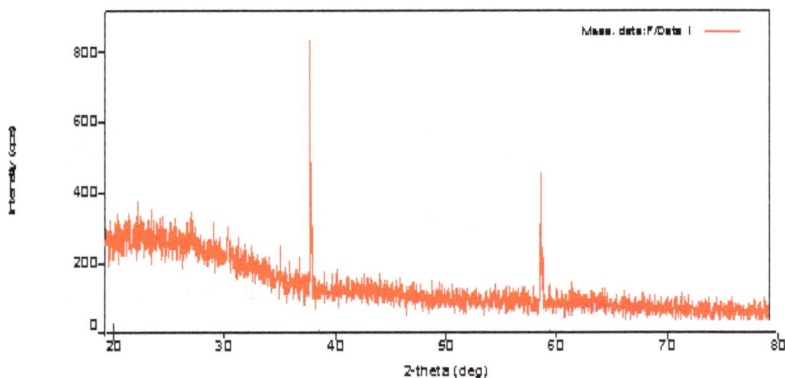

Figure 7(c): *X-ray diffraction of CuSe film from C-concentration (Double con).*

Table 7C: *Peak List.*

No.	2-theta (deg)	d (ang.)	Height (cps)	FWHM (deg)	Int. I (cps deg)	Int. W (deg)	Asym. factor
1	38.595	2.3309	709	0.055	47	0.066	1.0
2	59.410	1.55447	340	0.075	32	0.094	0.4

Table 8: *Particle size for different concentrations of precursor solution.*

S.N.	Sample Concentration	Particle Size
1	Half concentration (A)	25.495 nm
2	Normal concentration (B)	29.31 nm
3	Double concentration (C)	34.06 nm

The calculated average grain size using observed peaks for deposited copper selenide thin film is shown in Table 8. It is observed that the calculated average grain size for deposited copper selenide thin film increases with increase in concentration of precursor solution.

XRD shows the calculated average grain size using observed peaks for deposited copper selenide thin film is increases with deposition time and concentration of precursor solution but average grain size for deposited copper selenide thin film is decreases with higher temperature. From PDF card crystal class is found hexagonal.

5.   Scanning Electron Microscopy (SEM)

SEM monitors the formation and growth of thin films and nanostructures. Surface morphology and topography of thin films of deposited CuSe nanoparticles are investigated using SEM images.

Figure 8(a) shows the SEM images confirming surface morphology of copper selenide films for deposition time of 60 min. The surface of copper selenide has smoother contour with some edge lines between the grains of large size. It is seen that there is overlapping of grain and large number of small spherical grains are aggregated together to form big islands. Figure 8(b) shows SEM image of CuSe deposited thin film for deposition time of 120 min. In this case also the surface of copper selenide film has smooth contour with some edge lines between the grains of large size and there is overlapping of grain. From the micrograph it is clear that the film is composed of a compact structure of small densely packed nano crystals. Figure 8(c) shows SEM image of CuSe deposited thin film for deposition time of 180 min. It is observed that CuSe thin film is well covered to the substrate with the presence of grained surface particles. These results indicate that particle size of samples increase with increase in deposition time of CuSe photo electrodes.

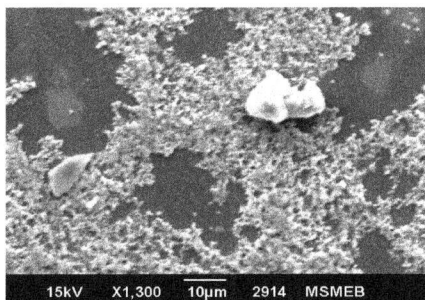

Figure 8(a): SEM Image (Deposition Time - 60 min).

Figure 8(b): SEM Image (Deposition Time - 120 min).

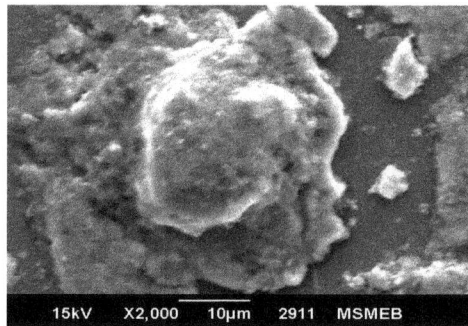

Figure 8(c): SEM Image (Deposition Time - 180 min).

Figure 9(a) shows the scanning electron micrograph of CuSe film deposited at temperature of 45°C with magnification X3,700. The surface of copper selenide has smooth contour with some edge lines between the grains of large size. It is seen that there is overlapping of grain. From the micrograph it is clear that the film is composed of a compact structure of packed nanocrystals. Figure 9(b) shows SEM image of CuSe deposited thin film for photo electrode deposited at 60°C temperature. The SEM images confirming surface morphology of copper selenide film at X2,300 magnification. The grains are well defined, spherical, of almost similar size, which were uniformly distributed over a smooth homogeneous background that may correspond to the amorphous phase of CuSe film. Figure 9(c) shows scanning electron micrographs of CuSe thin films deposited on glass substrate at X1,900 magnifications. In this case also the surface of copper selenide film has some edge lines between the grains of large size and there is overlapping of grain. From the micrograph it is clear that the film is composed of a compact structure of small densely packed nanocrystals and there is overlapping of grain.

Figure 9(a): SEM Image of deposited CuSe thin film (Temperature - 45°C).

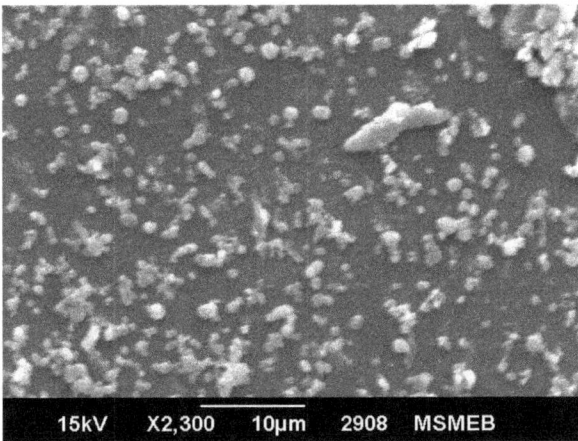

Figure 9(b): SEM Image of deposited CuSe thin film (Temperature - 60°C).

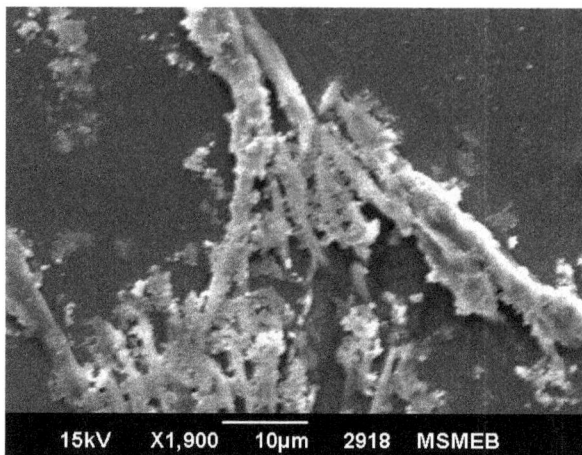

Figure 9(c): SEM Image of deposited CuSe thin film (Temperature - 90°C).

Figure 10(a) shows SEM image of CuSe deposited thin film on glass substrate for A-concentration (Half con.), Figure 10(b) shows a SEM image of CuSe deposited thin film on glass substrate for B-concentration (Normal con.) and Figure 10(c) shows the SEM image of CuSe deposited thin film on glass substrate for C-concentration (Double con.) of precursor solution of CuSe photo electrodes.

Figure 10(a): SEM image of CuSe thin film for A-concentration (Half con).

Figure 10(b): *SEM image of CuSe thin film for B-concentration (Normal con).*

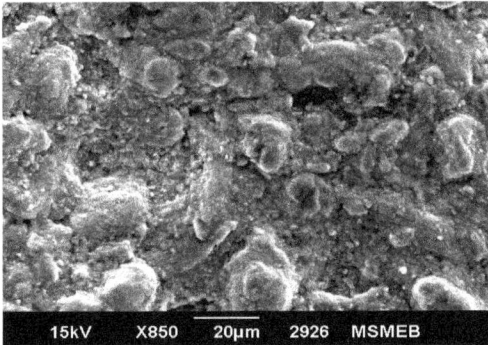

Figure 10(c): *SEM image of CuSe thin film for C-concentration (Double con) of precursor solution.*

Figure 10(a) shows the SEM images confirming surface morphology of copper selenide film at X3, 700 magnification. The grains are well defined, spherical, of almost similar size, which were uniformly distributed over a smooth background. Figure 10(b) shows the scanning electron microscopy micrograph of CuSe film with magnification X3, 700. In this case also the surface of copper selenide film has smooth contour with some edge lines between the grains of large size and there is overlapping of grain. From the micrograph it is clear that the film is composed of a compact structure of small densely packed nanocrystals. Figure 10(c) shows scanning electron micrographs of CuSe thin films deposited on glass substrate at X850 magnifications. It is observed that CuSe thin film is homogeneous and well covered to the substrate with the presence of grained surface particles. SEM images indicate overlapping of grains in all conditions of CuSe deposition.

6. Atomic Force Microscopy (AFM)

The atomic force microscope (AFM) is ideally suited for characterizing nanoparticles. Using the Atomic Force Microscope (AFM), individual particles and groups of particles can be resolved. It offers the capability of 3D visualization and both qualitative and quantitative information on many physical properties including size, morphology, surface texture and roughness. Statistical information, including size, surface area, and volume distributions, can be determined as well. The AFM offers visualization in three dimensions. Resolution in the vertical, or Z, axis is limited by the vibration environment of the instrument, where as resolution in the horizontal, or X-Y, axis is limited by the diameter of tip utilized for scanning.

For finding the results of Atomic Force Microscopy (AFM) studies of CuSe deposited thin film on glass plate, the AFM (NTMDT) instrument is used (Center of Nanoscience & Nanotechnology, Sathyabama University, Chennai). The goal is to determine the shape, size and size distribution of nanoparticles. CuSe thin films are prepared on glass substrate for deposition time, different temperature, and different concentration of solution. Samples are characterized by AFM in the dynamic mode.

(a) Atomic force microscopic image analysis of CuSe thin film for deposition time of 60 min

Figure 11(a) shows 2D and 3D atomic force microscopic images of CuSe thin film for deposition time of 60 min. There was agglomeration of particles in most of the cases as evident from the 3D image. The difference between the lowest and highest points on the surface is 40 nm for scans over 2.5x2.5 μm for the sample.

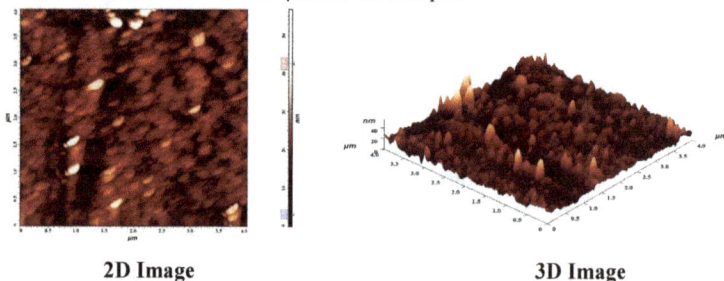

2D Image   3D Image

Figure 11(a): AFM Images of CuSe thin film for deposition time of 60 min.

(b) Atomic force microscopic image analysis of CuSe thin film for deposition time of 120 min

Figure 11(b) shows 2D atomic force microscopic image of CuSe thin film for deposition time of 120 min. There was agglomeration of particles in most of the cases as evident from the 3D image. The difference between the lowest and highest points on the surface is 120 nm, scans over 2.5x2.5 μm for the sample.

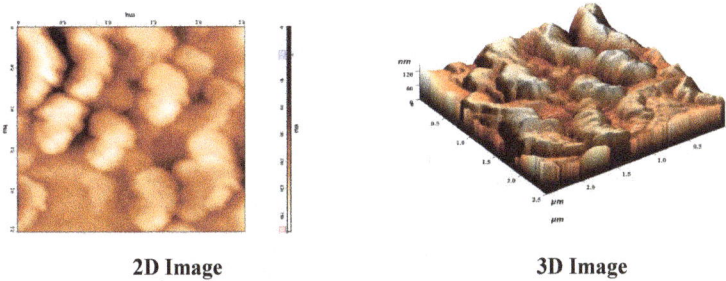

2D Image   3D Image

Figure 11(b): AFM Images of CuSe thin film for deposition time of 120 min.

(c) Atomic force microscopic image analysis of CuSe thin film for deposition time of 180 min

Figure 11(c) shows 2D and 3D atomic force microscopic image of CuSe thin film for deposition time of 180 min. The difference between the lowest and highest points on the surface is 160 nm, scans over 2.5x2.5 μm for the sample.

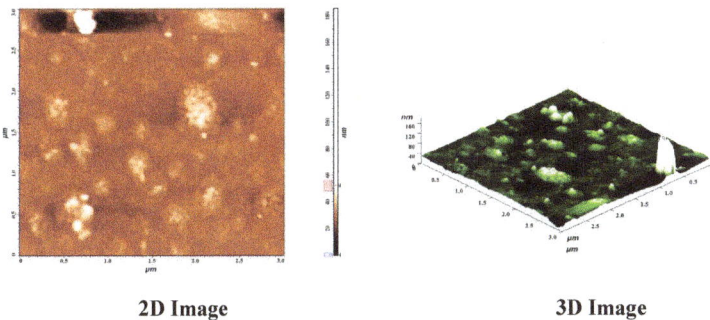

2D Image   3D Image

Figure 11(c): AFM Images of CuSe thin film for deposition time of 180 min.

This shows that as deposition time is increased the difference between highest and lowest points increases, which indicates that deposition is favered at higher elevated parts rather than the lower parts on the film and surface becomes more and more uneven.

(d)   Atomic force microscopic image analysis of CuSe thin film prepared at 45°C temperature

Figure 12(a) shows 2D and 3D atomic force microscopic image of CuSe thin film deposited at 45°C temperature. The difference between the lowest and highest points on the surface is 120 nm, scans over 2.5x2.5 μm for the sample. From the image it is clear that the film is uniform and the substrate surface is well covered by fine spherical or elliptical nature of the grains.

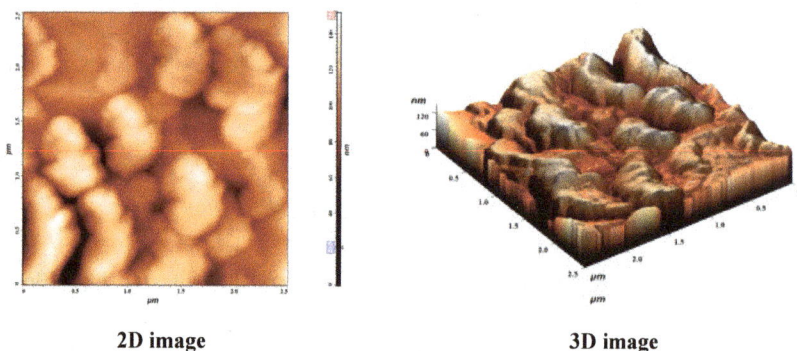

| 2D image | 3D image |

Figure 12(a): AFM Images of CuSe thin film deposited at 45°C temperature.

(e)   Atomic force microscopic image analysis of CuSe thin film prepared at 60°C temperature

Figure 12(b) shows 2D and 3D atomic force microscopic image of CuSe thin film deposited at 60°C temperature. The difference between the lowest and highest points on the surface is 100 nm, scans over 4x4 μm for the sample. From the image it is clear that the film is uniform and the substrate surface is well covered by fine spherical or elliptical nature of the grains. The observation reveals that the films are crystalline in nature.

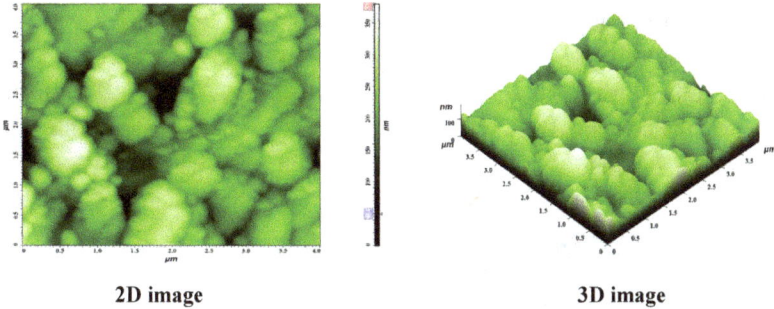

2D image   3D image

Figure 12(b): AFM Images of CuSe thin film deposited at 60°C temperature.

(f)   Atomic force microscopic image analysis of CuSe thin film prepared at 90°C temperature

Figure 12(c) shows 2D and 3D atomic force microscopic image of CuSe thin film deposited at 90°C temperature. The difference between the lowest and highest points on the surface is 40 nm, scans over 4x4 μm for the sample. From the image it is clear that the film is uniform and the substrate surface is well covered by fine spherical or elliptical nature of the grains. The observation reveals that the films are crystalline in nature.

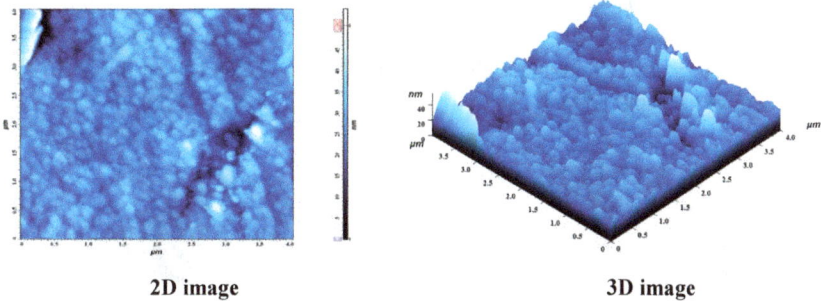

2D image   3D image

Figure 12(c): AFM Images of CuSe thin film deposited at 90°C temperature.

(g) Atomic force microscopic image analysis of CuSe thin film for A-concentration (Half con.) of precursor solution of CuSe photo electrodes

Figure 13(a) shows 2D and 3D atomic force microscopic images of CuSe thin film for A-concentration (Half con). There was agglomeration of particles in most of the cases as evident from the 3D image. The difference between the lowest and highest points on the surface is 40 nm for scans over 4x4 μm for the sample.

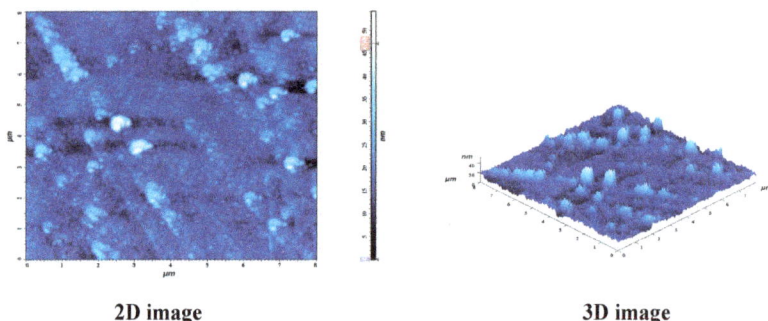

2D image   3D image

Figure 13(a): AFM Images of CuSe thin film d from A-concentration (Half con.) of precursor solution of CuSe photo electrodes.

(h) Atomic force microscopic image analysis of CuSe thin film for B-concentration (Normal con.) of precursor solution of CuSe photo electrodes

Figure 13(b) shows 2D and 3D atomic force microscopic images of CuSe thin film for B-concentration (Normal con.) of precursor solution of CuSe photo electrodes. The difference between the lowest and highest points on the surface is 120 nm for scans over 2.5x2.5 μm for the sample.

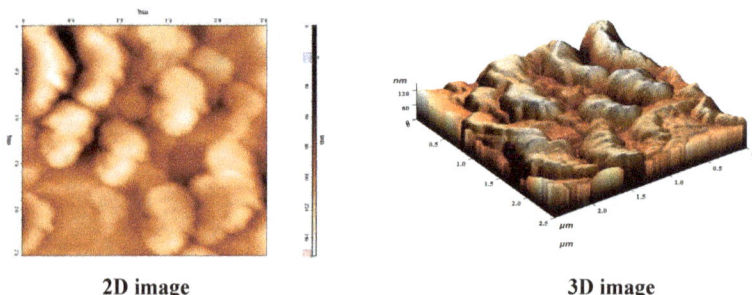

2D image   3D image

Figure 13(b): AFM Images of CuSe thin film deposited from B-concentration (Normal con.) of precursor solution of CuSe photo electrodes.

(i)   Atomic force microscopic image analysis of CuSe thin film for C-concentration (Double con.) of precursor solution of CuSe photo electrodes

Figure 13(c) shows 2D atomic force microscopic image of CuSe thin film for C-concentration (Double con). The difference between the lowest and highest points on the surface is 300 nm for scans over 4x4 μm for the sample.

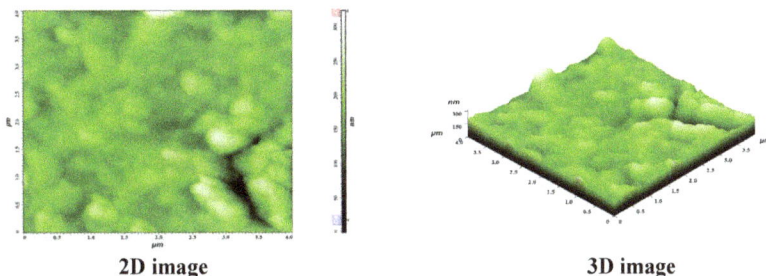

| 2D image | 3D image |

Figure 13(c): *AFM Images of CuSe thin film deposited from C-concentration (Double con.) of precursor solution of CuSe photo electrodes.*

Atomic Force Microscopy   (AFM) and Optical absorption methods show that thickness of the deposited copper selenide thin film increases with increase in deposition time and concentrations of precursor solution, but thickness of the deposited copper selenide thin film decreases with increase in temperature.

7.   Optical Absorption

Ultraviolet/Visible/Infrared (UV/Vis/IR) spectroscopy is a technique used to quantify the light that is absorbed and scattered by a sample. Copper selenide (CuSe) is well-known p-type semiconductors having potential applications in solar cells. We characterized the nanocrystalline CuSe thin Film by UV-Vis spectroscopy. UV-Vis spectroscopy involves photons in UV-Visible region. This means it uses light in the visible and adjacent near ultraviolet (UV) and near infrared (NIR) ranges (Rouessac, 1994). The absorption in the visible ranges directly affects the color of the chemicals involved. In this region of the electromagnetic spectrum, molecules undergo electronic transitions. Generally, wavelength of the maximum excitation absorption decreases as particle size decreases.

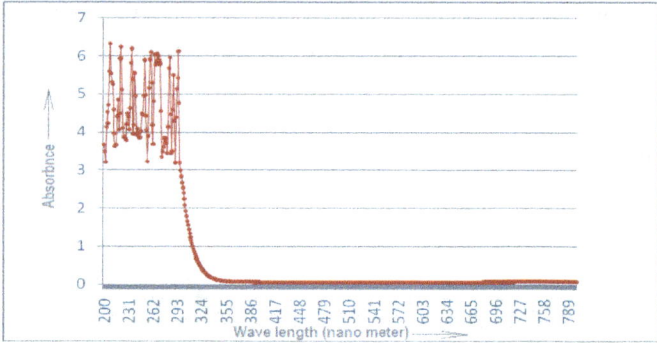

Figure 14(a): *Optical absorption spectra of CuSe film deposited for 60 min.*

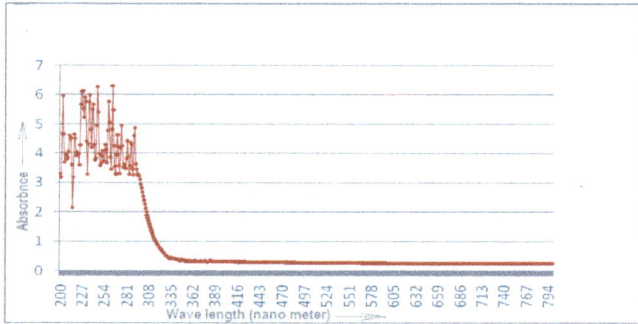

Figure 14(b): *Optical absorption spectra of CuSe film deposited for 120 min.*

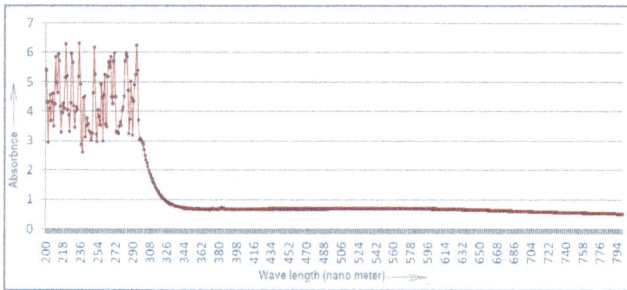

Figure 14(c): *Optical absorption spectra of CuSe film deposited for 180 min.*

Figure(s) 14(a), 14(b) and 14(c) show optical absorption spectra of deposited CuSe thin film for deposition time 60 min, 120 min and 180 min respectively. From UV-Vis spectral characteristics, absorption edge of the CuSe thin film for deposition time 60 min, 120 min and 180 min is located at wavelengths 312 nm, 317 nm and 323 nm respectively. The absorption spectrum of the nanocrystalline CuSe thin film clearly shows that the absorption edge shifts from 312 nm to 323 nm as the deposition time increases. The effective energy band gap (E_{nano}) is calculated as-

$$E_{nano} = h \times C / \lambda \tag{3}$$

where h = Plank's constant = 6.6261×10^{-27} erg/sec and C = Velocity of light = 2.9979×10^{10} cm/sec. The value of E_{nano} is found to be 3.98 eV for deposition time 60 min. Similarly the value of E_{nano} is found as 3.91 eV and 3.84 eV for deposition time 120 min and 180 min respectively. The increase in energy band gap ΔE_g is calculated as-

$$\Delta E_g = E_{nano} - E_{bulk} \tag{4}$$

Taking, $E_{bulk} = 2.2 eV$, the value of ΔE_g is found to be 1.78 eV for 60 min deposition time, 1.71 eV for 120 min deposition time and 1.64 eV for 180 min deposition time. From UV-Vis spectral characteristics particle size (r) is calculated from the relation-

$$\Delta E = h^2 \pi^2 / r^2 [1/m_e^* + 1/m_h^*] \tag{5}$$

or

$$r^2 = h^2 \pi^2 / \Delta E [1/m_e^* + 1/m_h^*] \tag{6}$$

taking, effective mass of electron of CuSe, $m_e^* = 0.445$ and effective mass of holes of CuSe, $m_h^* = 1.22$.

Table 9: Effect of deposition time on band width and particle size

S.N.	Deposition Time	Absorption Edge (λ)	E_{nano}	ΔE, Increase in Band Width	Particle Size
1	60 min	312 nm	3.98	1.78	27.3 nm
2	120 min	317 nm	3.91	1.71	27.88 nm
3	180 min	323 nm	3.84	1.64	28.47 nm

From Table 9, it is observed that the increase in band width (ΔE) of deposited copper selenide decreases with increase in deposition time and hence calculated value of particle size (r) for deposited copper selenide thin film increases with increase in deposition time.

Figure 15 shows the optical absorption spectra of CuSe thin film prepared at different temperatures 45°C, 60°C and 90°C by UV-spectroscopy.

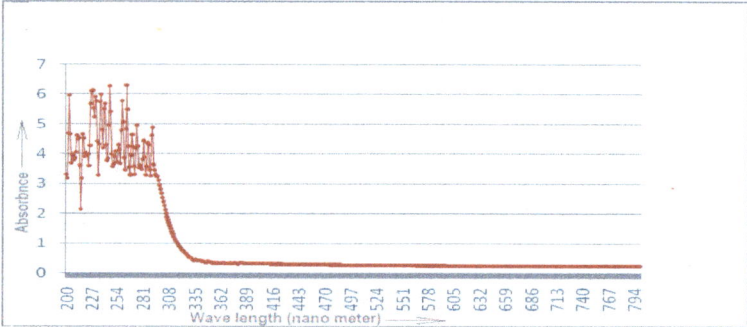

Figure 15(a): *Optical absorption spectra of CuSe film deposited at temperature 45°C.*

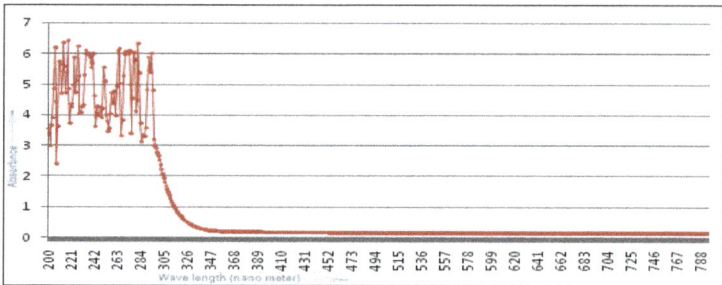

Figure 15(b): *Optical absorption spectra of CuSe film deposited at temperature 60°C.*

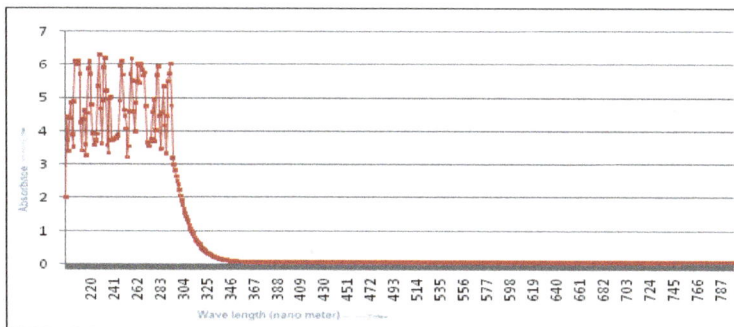

Figure 15(c): Optical absorption spectra of CuSe film deposited at temperature 90°C.

The UV-spectral characteristics of the CuSe thin film photo electrodes deposited at different temperatures 45°C, 60°C and 90°C show that the absorption edge is located at wavelengths 317 nm, 314 nm and 312 nm respectively. The absorption spectrum of the nanocrystalline CuSe thin film clearly shows that the absorption edge shifts from 317 nm to 312 nm as the temperature during deposition increases.

The value of Enano is found to be 3.91 eV for deposited CuSe thin film photo electrode prepared at temperature 45°C. Similarly the value of E_{nano} is found to be 3.95 eV and 3.98 eV for deposited CuSe thin film photo electrode prepared at temperatures 60°C and 90°C respectively. The increase in energy band gap (ΔE_g) is calculated from relation (4).

Table 10: Effect of temperature on band width and Particle size

S.N.	Temperature	First Edge (λ)	E_{nano}	ΔE, Increase in Band Width	Particle Size
1	45°C	317 nm	3.91	1.71	27.88 nm
2	60°C	314 nm	3.95	1.75	27.5 nm
3	90°C	312 nm	3.98	1.78	27.33 nm

From Table -10, it is observed that the calculated value of particle size (r) for deposited copper selenide thin film decreases with increase in temperature, since ΔE i.e. increase in band width of deposited copper selenide increases with increase in temperature.

Fig.16 shows the optical absorption spectra of CuSe thin film for different concentrations of precursor solution by UV–vis spectroscopy.

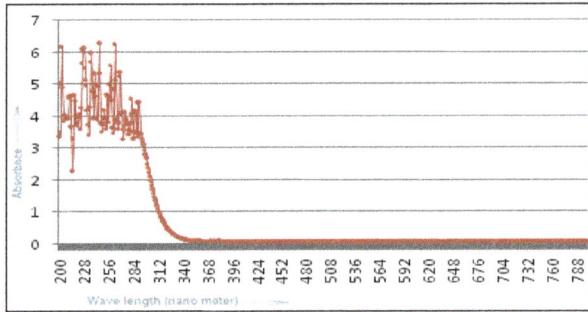

Figure 16(a): Optical absorption spectra of CuSe film deposited from A-concentration.

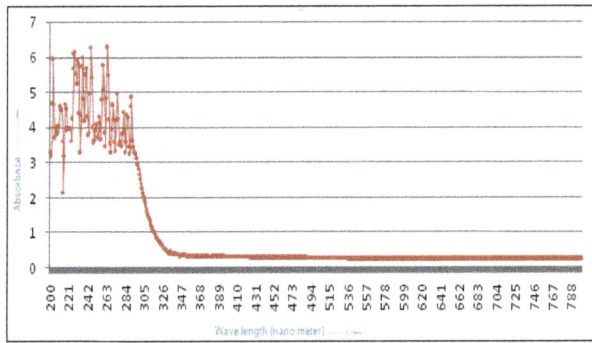

Fige 16(b): Optical absorption spectra of CuSe film depoaited from B-concentration.

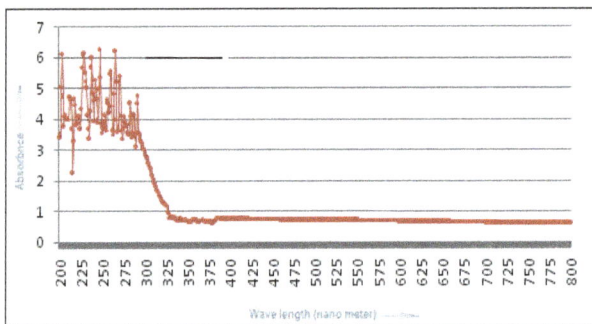

Figure 16(c): Optical absorption spectra of CuSe film deposited from C-concentration.

The UV-spectral characteristics absorption peaks of the CuSe thin film for A-concentration (Half con.), B-concentration (Normal con.) and C-concentration (Double con.) of precursor solution of CuSe photo electrodes are located at wavelengths 313 nm, 317 nm and 326 nm. The absorption spectrum of the nanocrystalline CuSe thin film clearly shows that the absorption edge shifts from 313 nm to 326 nm as the concentration increases.

We observed a little increase in the absorbance for the annealed films in the visible range that decreases in ultraviolet range. The effective energy band gap (E_{nano}) is calculated using the relation (3).

Table 11: *Effect of different concentrations of precursor solution on band width and particle size*

S.N.	Sample Concentration	Absorption Edge (λ)	E_{nano}	ΔE, Increase in Band Width	Particle Size
1	Half concentration (A)	313 nm.	3.96eV	1.76	27.4 nm.
2	Normal concentration (B)	317 nm.	3.91	1.71	27.88 nm.
3	Double concentration (C)	326 nm.	3.80	1.60	28.83 nm

From Table 11, it is observed that the increase in band width (ΔE) of deposited copper selenide decreases with increase in concentrations of precursor solution and hence calculated value of particle size (r) for deposited copper selenide thin film increases with increase in concentrations of precursor solution.

Optical absorptions result indicates that ΔE, i.e. increase in band width of deposited copper selenide increases with increase in temperature and decreases with increase in deposition time and concentrations of precursor solution. Average grain size for all deposited copper selenide thin films is found in range of ~ 25 to 35 nm. Thus CuSe thin films are nanocrystalline.

8.   Photoeletrochemical Cell Studies

Copper selenide is studied with great interest during the past decades because of its potential application in the fabrication of photovoltaic devices. Their photoelectrochemical performance was investigated in standard two electrode configuration with redox electrolyte, when the nanocrystalline CuSe films are deposited for different deposition time, at different temperature, and different concentration of precursor solution.

8.1   CuSe Nanocrystalline Photoelectrode Deposited for Different Durations

(a)   Current-Voltage (I-V) Characteristics

Figure 17 shows I-V characteristics of PEC cells having photo electrodes prepared for different deposition time.

Observations show that as load is reduced there is increase in output current and output voltage decreases. Copper selenide is studied with great interest during the past decades because of its potential application in the fabrication of photovoltaic devices. I-V characteristics of solar cell are plotted for four different electrodes. The deposition time of CuSe for each anode is different, viz. 60 min., 90 min, 120 min and 180 min.

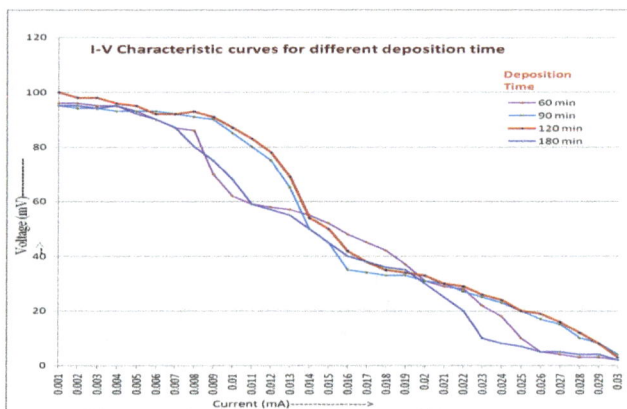

Figure 17: *I-V characteristics of PEC solar cells (for photo electrodes deposited for different deposition time).*

(b)   Photovoltaic Output Characteristics

From I-V curve it can be concluded that with decrease in resistance, there is increase in current and the terminal potential of solar cell decreases. It was observed that the solar cell parameters are influenced by deposition time used for preparation of the photo electrodes. It is seen that short circuit current (Isc), open circuit voltage (Voc). Fill factor (FF) and efficiency (η) change with increase in the deposition time. These parameters for various deposition times are given in Table 12.

Table 12: *Solar cell parameters.*

Deposition time (min)	Short circuit current Isc (mA)	Open circuit voltage Voc (mV)	Fill Factor	Efficiency, η (%)
60	0.046	98	0.17303	1.128
90	0.043	98.5	0.22127	1.272
120	0.042	99.5	0.24097	1.457
180	0.036	100	0.20475	1.034

Figure 18(a) shows variation of solar cells parameters Voc and Isc with different deposition time of photo electrodes. From the results it is observed that short circuit current (Isc) is maximum for 60 min deposition and decreases by increasing deposition time, where as it is just reveres in case of open circuit voltage (Voc), which is minimum for 60 min deposition and increases by increasing deposition time.

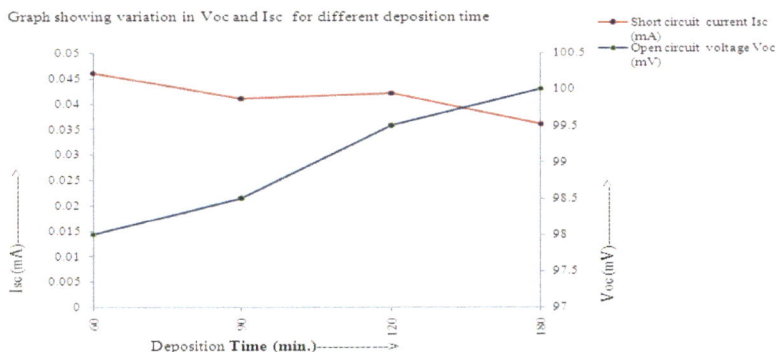

Figure 18(a): *Variation of Voc and Isc for different deposition time of photo electrodes.*

Figure 18(b) shows variation of solar cell parameters fill factor and efficiency with different deposition time of photo electrodes. From these results we see that solar cell efficiency as well as fill factor is maximum when electrodes are prepared with 120 minute deposition time.

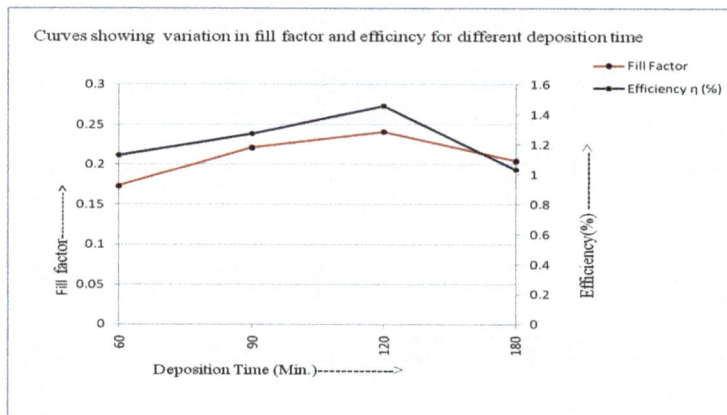

Curves showing variation in fill factor and efficincy for different deposition time

Figure 18(b): *Variation of Fill factor and efficiency with deposition time of photo electrodes.*

It is observed that the solar cell is more efficient which uses photo electrode with deposition time 120 minute. Observation shows copper selenide electrode in solar cell is suitable p-type candidate.

8.2   CuSe Nanocrystalline Photoelectrode deposited at Different Temperatures

(a)   Current-Voltage (I-V) Characteristics

Figure 19 shows the I-V characteristics of annealed copper selenide photoelectrodes deposited at different temperature in dark and under light illumination with 315 lux intensity. An increase in the current is observed when films are illuminated. The photo response upon illumination indicates that the films are sensitive towards light.

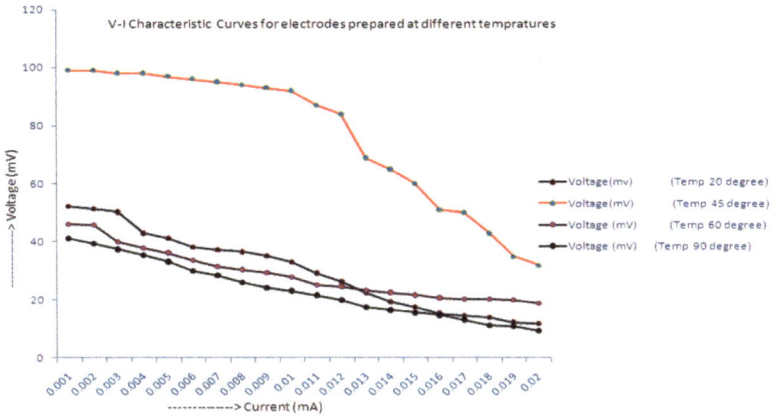

Figure 19: *I-V characteristics of PEC Solar cells (photoelectrodes deposited at different temperatures).*

Figure 19 shows the I-V characteristics of a solar cell that uses a photo anode with copper selenide deposited on it. The characteristics are plotted for four anodes which were prepared at different temperatures viz 20°C, 45°C, 60°C & 90°C. Observing I-V curve it can be concluded that with decrease in resistance, the current increases and the terminal potential of the solar cell decreases. It is observed that the photo current due to the anode which is prepared at temperature of 20°C is lower than with an anode which is prepared at 45°C, it is also seen that the solar cell is more efficient if it uses an anode prepared at 45°C.

(b) Photovoltaic Output Characteristics

It is observed that the solar cell parameters: short circuit current (Isc), open circuit voltage (Voc), fill factor (FF) and efficiency, η (%) are influenced by different temperature used for preparation of photo electrodes. These parameters are given in Table 13.

Table 13: Solar Cell Parameters.

Temperature (Degree Celsius)	Short Circuit Current, Isc (mA)	Open Circuit Voltage, Voc (mV)	Fill Factor	Efficiency (%)
20	0.06	53.1	0.10389203	0.478565749
45	0.042	99.5	0.24120603	1.457384515
60	0.08	47.2	0.10010593	0.546519193
90	0.05	42.5	0.11341176	0.348442131

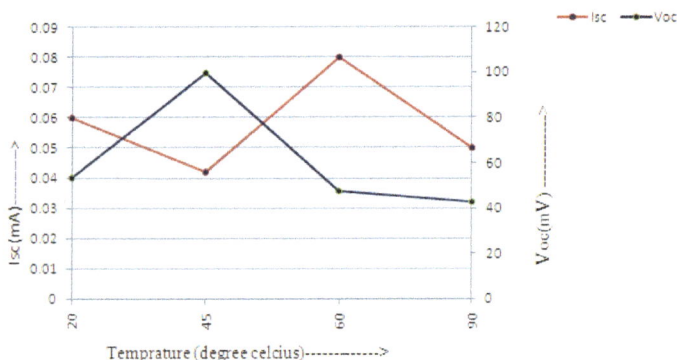

Figure 20(a): *Variation of Voc and Isc for photoelectrodes deposited at different temperatures.*

Curves showing variation in Fill factor and efficiency for different Temprature

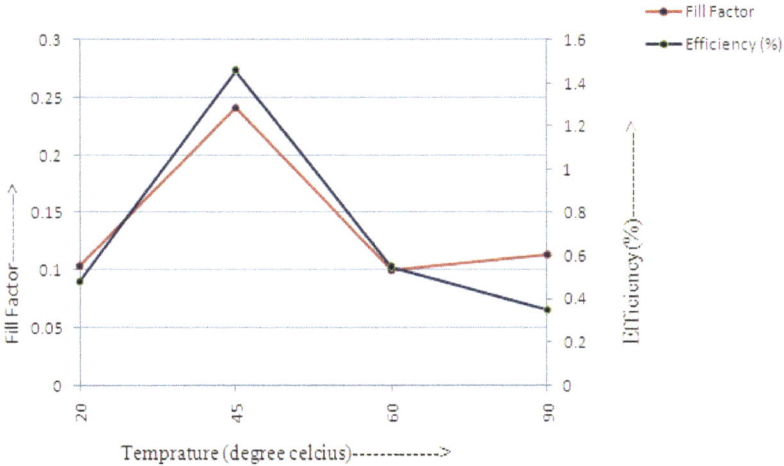

Figure 20(b): *Variation of fill factor and efficiency for photoelectrodes deposited at different temperatures.*

Figure 20(a) shows variation of solar cells parameters Voc and Isc for photo electrodes deposited at different temperatures. From the results it is observed that open circuit voltage (Voc) is highest when photo electrode deposition is done at 45°C temperature and short circuit current (Isc) is highest when photo electrode deposition is done at 60°C temperature.

Figure 20(b) shows variation of solar cell parameters fill factor and efficiency for photo electrodes deposited at different temperature. From these results we see that solar cell efficiency as well as fill factor is maximum when photo electrode deposition is done at 45°C temperature.

8.3   CuSe Nanocrystalline Photoelectrode Deposited from Various Concentration of Precursor Solutions

(a)   Current-Voltage (I-V) Characteristics

Figure 21 shows the I-V characteristics of solar cell that uses photo anode with copper selenide deposited on it. The characteristics are plotted for three different anodes. The concentration of CuSe for each anode is kept different, viz. half, normal and double concentration.

Observing I-V curve it can be concluded that with change in resistance, increase in current the terminal potential of solar cell decreases. The different behavior of I-V curve is obtain because the roughness of anode surfaces changes with deposition time and the smoothness of surface increase with increases in deposition time. It is observed that the solar cell is more efficient which uses anode with double concentration.

I-V Characteristic curves for different concentrations

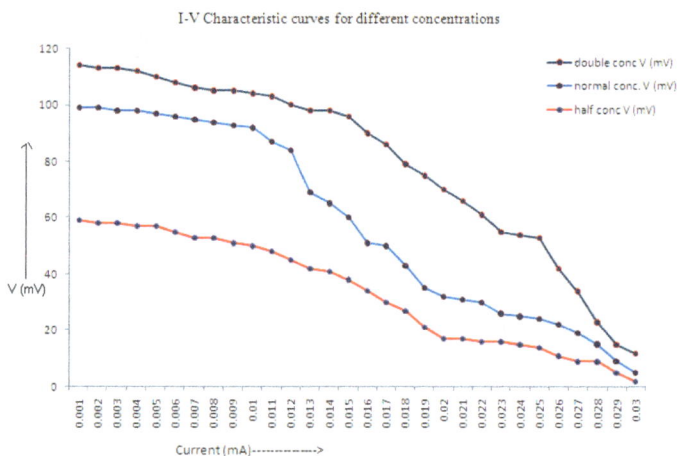

Figure 21: I-V characteristics of PEC solar cells for photo electrodes prepared from different concentrations of precursor solution.

(b) Photovoltaic Output Characteristics

Observation shows copper selenide electrode in solar cell is suitable p-type candidate. It is observed that the solar cell parameters: short circuit current (Isc), open circuit voltage (Voc), fill factor (FF) and efficiency, η (%) are influenced by concentration of precursors used for preparation of photo electrodes.

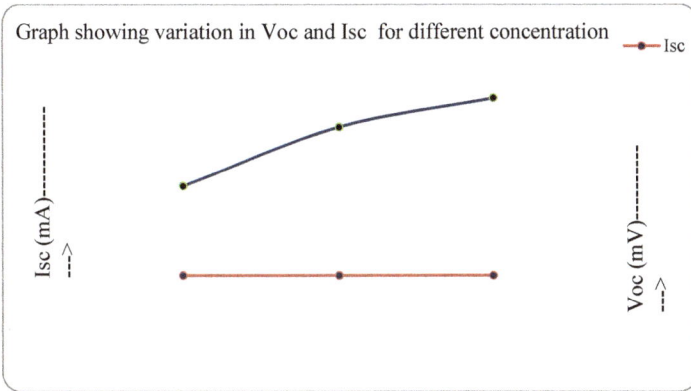

Figure 22(a): *Variation of Voc and Isc for different precursor concentrations.*

Figure 22(a) shows variation of solar cells parameters Voc and Isc with different deposition time of photo electrodes. From the results it may be observed that performance of solar cell is best for double concentration.

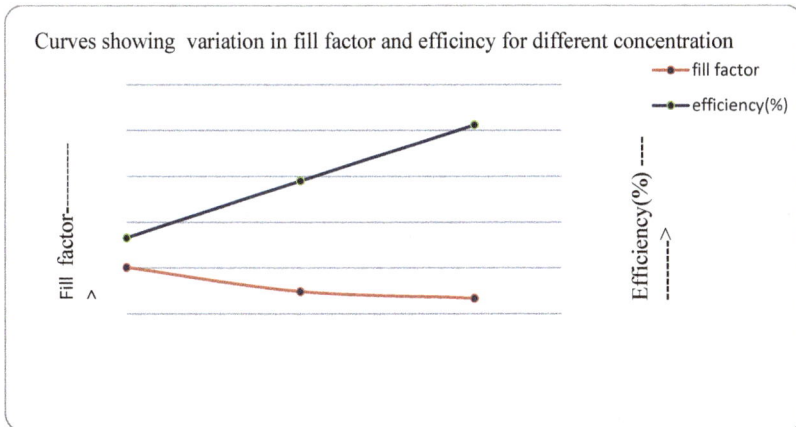

Figure 22(b): *Variation of fill factor and efficiency for different precursor concentration.*

Figure 22(b) shows variation of solar cell parameters fill factor and efficiency with different deposition time of CuSe photo electrodes. The value of fill factor and efficiency increases with increases in concentration because the amount of deposited CuSe increases. Solar cell efficiency as well as fill factor is maximum when electrodes are prepared with double concentration.

9. Conclusions

The chemically synthesized copper selenide photoelectrode have been introduced which exhibits p-type photoconductivity. The different behavior of I-V curve for different deposition time is obtained because the roughness of CuSe photoelectrode surfaces changes with deposition time. For highest deposition time 180 min, smoothness of surface increases more and surface area for current flow decreases so the result is a decrease in current. The experiment indicates that the best results are obtained with deposition of CuSe for 120 min, which has maximum roughness. The photoeletrochemical cell studies of CuSe photo electrodes deposited at different temperature show that the efficiency of solar cell is best for deposited CuSe thin film photoelectrode prepared at temperature 45°C. Similarly, the photoeletrochemical cell studies of CuSe photoelectrodes deposited from different concentration of precursor solution show that the efficiency of solar cell increases with increase in concentration of precursor solution. The photoeletrochemical cell studies of CuSe photoelectrodes conclude that the best result is obtained when CuSe photo electrodes are deposited from concentration- C (Double) of precursor solution, at temperature 45°C and for 120 min deposition time. Thus copper selenide can be considered as a potential candidate for solar cell.

XRD shows the average grain size using observed peaks for deposited copper selenide thin film increases with deposition time and concentration of precursor solution but average grain size decreases with temperature. From PDF card, crystal class is found to be hexagonal. SEM images indicate overlapping of grains in all cases.

Atomic Force Microscopy (AFM) and optical absorption methods show that thickness of the deposited copper selenide thin film increases with increase in deposition time and concentrations of precursor solution, but it decreases with increase in temperature.

Optical absorptions result indicate that ΔE i.e. increase in band width of deposited copper selenide increases with increase in temperature and decreases with increase in deposition time and concentrations of precursor solution. Average grain size for all deposited copper selenide thin films is found in range of ~ 25 to 35 nm. Thus CuSe thin films are nanocrystalline.

A photoelectrochemical study confirms p-type conductivity and thus copper selenide can be considered as a potential candidate for solar cells. In preparation of photo electrodes, deposition time and temperature plays an important role. The investigation may be useful in obtaining efficient, stable and low cost solar cell to compete with the existing technology. This may help to overcome the energy crisis due to exhausting fossil fuels.

References

[1]   A Kathaligam, Bull. of the Indian Association of Physics Teacher, vol. 22. No. 8, page 268 2006.

[2]   B. A. MANSOUR, I. K. EL ZAWAWI, M. KAMAL,, T. A. HAMEED, Journal of Ovonic Research Vol. 6, No. 5, September-October, p. 193 - 200, 2010.

[3]   Di Liu and Prashant V. Kamat American chemical society J. physics, chem., 97, 10769-10773, 1993.

[4]   Fritz Allhoff, Patrick Lin, and Daniel Moore; What Is Nanotechnology and Why Does It Matter? From Science to Ethics, A John Wiley & Sons, Ltd., Publication, 2010. https://doi.org/10.1002/9781444317992

[5]   H. M. Pathan, C. D. Lokhande, D. P. Amalnerkar, Appl. Sur. Sc. 211, 48-56, 2003. https://doi.org/10.1016/S0169-4332(03)00046-1

[6]   J. E. Dickman and A. F. Hepp, D. L. Moral and C. S. Ferekides, J. R. Tuttle, D. J. Hoflman, N. G. Dhere; NASA/TM, AIAA-5922, 2003.

[7]   Kavita Gour, Preeti Pathak, M. Ramrakhiani and P. Mor "International Journal of Electrical, Electronics & Computing Technology ISSN 2229-3027 Volume 8 (I) Issue 7 March-May 2013.

[8]   Kavita Gour, Preeti Pathak, M. Ramrakhiani and P. Mor, COSMIC International Journal, Vol 5 April- June 2014.

[9]   Mohd Fairul, Sharin Abdul Razak & Zulkaranain Zainal The Malaysian journal of analytical sciences Vol.11 no. 1324-330, 2007.

[10]   S. Kashida, J. Akai, J. Phys. C. Solid State Phys. 21, 75329, 1998.

[11]   S. R. Gosavi, N. G. Despande, Y. G. Gudage, Ramphal Sharma, ELSEVIER Journal of alloy and compounds 448, 344-348, 2008. https://doi.org/10.1016/j.jallcom.2007.03.068

[12]   Swapnil B. Ambade, R.S. Mane S.S. Kale, S.H. Sonawane Arif V. Shaikh, Sung-Hwan Han Sc. Direct Elsevier, applied surface science 253, 2123-2126, 2006.

[13] W. R. Fahrner, Nanotechnology and Nanoelectronics: Materials, Devices, Measurement Techniques, ISBN 3-540-22452-1 Springer Berlin Heidelberg New York 2005. https://doi.org/10.1007/b137771

[14] Yunxiang Hu, mohammad Afzal, Mohammad A. Malik, Paul o'Brien Science direct journal of crystal growth 297, 61-65, 2006.

[15] Zulkaranain Zainal, Saravanan Nagalingam, Tan Chin Loo, Elsevier, Science direct, materials letters 59, 1391-139 2005.

[16] http://www.worldscibooks.com/physics/p276.html. The Physics of Solar Cells © Imperial College Press.

Chapter 6

Photovoltaic response of nanocrystalline cadmium telluride in photoelectrochemical cells

Preeti Pathak[1], Kavita Gour [1], M.Ramrakhiani[2], P. Mor[2]

[1]Mata Gujri Mahila Mahavidyalaya, Jabalpur.(MP), India

[2]Department of Physics and Electronics, Rani Durgawati Vishwavidyalaya, Jabalpur .(MP), India

preetipathak74@yahoo.co.in

Abstract

Among various techniques for preparation of nanometer size materials, chemical methods offer better orientation, which are less expensive, non-polluting and easy to incorporate suitable doping materials for altering the film properties. These processes enable formation of thin films onto glass substrates in addition to the conducting films (substrates) which can potentially lead to a new generation of photovoltaic devices that are light in weight, foldable, flexible and mouldable. The thin films of cadmium telluride have been prepared in this work by direct chemical deposition method and annealed at 200°C for 2 hrs for crystallinity improvement. The preparation of CdTe film photoelectrodes for different deposition time, at different temperature and from different concentrations of precursor solution are discussed here. Effect of preparation parameters on physical, structural, microstructural, and electrical properties of these films and their photovoltaic response in a PEC cell have been investigated. The thin films of CdTe on glass substrate were used for characterization by means of X-ray diffraction (XRD), scanning electron microscopy (SEM), atomic force microscopy (AFM) and absorption spectroscopy. The nanocrystalline CdTe thin film on conducting glass substrates were used as photoelectrodes in PEC cells and the photovoltaic response has been studied optimising the preparation parameters.

Keywords

Cadmium Telluride Thin Films, Chemical Bath Deposition (CBD), Photoelectrochemical Cells, Photovoltaic Effect, XRD, UV-Vis Absorption

Contents

1.   Introduction...187

1.1   Photovoltaic effect...188

1.2   Nanotechnology used in solar cells...189

1.3   Thin film technology...189

1.4   CdTe thin film solar cells 189

2.   Synthesis of nanocrystalline CdTe thin films...................................191

2.1   For various deposition times..192

2.2   For various temperatures during deposition 195

2.3.   For various concentrations of precursors 195

3.   Photovoltaic response of CdTe films in photoelectrochemical cell...196

3.1   For various deposition times..197

3.2.   For various temperatures during deposition 199

3.3   For various concentrations of precursors 202

4.   Characterization of nanocrystalline CdTe thin films 204

4.1   Determination of film thickness 204

4.2   Structural studies using XRD...207

4.3   Microstructural properties using SEM and AFM...............................217

4.3.1 Scanning Electron Microscopy...217

4.3.2 Atomic force microscopy...221

4.4   Optical properties using UV-spectroscopy...228

5.   Conclusions...235

References 236

1. Introduction

Rapid growth of the world population, from 7 billion today towards 10 billion by the Middle of the 21st century, is going to drastically increase the energy demand. Conventional fossil fuels are getting expensive due to extraction difficulties and hence the world is going to face a widening gap between supply and demand. Colossal use of fossil fuel has created noticeable pollution problems contributing to climate change and health hazards. A major prerequisite for the modern style of life is existence of sustainable and inexpensive energy sources. Without affordable electricity, oil, and gas resources, we simply would not be able to exploit most of the advantages provided to us by new technologies [1]. There are two solutions to this problem; the use of conventional energy with improved efficiency and rapid introduction of clean energy technologies. The conventional fossil energy resources will contribute to irreversible climate changes in near the future. In order to overcome the energy deficit in the future, a number of alternative energy sources are considered. Most common types include nuclear, hydro, solar, wind, geothermal energy, and biomass fuels. Each of these options has certain advantages and disadvantages over the others and it is very likely that a combination of alternatives rather than a single source will be utilized in future global energy production.

Nuclear energy, even though a well-developed, an economically efficient and a controllable means of power production, is associated with a number of health, safety and security concerns related to nuclear waste management, natural disasters, terrorist attacks, etc. Hydro, wind and geothermal power are renewable energy sources but are limited by geographical conditions and thus are not capable of expanding enough to provide a sufficient energy supply by themselves. Most of the world's locations with hydro and geothermal potentials are already being exploited, while wind energy is associated with high capital costs [2]. Biomass fuels are accompanied with greenhouse emissions during their production and the process of their extraction is very expensive. Solar energy source is our only long term natural source of energy.

Solar energy is renewable and carbon-free and is the most abundant energy source. Currently; solar cells are not economically competitive with fossil fuels due to either their relatively low power conversion efficiencies or high manufacturing costs. However, conversion efficiencies of many commercialized solar devices are still well below their theoretical limits, thus leaving room for improvement, while different approaches to reduce manufacturing costs are being investigated. Therefore, solar energy has a significant potential to provide a sustainable, economically feasible, safe and environmentally friendly energy supply. Photovoltaic (PV) solar energy is at the top of the renewable energy list and capable of providing "Energy for all". But the initial cost of the manufacturing remains high at present, and therefore the market penetration is slow.

Therefore, the worldwide PV community is engaged in an active research programme to reduce this cost, in order to accelerate the take-up of solar energy applications [3].

Solar cells are considered as potential energy source for our future. The literature describes the physics of photovoltaic (PV) effect and characteristic of solar cell. Photocells are already available. But the efficiency of existing photo cell is about 10% to 20%. In order to enhance this efficiency, we need to use new chemicals and techniques to develop solar cell. The relevance of the study is to find photovoltaic effect using different chemicals to obtain a better and improved photocell. The development of thin film solar cell is an active area of research at this time. Much attention has been paid to the development of low cost high efficiency thin film solar cell.

1.1 Photovoltaic effect

The photovoltaic effect is the creation of voltage or electric current in a material upon exposure to light. Photovoltaic is marriage of two words: — photo from Greek roots, meaning light, and — voltaic from —volt, which is the unit used to measure electric potential at a given point. Thus, the term photovoltaic literally means — electricity from light. Photovoltaic is the technology that allows producing electricity by direct conversion of sunlight without the use of fuel and no moving parts are involved. It is possible because of the use of a cell to convert solar radiation into electricity. The cell consists of one or two layers of a semi-conducting material. When light shines on the cell it creates an electric field across the layer, causing electricity flow. The greater the intensity of the light, the greater is the flow of electricity [4].

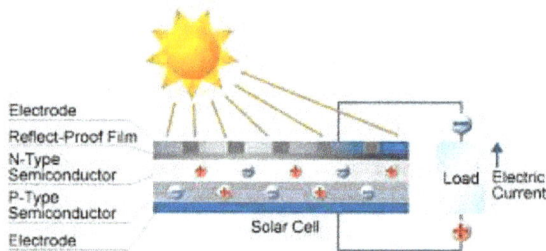

***Figure 1** : Photovoltaic Effect.*

Photoelectrochemical cells are considered as potential energy source. The photoelectrochemical (PEC) cell is an attractive means of converting solar energy into electricity. It is considered a major candidate for obtaining energy from the sun, since it can convert sunlight directly into electricity. It can provide nearly permanent power and is virtually free of pollution.

1.2 Nanotechnology used in solar cells

In recent years nanoscience has shown itself to be one of the most exciting areas in science, with experimental development being driven by pressing demands for new technological applications. It is highly multidisciplinary research field and the experimental and theoretical challenges for researchers in the physical sciences are substantial. Nowadays, scientists and researchers have been developing new kinds of nano-materials which could be used for forensic science, biology, electronic technology, environmental science, computer manufacturing, sports facility production as well as food industries in addition to PV. The term 'nanotechnology' can be traced back to 1974. It was first used by Norio Taniguchi in a paper entitled — 'On the Basic Concept of Nano-Technology'. Nanotechnology literally means any technology performed on a nanoscale that has applications in the real world. The new and emerging technology is "nanocrystalline thin film technology". The significant advantage of thin films is that these exist in solid state and solids tend to be more compact, easily cooled. Thin film converts light into electricity in a tuneable manner, depending on the thickness of the thin film.

1.3 Thin film technology

Thin film technology is an ever growing field in the physical & chemical sciences which are confluence of materials science, surface science, applied physics and applied chemistry. Thin film materials are the key elements of continued technological advances made in the fields of optoelectronic, photonic, and magnetic devices.

The processing of materials into thin films allows easy integration into various types of devices. The properties of material significantly differ when analyzed in the form of thin films. Thin film is defined as a low dimensional material created by condensing, one by one, atomic/molecular/ionic species of matter. The thickness is typically less than several microns. A thin - less than about one micron (10,000 A0, 1000 nm) film - layer of material on a surface with no substrate is called foil.

1.4 CdTe thin film solar cells

When considering biological applications, Cadmium Telluride (CdTe), is a notorious name when it is caught on the first sight due to its toxicity, but only so if ingested, its dust inhaled, or it is handled inappropriately. If it is properly and securely encapsulated, CdTe may be rendered harmless. Nowadays, it became a very useful material in the thin film solar cell industry, or in infrared optical material for optical windows and lenses. Bulk

CdTe is transparent in the infrared wavelength, from close to its band gap energy which is approximately 1.44eV at 300K (i.e. 860 nm) already in the infrared region, to the wavelength greater than 20 μm. As it has been presented that if the size of the bulk CdTe material shrinks to nanometer scale, normally 2 to 5 nm, the bandgap energy of the material will increase, due to quantum confinement effect, meaning the fluorescence peak will shift towards the infrared region or even visible range. This will open a new gate of application for this magical semiconductor material to be used in several areas which require small things to penetrate. [5]

Cadmium telluride (CdTe) is a stable crystalline compound formed from cadmium and tellurium. It is mainly used as the semiconducting material in cadmium telluride photovoltaics and an infrared optical window. Cadmium telluride (CdTe) photovoltaic describes a photovoltaic (PV) technology that is based on the use of cadmium telluride, a thin semiconductor layer designed to absorb and convert sunlight into electricity. CdTe is used to make thin film solar cells, accounting for about 8% of all solar cells installed in 2011. They are among the lowest-cost types of solar cell [6]. Of the II-IV semiconducting photovoltaic materials available, cadmium telluride has good optical performance across a wide range of temperatures and has provided adequate mechanical robustness to be used as a substrate material. Compared to the limited selection of alternative materials, CdTe has a high resistance, low moisture sensitivity and is available at a reasonable price. It has a cubic zinc blend structure with a lattice parameter of 6.481 Å and a density of 6.2 g/cm^3. It has a thermal coefficient of expansion of 4.9 x 10^{-6} K^{-1} and electron affinity of 4.3eV. It has a melting point of 1040°C. P type CdTe has a very high work function of 5.5eV. One special property of CdTe is that it can be doped both as p-type as well as an n-type material.

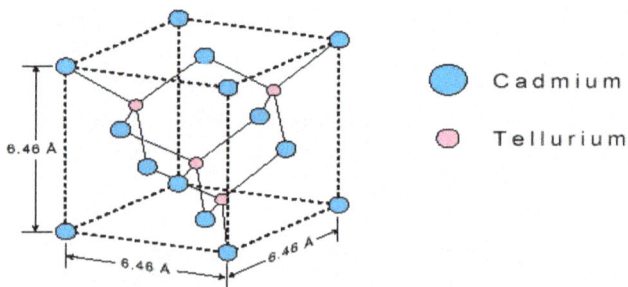

Figure 2: *Atomic structure of Cadmium telluride (CdTe).*

The choice of the material for solar cells depends upon the potential conversion efficiency expected from the devices made from them. It is found that for maximum conversion efficiency, the bandgap of the material should be approximately 1.5 eV. CdTe is a direct band gap semiconductor material with an energy gap of 1.45 eV and very high absorption coefficient for visible light of over 10^4 cm^{-1} and hence only a micron thick CdTe can absorb over 90% of photons with energy greater than 1.45eV. Therefore it is well suited for efficient conversion of solar light into electricity and is the leading material for solar cell fabrication. CdTe is stable up to 500 °C.

In the present work Cadmium Telluride CdTe nanocrystalline films have been grown on glass and indium doped tin oxide substrates using chemical method. The PV effect using nanocrystalline CdTe thin film as photo electrode in PEC cell have been studied. The films are characterised by XRD, SEM, AFM and VU-VIS absorption.

2.   Synthesis of nanocrystalline CdTe thin films

The nanocrystalline CdTe thin films have been deposited on plane as well as ITO coated glass substrates using the chemical bath deposition (CBD) technique. This is also known as controlled precipitation or solution growth method, or simply chemical deposition, recently emerged as a method for the deposition of metal chalcogenide thin films. In the chemical bath deposition metal chalcogenide thin film is formed due to the substrates maintained in contact with dilute chemical baths containing the metal and chalcogenide torts. The film formation takes place when the ionic product exceeds the solubility product. Lokhande [7] has published a review article with an emphasis on describing the deposition of the various semiconductors. A review by Mane and Lokhande [8] reported the effect of a number of deposition parameters on structural, optical, electrical and morphological properties of chemically deposited PbS, HgS, $Pb_{1-x}.Hg_xS$ and a number of binary and ternary thin films. Recently Hodes [9] wrote a book entitled —Chemical Solution Deposition of Semiconductor Films.

The chemical bath deposition has numerous advantages such as:

• The method does not require sophisticated instrumentation.

• It is ideally suited for large area deposition and substrate of accessible as well as non-accessible nature.

• Electrical conductivity of the substrate material is not an essential criterion.

• The deposition is easy even at low temperature and avoids the oxidation or corrosion of the metallic substrate. An intimate contact between the reacting solution and the substrate material permits for pinhole free and uniform deposits on the substrate of film.

- Stiochiometry of the deposit can easily be maintained since the basic building blocks are ions instead of atoms.

- Mixed/doped film structures could be obtained by merely adding the mixent/dopent solution directly into the reaction bath

- The process of film growth is slow at low temperatures that suppress the agglomeration of particles

- The method can be used to deposit a large number of metal chalcogenides.

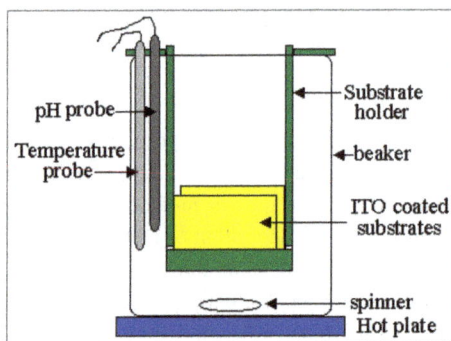

Figure 3: *Chemical bath deposition (CBD).*

The CdTe nanocrystalline films have been synthesised by varying the three different preparation conditions – (i) By varying deposition time keeping temperature constant at fixed concentration, (ii) changing temperature of bath keeping same deposition time and fixed concentration and (iii) at constant temperature and deposition time varying the concentration of precursor solutions. The details are described below.

2.1   For various deposition times

Nanocrystalline CdTe thin films are synthesized by the wet chemical route at pH 11.2 using cadmium chloride and potassium telluride as starting materials. All chemicals were of analytical grade, and used as received without further purification. De-ionised water is used throughout the experiments. The reaction vessel is filled with the composition of solution: [0.1 M 20 ml $CdCl_2$ + 0.1M 20 ml K_2Te], buffer solution is used to maintain 11.2 pH. The reaction is carried out by refluxing the mixture of starting materials at

900°C for 5 hrs. Water bath with reflux assembly is used for this purpose. Addition of 2-propanol produced suspension of CdTe nanocrystals named as solution A [10].

Chemical reaction:

$$CdCl_2 + K_2Te \longrightarrow CdTe + 2KCl$$

Figure 4: *View of water bath set at 90°C with reflux assembly.*

Figure 4 shows experimental setup photographs of water bath with reflux assembly filled with cadmium chloride and potassium telluride as starting materials for preparation of CdTe nanocrystalline solution.

Nanocrystalline thin films of cadmium telluride have been grown on glass and indium tin oxide (ITO) substrates using chemical bath method. (ITO) coated glass slides and plane glass slides of dimension 75×25×1.35 mm were procured from Nuchem International Company Jabalpur. The substrate cleaning is very important in the deposition of thin films. Glass slides are washed using soap solution and subsequently kept in hot chromic acid and then cleaned with deionised water followed by rinsing in acetone. Finally, the slides are wiped with soft cotton and stored in a hot oven [11].

Figure 5 : *Experimental setup photograph of deposition bath assembly with 12V battery, thermostat hot plate, digital multimeter, slide holder connected with DC motor*

A reaction vessel containing solution A used in the experiment is connected to auto-thermostat hot plate to maintain and control the accurate temperature of solution. The glass and ITO substrate are fixed to the circular holder and dipped in solution A (shown in Figure 4) with ITO side facing towards the solution and allowed to rotate with a speed of 25 rpm, in order to obtain CdTe thin films. The deposition is done for different deposition times [12]. The temperature and concentration of precursor solution and pH of the solution are tabulated in Table 1.

Table 1- Parameters used for deposition of nanocrystalline CdTe thin films [13].

Deposition parameter	Optimum value / item
Deposition time	60 min /120min/180min
pH	11.2
Concentration of precursor	Normal B
Deposition temperature	45°C
Solvent	Deionised water

The thermostat is set to a temperature at 45 ^{0}C and deposition is done on both glass and ITO slide for 60, 120, and 180 min with constant stirring of the solution throughout the experiment.

2.2 For various temperatures during deposition

We apply same synthesis process here, but with changed preparation conditions. The deposition is done at different temperature of the precursor solution. The deposition time and concentration of precursor solution and pH of the solution are tabulated in Table 2. The thermostat was set to temperatures at **45^{0}C, 60°C, 90°C** and deposition was done on both glass and ITO slide with constant stirring of the solution throughout the experiment.

Table 2- Parameters used for deposition of nanocrystalline CdTe thin films.

Deposition parameter	Optimum value / item
Deposition time	120 min
pH	11.2
Concentration of precursor solution	Normal B
Deposition temperature	45°C/60°C/90°C
Solvent	Deionised water

2.3. For various concentrations of precursors

Again the synthesis steps remain same with changed parameters. The deposition is done with different concentration of precursor solution i.e. half concentration (named-A), normal concentration (named-B) and double concentration (named-C). Different concentration of precursor solutions is tabulated in Table 3

Table 3- Parameters used for deposition of nanocrystalline CdTe thin films.

Deposition parameter	Optimum value / item
Deposition time	120 min
pH	11.2
Concentration of precursor solution	Half A/Normal B /Double C
Deposition temperature	45°C
Solvent	Deionized water

Table 4- Different Concentration of precursor solution.

Chemicals	Half concentration A	Normal concentration B	Double concentration C
$CdCl_2$	.05 Mole (20 ml)	0.1Mole (20 ml)	.2 Mole (20 ml)
K_2Te	.05 Mole (20 ml	0.1Mole (20 ml)	.2 Mole (20 ml)

The thermostat is set to a temperature at 45°C and deposition is done on both glass and ITO slide with constant stirring of the solution throughout the experiment. The deposition time, temperature of deposition and pH of the solution are tabulated in Table 3.

The deposited samples of CdTe thin film are annealed at 200°C for 2 hrs in oven for crystallinity improvement. Good transparent films are obtained onto both glass and ITO substrates. The thin films of CdTe on ITO coated glass substrate are used for study of photovoltaic effect in PEC cell. The thin films of CdTe on plane glass substrate are examined for their structural, surface morphological and optical characteristics by means of X-ray diffraction (XRD), scanning electron microscopy (SEM), atomic force microscopy (AFM), and UV spectrophotometer techniques respectively.

3. Photovoltaic response of CdTe films in photoelectrochemical cell

The PEC cell with a configuration of CdTe / [1M Sulpher (S) + 1M Sodium Hydroxide (NaOH) + 1M Sodium sulphide Na_2S] / graphite is used to check the type of conductivity exhibited by the CdTe thin films. The polarity of the dark voltage is positive toward the CdTe photoelectrode and negative toward the graphite electrode for all samples showing p-type semiconducting behaviour. An increase in the current is observed when films are illuminated. The photo-response upon illumination indicates that the films are sensitive towards light.

The configuration of PEC cell is a single glass vessel coated with black paint to avoid extra light, using CdTe nanocrystalline film on ITO substrate as photo-electrode, graphite as counter electrode and polysulphide solution [using 1M Sulphur (S) + 1M Sodium Hydroxide (NaOH) + 1M Sodium Sulphide Na2S] as electrolyte (shown in Figure 6a). The distance between working electrode and counter electrode is kept 2 cm. In this experiment the tungsten filament light source is kept at distance of 10 cm from beaker which produced the intensity of 315 lux as measured by standard lux meter (Scientific Laboratory Instruments, Roorkee, India). A potentiometer is connected with the PEC cell and voltage across the cell and current through the circuit are measured with the help of multimeters. The circuit is shown in Figure 6b. The resistance is varied and the different values of voltage and current are measured and I-V characteristics are obtained. The photovoltaic response of the nanocrystalline CdTe thin films on ITO coated glass substrate was studied for all the three different preparations conditioned [14].

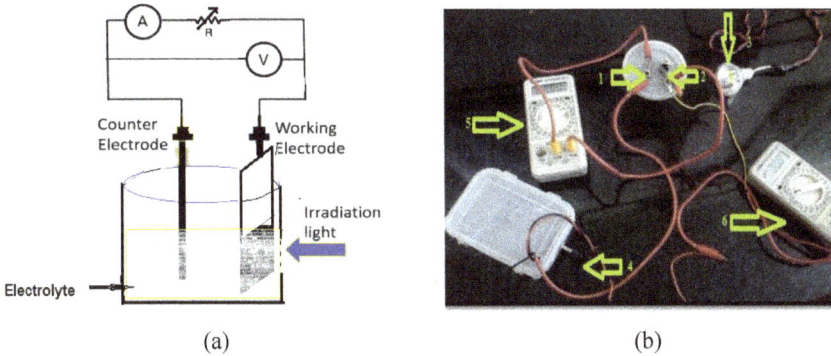

(a)   (b)

Figure 6: a) Circuit diagram and *b)* Photograph of PEC Assembly with single glass vessel.

(1) CdTe nanocrystalline film on ITO substrate as photo electrode,(2) graphite as counter electrode and polysulphide solution as electrolyte,(3) tungsten light source,(4) potentiometer, (5,6) Digital multimeter.

3.1 For various deposition times

The **I–V** characteristics are plotted for three electrodes prepared with different deposition time viz 60, 120 and 180 Minutes. Observing I-V curve it can be seen that with increase in resistance, current decreases and the terminal potential of solar cell increases.

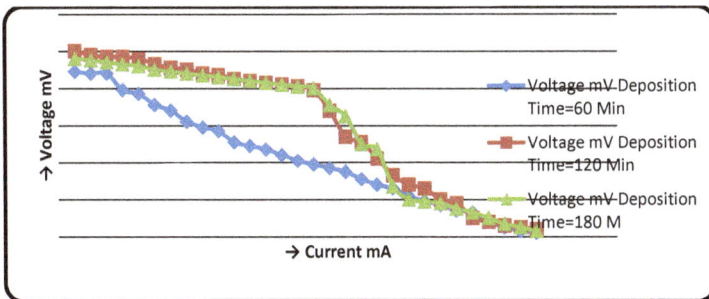

Figure 7: I–V characteristics of cadmium telluride photoelectrodes for different deposition time.

Table 5- Solar cell parameter for films deposited for different time duration.

S. No.	Sample Deposition Time (in Min	Temp (^{0}C)	Short Circuit Current Isc (mA)	Open Circuit Voltage Voc (mV	Fill Factor ff (%)	Efficiency η (%)	Max. Power (mw)	Series resistance R_S Ω	Shunt resistance R_{SH} kΩ
1	60	45	0.035	95	.189	0.91	0.63	33.33	2.5
2	120	45	0.034	110	.338	1.82	1.26	66.6	2.75
3	180	45	0.031	115	.366	1.85	1.28	100	3.0

From the I–V characteristics curves, solar cell parameters - short circuit current (Isc), open circuit voltage (Voc), Fill factor (FF), efficiency (η%), series resistance (R_S) and shunt resistance (R_{SH}) are determined. These are given in Table 5. It is observed that the solar cell parameters are influenced with deposition time.

Figure 8 shows variation of solar cells parameters - Voc and Isc with different deposition time of photo-electrodes. From the results it is observed that Isc decreases where as Voc increases by increasing deposition time. Figure 9 shows variation of solar cell parameters - fill factor and efficiency with different deposition time of photo-electrodes. The value of fill factor and efficiency increases with increase in deposition time (i.e. 60 min, 120 min upto180 min). From these results we also see that solar cell efficiency as well as fill factor is optimum when electrodes are prepared by 180 minute deposition time. It is observed that the performance is poor for 60 min deposited film and improves for thicker CdTe film. It is nearly same for 120 min and 180 min deposition time. The PEC performance of CdTe photoelectrode is improved with increase in deposition time.

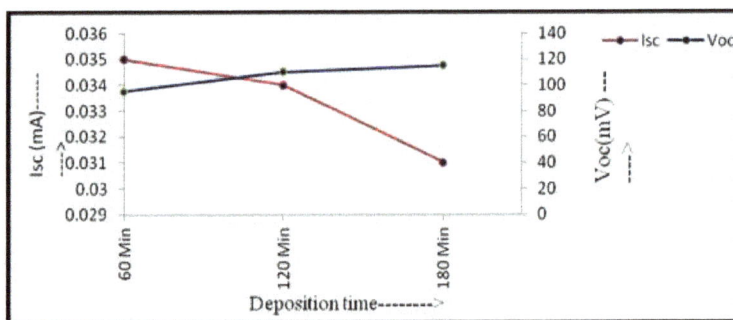

Figure 8: *Variation of Voc and Isc with different deposition time of photo-electrodes.*

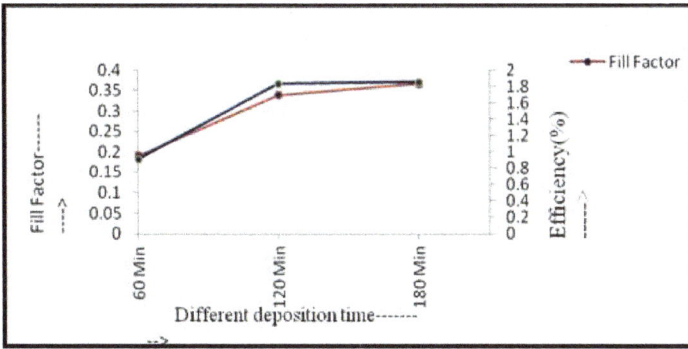

Figure 9: *Variation of Fill factor and efficiency with different deposition time of photo-electrodes.*

The value of series resistance R_S and shunt resistance R_{SH} were evaluated from the slope of the I-V plots using relation [15] →

$$[dI/dV]_{I=0} = 1/R_S \qquad (1)$$

$$[dI/dV]_{V=0} = 1/R_{SH} \qquad (2)$$

The different values of R_S and R_{SH} are obtained from I-V curve. The value of R_S and R_{SH} increases as CdTe thin film photo-electrode deposition time increases (as 60 Min, 120 Min, 180 Min). This variation is tabulated in table 5. The R_S value reflects the mobility of specific charge carrier in the CdTe. The mobility is affected by space charges and traps or other barriers (hopping) and it decreases with increase in thickness of emissive layer, because the layers exhibit larger resistivity than the two electrodes. With the increasing of deposition time, the R_S value is increased from **33.33** Ω to **100** Ω in this cell, which means that carrier mobility is decreased with deposition time. Generally, R_{SH} is determined by leakage due to recombination of charge carriers [16, 17]. R_{SH} is increased from **2.5** kΩ to **3** kΩ with deposition time, indicating that the leakage is reduced. The reduction of the carrier mobility will reduce the possibility of the carrier's recombination before carriers are collected by electrodes, and this will increase the R_{SH} value.

3.2. For various temperatures during deposition

The circuit diagram of PEC cell is same as shown in Figure 6 for obtaining I-V characteristics. In this experiment the light source is kept at distance of **10** centimeter from beaker which resulted the intensity of **315 lux** as in an earlier case as measured by standard lux meter. The resistance is varied and the different values of voltage and current are measured.

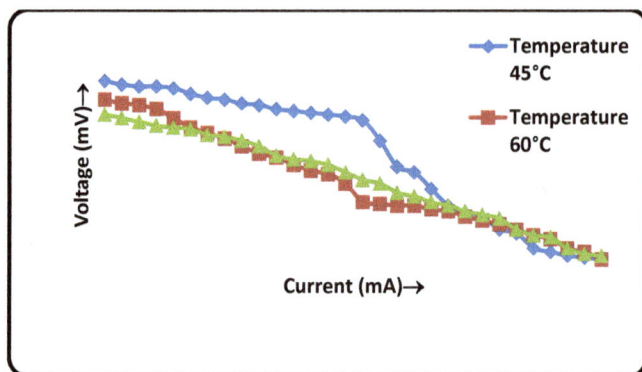

Figure 10: *I–V characteristics of cadmium telluride photoelectrodes deposited at different temperatures.*

Table 6- solar cell parameter for films deposited at deferent temperatures.

Sample temp. of precursor solution	Depositi-on Time	Short Circuit Current Isc (mA)	Open Circuit Voltage Voc (mV)	Fill Factor ff (%)	Efficiency η (%)	Max. Power (mW)	Series resistance R_S Ω	Shunt resistance R_{SH} kΩ
1 45^0C	120 min	.034	110	.338	1.82	1.254	103.44	4.33
2 60^0C	120 min	.036	92	.267	1.278	0.884	68.965	3.33
3 90^0C	120 min	.038	83	.253	1.155	0.799	34.482	3.00

The **I–V** characteristics (Figure 10) are plotted for three electrodes prepared at temperatures viz at **45^0C, 60°C, 90° C**. From the I–V characteristics curves solar cell parameters short circuit current (Isc), open circuit voltage (Voc).Fill factor (FF) efficiency (η%) Series resistance (R_S) and Shunt resistance (R_{SH}) are determined. These parameters are calculated and given in Table 6. It is observed that the solar cell parameters are influenced with temperature of precursor solution.

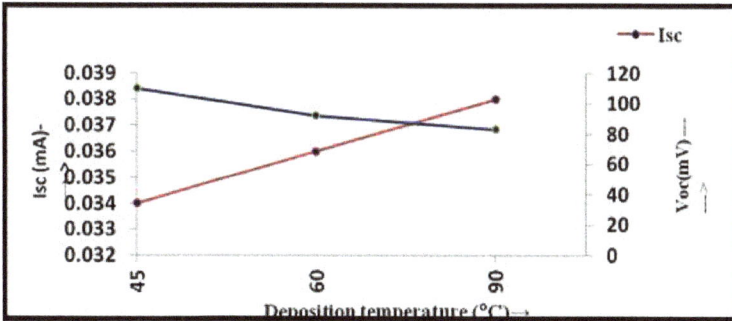

Figure 11: *Variation of Voc and Isc for different temperature during deposition.*

Figure 11 shows Variation of solar cells parameters Voc and Isc for different temperature of precursor solution during deposition of photo-electrodes. From the results it may be observed that Isc increases where as Voc decreases by increasing temperature. Figure 12 shows variation of solar cell parameters - fill factor and efficiency for different temperature of precursor solution during deposition of photo-electrodes. The values of fill factor and efficiency decrease with increase in temperature (i.e. at **45^{0}C, 60°C, 90°C.**) during deposition. From these results we also see that solar cell efficiency as well as fill factor is optimum when electrodes are prepared at **45^{0}C**. It is found that the PEC performance of CdTe photoelectrode becomes poor with increase in temperature during deposition of CdTe film.

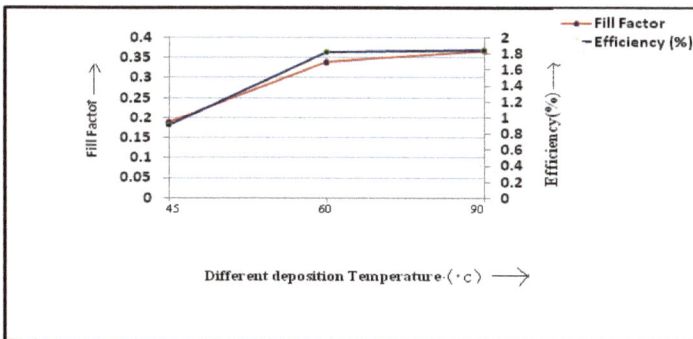

Figure 12: *Variation of Fill factor and efficiency for different temperature during deposition.*

The different values of R_S and R_{SH} are obtained from I-V curve. With the increasing of deposition temperature, the R_S value is decreased from **103.44** Ω to **34.482** Ω in this cell, which means that carrier mobility is increased with temperature. R_{SH} is decreased from **4.33** **kΩ** to **3** **kΩ** with temperature, indicating that the leakage is increased. The improvement of the carrier mobility will increase the possibility of the carrier's recombination before carriers are collected by electrodes, and this will decrease the R_{SH} value.

It is observed that the solar cell is more efficient and photovoltaic response is optimum for film deposited at 45°C temperature, deposition time is120 Min.

3.3   For various concentrations of precursors

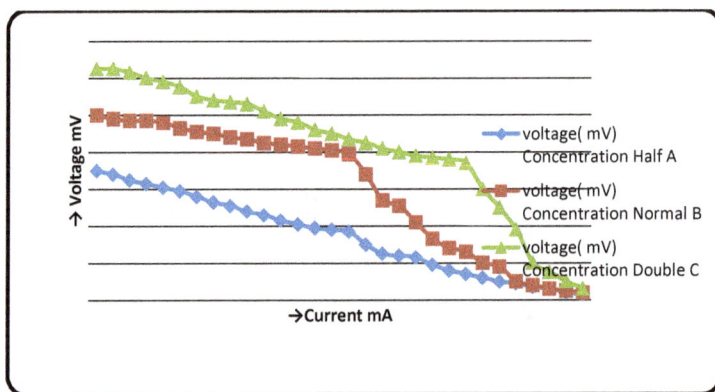

Figure 13: *I–V characteristics of Cadmium Telluride photoelectrodes from different concentration of precursor solution.*

Here also the **I–V** characteristics are plotted for three electrodes prepared from different concentration of precursor solution viz, half concentration (named-A), normal concentration (named-B) and double concentration (named-C) as described earlier and shown in Figure 13. Observing I-V curve it can be seen that with increase in resistance, current decreases and the terminal potential of solar cell increases.

Table 7- Solar cell parameter for films deposited from different concentration of precursor solution.

Sample (Conc. of precursor)		Tem p (^{o}C)	Depo sition Time (min)	Isc (mA)	Voc (mV)	Fill Factor (%)	Effici ency η(%)	Max. Power (mw)	Series resistance R_S Ω	Shunt resistance R_{SH} kΩ
A	Half	45	120	0.032	75	.247	0.855	0.592	31.25	3.5
B	Normal	45	120	0.034	110	.338	1.83	1.264	32.258	4.0
C	Double	45	120	0.038	126	.356	2.46	1.702	33.33	5.0

From the I–V curves solar cell parameters short circuit current (Isc), open circuit voltage (Voc).Fill factor (FF) efficiency (η%) series resistance (R_S) and shunt resistance (R_{SH}) are determined; these are given in table 7. It is observed that the solar cell parameters are influenced by concentration of precursor solution.

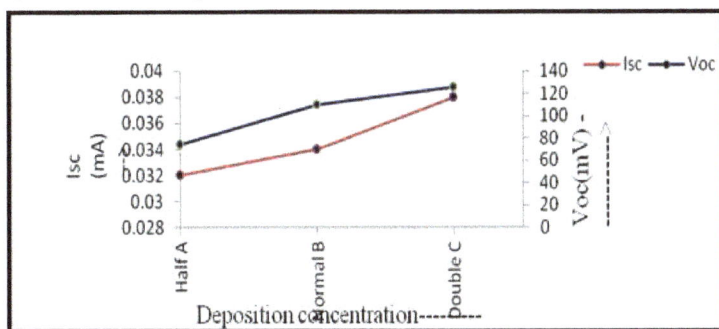

Figure 14: *Variation of Voc and Isc for different concentration of precursor solution.*

Figure14 shows variation of solar cells parameters - Voc and Isc for different concentration of precursor solution. From the results it may be observed that Isc and Voc increases by increasing concentration of precursor solution. It is observed that open circuit voltage (Voc) and short circuit current (Isc) are highest when photo electrode deposition is done at 45°C temperature and double concentration of precursor solution (named C).

Figure 15 shows variation of solar cell parameters - fill factor and efficiency for different concentration of precursor solution. The value of fill factor and efficiency increases with increase in concentration of precursor solution viz, half concentration (named-A), normal concentration (named-B) and double concentration (named-C). From these results we also see that solar cell efficiency as well as fill factor is optimum when electrodes are prepared with double concentration (named-C) of precursor solution.[18] [19]

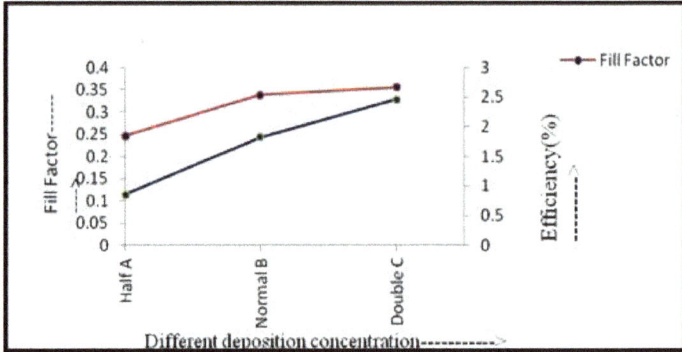

Figure 15: Variation of Fill factor and efficiency for different concentration of precursor solution.

The different values of R_S and R_{SH} are obtained from I-V curve. The values of R_S and R_{SH} increases as CdTe thin film photo-electrode prepared from higher concentration of precursor solution. This variation is tabulated in Table 7.

It is observed that the solar cell is more efficient and photovoltaic response is highest for film deposited at 45°C temperature, for 120 minutes from double concentration of precursor solution (named C).

4. Characterization of nanocrystalline CdTe thin films

The thin films of CdTe on plane glass substrate were used to study their properties. The thickness of the films was obtained by the gravimetric method. For their structural, surface morphological and optical characteristics, X-ray diffraction (XRD), scanning electron microscopy (SEM), atomic force microscopy (AFM), and UV spectrophotometer techniques were used.

4.1 Determination of film thickness

The thickness of photelectrode has a strong impact on the performance of photoeletrochemical cell [14]. The thickness, sample weight and sample area are related as:

$t = M/A \cdot \rho$ ----------------- (3) [20]

Where:

M = weight of the sample in *gm*,

A = area of the sample in cm^2

ρ = materials density in gm/cm^{-3}

The thickness of the CdTe film was determined by a weight difference method [21-24] or gravimetric method .The weight of substrate was obtained employing sensitive electronic microbalance before and after the deposition. The deposition has been done on [1×1.5] cm^2 area and bulk density of CdTe is taken as (6.02gm/cm^3).

a) Thickness of CdTe thin film deposited for various deposition times

The computed values of thickness of CdTe films for different deposition time are given in Table 8. Figure 16(A) shows the variation of thickness of the film for various deposition times. It is observed from Table 8 and figure 16 (A) that in chemical deposition, thickness increases with duration of deposition [25].

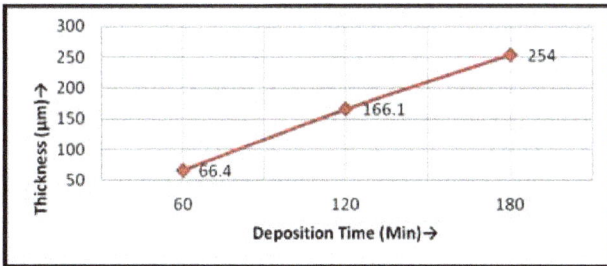

Figure 16(A): *Thickness of the films as a function of deposition time.*

Table 8- The thickness variation with deposition time.

S.No.	Deposition Time	Thickness (μm)
1	60 Min	66.4
2	120 Min	166.1
3	180 Min	254

b) Thickness of CdTe thin film deposited at various temperatures

Figure 16 (B) shows the variation of thickness with deposition at different temperature. It is observed from Table 9 and Figure 16 (B) that the thickness of the deposited cadmium telluride thin film decreases with increase in temperature of precursor solution.

Table 9- The thickness variation with deposition at different temperatures.

S.No.	Temperature	Thickness (μm)
1	45°C	166.1
2	60° C	132.8
3	90° C	110.7

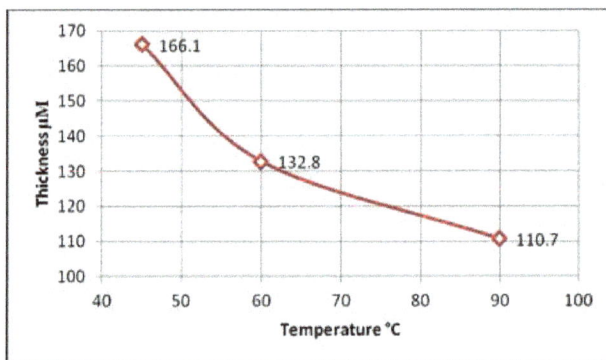

Figure 16 (B): *Thickness of the films as a function of deposition temperature.*

c) Thickness of CdTe thin films deposited from various concentrations of precursors

Figure 16 (C) shows the graph of thickness of the film for various concentration of precursor solution. It is observed from Table 10 and Figure 16 (C) that in chemical deposition, thickness increases with concentration of precursor solution.

Table 10- The thickness variation with concentration of precursor solution.

S.No.	Concentration of precursor solution	Thickness (μm)
1	Half concentration named A	66.4
2	Normal concentration named B	166.1
3	Double concentration named C	232.5

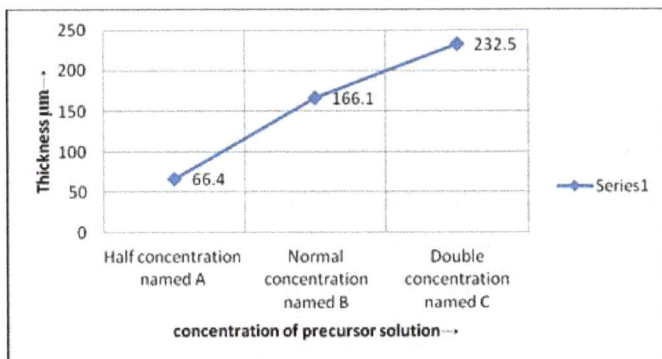

Figure 16 (C): *Thicknesses of the films as a function of concentration of precursor solution.*

4.2 Structural studies using XRD

The X-Ray Diffractometer (Epifluorescence Microscopy Model, Rigaku, NIKON, Japan) is used for the X-ray diffraction studies of CdTe thin film deposited on glass plate. JCPDS card number 9007154 is used for comparing the standard values with the experimental data. From the XRD profile, the inter-planer spacing and Miller indices (hkl) are calculated using the Bragg's relation and 2θ *is the angle between incident and the scattered X ray. The wavelength of the X-ray is 1.540559 A°* (given by Centre for Nanoscience and Nanotechnology Sathyabama University Chennai). From the XRD profile a narrow peak in CdTe thin film can confirms grain size growth.

Table 11 - Measurement conditions.

X-Ray	30 kV , 100 mA	Scan speed / Duration time	4.0000 deg/min
Goniometer	SmartLab	Step width	0.0200 deg
Attachment	Standard	Scan axis	Theta/2-Theta
Filter	Cu K-beta	Scan range	10.0000 -
CBO	BB	Incident slit	2/3deg
Diffracted	None	Length limiting slit	10.0mm
Detector	SC-70	Receiving slit #1	2/3deg
Scan mode	CONTINUOUS	Receiving slit #2	0.150mm

The X-ray diffraction patterns of nanocrystalline CdTe thin films are shown in Figures 17, 18 and 19. The average particle size of CdTe thin film was calculated by using Scherrer's equation.

Scherrer's formula:

$$D = C\lambda / \beta Cos(\theta) \quad [26]$$

Where λ is wavelength (Å),

β is FWHM (radians) corrected for instrument broadening,

θ is Bragg angle (Diffraction peak position)

C is a crystal shape factor from 0.9~1.

D = Average crystallite size

a)   XRD of nanocrystalline CdTe thin film with deposition time variation:-

Figure 17 [A], [B], [C] shows the X-ray diffraction pattern of a chemically deposited cadmium telluride thin film deposited on glass substrates at 45^0C temperature for deposition time is 60, 120 and 180 minutes. From JCPDS card number 9007154, crystal class is predominantly tetragonal and reasonably crystalline and this is also used for comparing the standard values with the experimental data. The value of cell parameter is found to be, a=b=4.0440, c=2.6190 and $\alpha = \beta = \gamma = 90^0$[27]. The diffractogram of the deposited cadmium telluride thin films seems to exhibit nanocrystalline nature.   A narrowed peak (horizontal) width in annealed film confirms grain size growth. The grain size (D) is calculated using the Scherrer's formula from the full-width at half maximum (FWHM)-β of corresponding peak of the XRD pattern. The calculated average grain size using observed peaks for deposited cadmium telluride thin film were (4.69 nm, 25.32nm, 26.09nm) for deposition time 60, 120 and 180 minutes respectively .

Figure 17(A): *X-ray diffraction pattern of CdTe thin film (Deposition time- 60 min.).*

Table 12(a)- Peak list.

No.	2theta (deg)	D (ang.)	Height (cps)	FWHM (deg)	Int. I (cps deg)	Int. W(deg)	Asym. factor
1	12.715	6.956	587	0.427	390	0.66	1.4
2	28.025	3.181	277	0.295	104	0.38	0.8
3	40.19	2.242	66	0.30	21.4	0.32	1.0

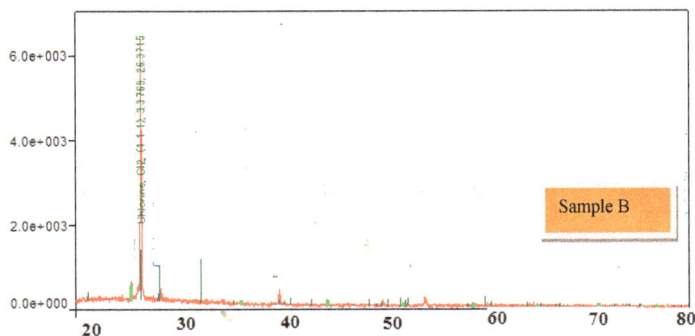

Figure 17 (B): X-ray diffraction pattern of CdTe thin film (Deposition time- 120 min.).

Table 12(b-) Peak list.

No.	2theta (deg)	d(ang)	Height (cps)	FWHM (deg)	Int. I (cps deg)	Int.W (deg)	Asym. factor
1	25.398	3.504	387	0.064	39	0.101	0.8
2	26.348	3.379	5784	0.068	592	0.102	1.2
3	28.336	3.147	282	0.057	22	0.077	3
4	39.985	2.253	369	0.062	37	0.101	1.6
5	50.140	1.817	156	0.053	8.7	0.06	1.0
6	54.238	1.689	190	0.088	28	0.15	1.9

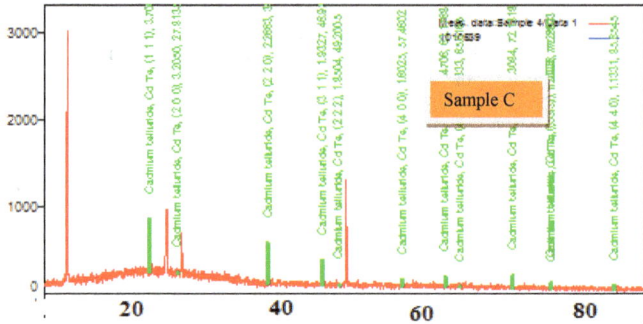

Figure 17 (C): *X-ray diffraction pattern of CdTe thin film (Deposition time- 180 min.).*

Table 12(c)- Peak list.

No.	2theta (deg)	D (ang.)	Height (cps)	FWHM (deg)	Int.I (cps deg)	Int.W (deg)	Asym. factor
1	12.983	6.813	1180	0.060	87	0.074	3.1
2	13.086	6.760	2107	0.089	231	0.110	2.3
3	26.360	3.37834)	561	0.091	105	0.19	3
4	28.336	3.14707)	479	0.055	30	0.063	2.0
5	50.148	1.81765(8)	1340	0.055	88	0.065	0.84

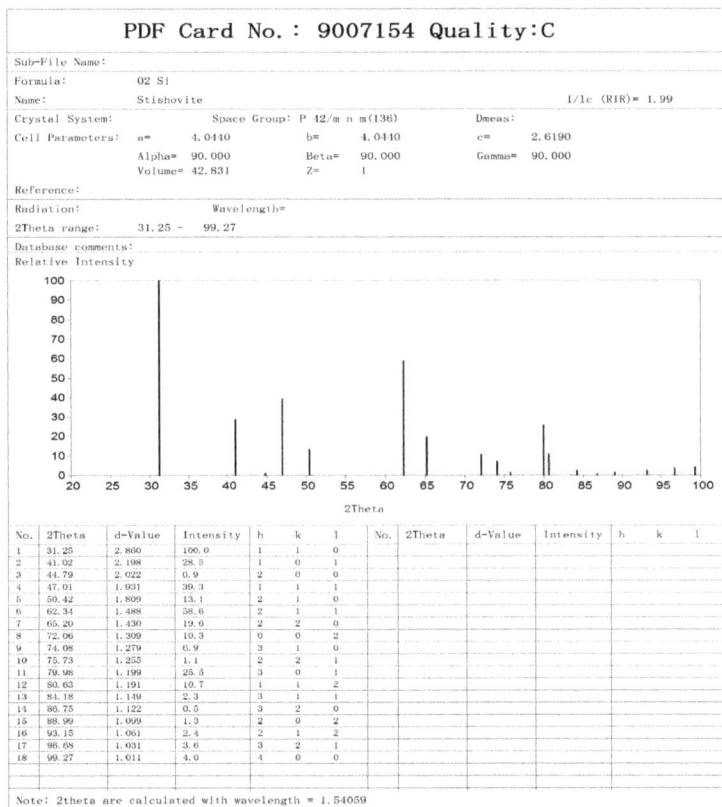

PDF Card No. : 9007154 Quality:C

Sub-File Name:							
Formula:	O2 Si						
Name:	Stishovite						1/Ic (RIR)= 1.99
Crystal System:		Space Group: P 42/m n m(136)			Dmeas:		
Cell Parameters:	a= 4.0440		b= 4.0440		c= 2.6190		
	Alpha= 90.000		Beta= 90.000		Gamma= 90.000		
	Volume= 42.831		Z= 1				
Reference:							
Radiation:		Wavelength=					
2Theta range:	31.25 - 99.27						
Database comments:							

Relative Intensity

No.	2Theta	d-Value	Intensity	h	k	l	No.	2Theta	d-Value	Intensity	h	k	l
1	31.25	2.860	100.0	1	1	0							
2	41.02	2.198	28.8	1	0	1							
3	44.79	2.022	0.9	2	0	0							
4	47.01	1.931	39.3	1	1	1							
5	50.42	1.809	13.1	2	1	0							
6	62.34	1.488	38.6	2	1	1							
7	65.20	1.430	19.6	2	2	0							
8	72.06	1.309	10.3	0	0	2							
9	74.08	1.279	6.9	3	1	0							
10	75.73	1.255	1.1	2	2	1							
11	79.98	1.199	25.5	3	0	1							
12	80.63	1.191	10.7	1	1	2							
13	84.18	1.149	2.3	3	1	1							
14	86.75	1.122	0.5	3	2	0							
15	88.99	1.099	1.3	2	0	2							
16	93.15	1.061	2.4	2	1	2							
17	96.68	1.031	3.6	3	2	1							
18	99.27	1.011	4.0	4	0	0							

Note: 2theta are calculated with wavelength = 1.54059

2013-Sep-02 12:40:39 Page-1/1

Figure 18

The calculated average crystal size using observed peaks for deposited CdTe thin film is given in Table 13.

It is observed from Table 13 that the calculated average crystal size for deposited CdTe thin film increases with increase in deposition time. After 120 min there is very little change in particle size.

Table 13- Effect of deposition time on, Particle Size

S.No.	Deposition Time	Particle Size
1	60 Min	4.69 nm
2	120 Min	25.32 nm
3	180 Min	26.09nm

b) XRD of nanocrystalline CdTe thin film with deposition temperature variation:-

Figure 19 [A], [B], [C] shows the X-ray diffraction pattern of a chemically deposited cadmium telluride thin film deposited on glass substrates at different temperature of precursor solution viz 45°C, 60°C and 90°C and deposition time is 120 min. Similar crystal structure, planes and cell parameters are obtained.

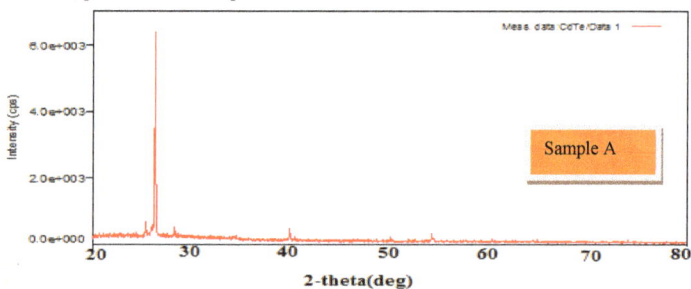

Figure 19 (A): *X-ray diffraction pattern of CdTe thin film (Temperature during deposition 45°C).*

Table 14(a)- Peak list.

No.	2-theta(deg)	d (ang.)	Height(cps)	FWHM(deg)	Int. I (cps deg)	Int. W(deg)	Asym. factor
1	25.398	3.50406)	387	0.064	39	0.101	0.8
2	26.348	3.37984)	5784	0.068	592	0.102	1.2
3	28.336	3.14719)	282	0.057	22	0.077	3
4	39.985	2.25305)	369	0.062	37	0.101	1.6
5	50.140	1.81794)	156	0.053	8.7	0.06	1.0
6	54.238	1.68983)	190	0.088	28	0.15	1.9

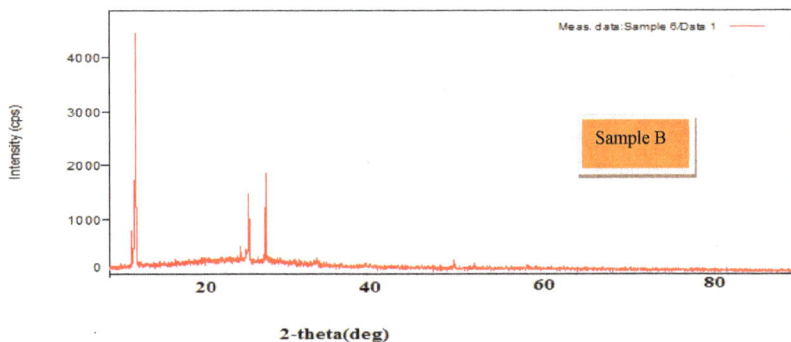

Figure19 (B): *X-ray diffraction pattern of CdTe thin film (Temperature during deposition 60°C.).*

Table 14(b)- Peak list.

No.	2-theta(deg)	d(ang.)	Height(cps)	FWHM(deg)	Int. I (cps deg)	Int. W(deg)	Asym. Factor
1	12.611	7.014	658	0.041	44	0.067	0.38
2	13.080	6.763	3325	0.074	434	0.130	1.4
3	26.367	3.37759	1052	0.070	125	0.119	4
4	28.324	3.14833	1560	0.056	113	0.072	1.1

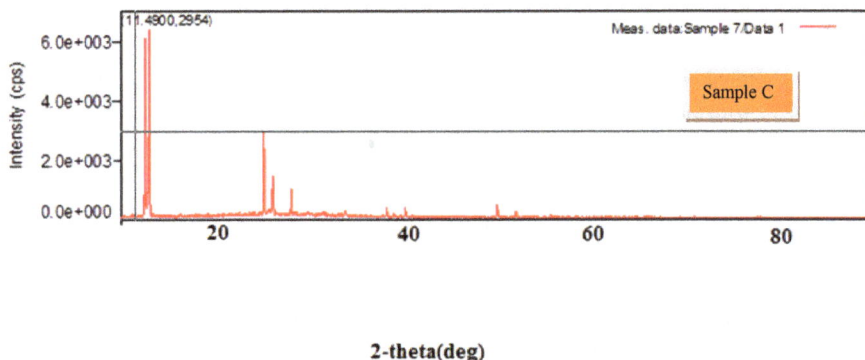

Figure 19(C): *X-ray diffraction pattern of CdTe thin film (Temperature during deposition 90°C).*

213

Table 14(c)- Peak list

No.	2-theta(deg)	d(ang.)	Height (cps)	FWHM (deg)	Int. I (cps deg)	Int. W(deg)	Asym. factor
1	12.6072	7.0157	4562	0.058	350	0.077	2.9
2	13.046	6.781	4752	0.082	603	0.127	1.7
3	25.381	3.5064	2690	0.052	179	0.067	1.3
4	26.319	3.3836	1072	0.084	128	0.120	1.1
5	28.290	3.1521	963	0.048	51	0.053	0.8
6	38.483	2.3374	288	0.050	22.4	0.078	0.8
7	40.471	2.2271	395	0.039	22.4	0.057	1.3
8	50.128	1.8183	422	0.052	32.8	0.078	2.0
9	52.164	1.7520411)	182	0.080	16.6	0.09	4.0

The average grain size has been estimated from the width of the peaks. The calculated average grain size using observed peak for deposited CdTe thin film is shown in Table 15.

Table 15- Effect of temperature of precursor solution on Particle Size.

S.No.	Temperature of precursor solution	Particle Size
1	45°C	25.32 nm
2	60°C	20.25 nm
3	90°C	16.90 nm

It is observed that the calculated average grain size for deposited CdTe thin film decreases with increase in temperature during deposition of films.

c) XRD of nanocrystalline CdTe thin film with precursor concentration variation:-

Figure 20 [A], [B], [C] shows the X-ray diffraction pattern of a chemically deposited cadmium telluride thin film deposited on glass substrates at different concentration of precursor solution viz half concentration (named-A), normal concentration (named-B) and double concentration (named-C) respectively, and deposition time is 120 minutes. Similar crystal structure, planes and cell parameters are obtained.

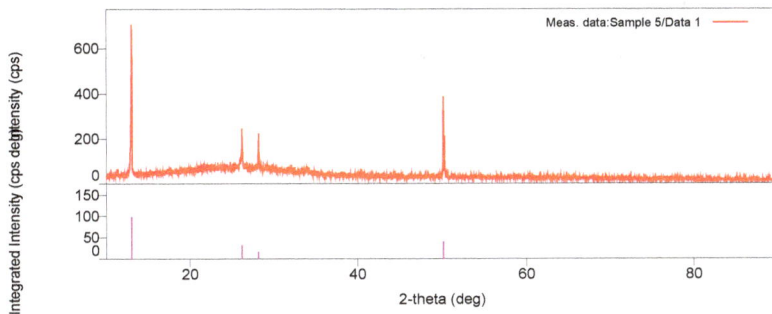

Figure 20 (A): *X-ray diffraction pattern of CdTe thin film (precursor concentration A - half).*

Table 16(a)- Peak list.

No.	2-theta (deg)	d(ang.)	Height (cps)	FWHM (deg)	Int. I (cps deg)	Int. W (deg)	Asym. Factor
1	13.080	6.763	518	0.074	66	0.127	1.4
2	26.349	3.3797	154	0.083	19.9	0.130	2.3
3	28.315	3.1493	139	0.049	9.3	0.067	0.8
4	50.138	1.8179811	367	0.055	25.8	0.070	0.8

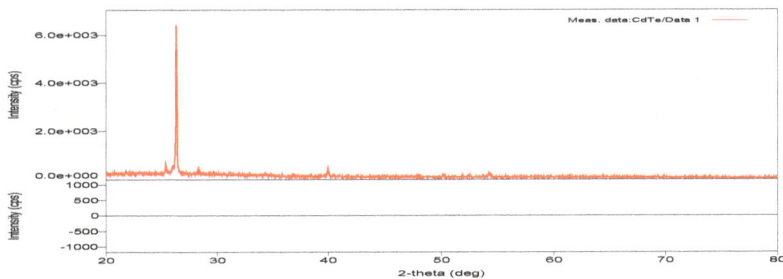

Figure 20(B): *X-ray diffraction pattern of CdTe thin film (precursor concentration B - normal).*

Table 16(b-) Peak list.

No.	2-theta (deg)	d(ang.)	Height(cps)	FWHM(deg)	Int. I (cps deg)	Int. W (deg)	Asym. factor
1	25.398	3.5040	387	0.064	39	0.101	0.8
2	26.348	3.3798	5784	0.068	592	0.102	1.2
3	28.336	3.1471	282	0.057	22	0.077	3
4	39.985	2.2530	369	0.062	37	0.101	1.6
5	50.140	1.8179	156	0.053	8.7	0.06	1.0
6	54.238	1.6898	190	0.088	28	0.15	1.9

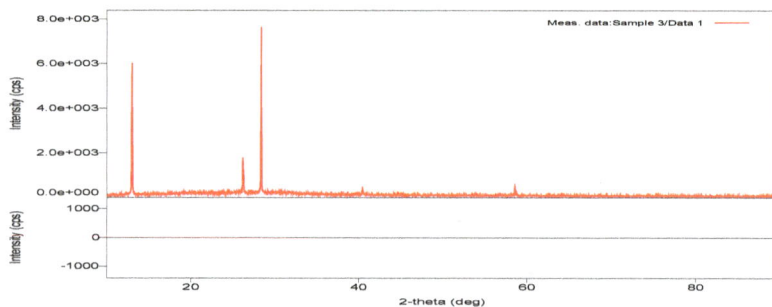

Figure 20 (C): *X-ray diffraction pattern of CdTe thin film ((precursor concentration C – double).*

Table (16(c)- Peak list.

No.	2-theta (deg)	d (ang.)	Height (cps)	FWHM (deg)	Int. I (cps deg)		Int. W (deg)	Asym. Factor
1	25.398	3.5040	387	0.064	39		0.101	0.8
2	26.348	3.3798	5784	0.068	592		0.102	1.2
3	28.336	3.1471	282	0.057	22		0.077	3
4	39.985	2.2530	369	0.062	37		0.101	1.6
5	50.140	1.8179	156	0.053	8.7		0.06	1.0
6	54.238	1.6898	190	0.088	28		0.15	1.9

The average grain size has been estimated from the width of the peaks. The calculated average particle size using observed peaks for deposited CdTe thin film is shown in table 17.

Table 17- Effect of concentration of precursor solution on Particle Size.

S.No.	Concentration of precursor solution	Particle Size
1	Half Concentration A	17.24 nm
2	Normal Concentration B	25.32 nm
3	Double Concentration C	30.30 nm

It is observed from Table 17 that the calculated average grain size for deposited CdTe thin film increases with increase in concentration of precursor solution.

4.3   Microstructural properties using SEM and AFM

4.3.1 Scanning Electron Microscopy

Scanning electron microscopy is a convenient technique for surface microstructure of thin films. SEM monitors the formation and growth of thin films and nanostructures. Scanning electron microscope (SEM) was used for the morphological study of prepared CdTe thin films.

a)   SEM study of nanocrystalline CdTe thin film for deposition time variation

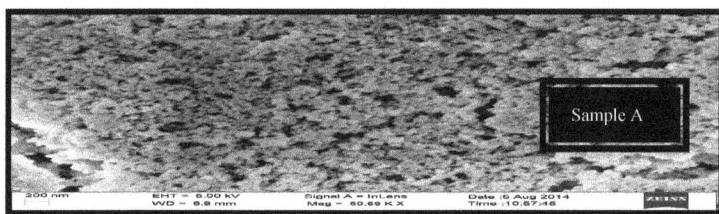

Figure 21 [A]: SEM image of CdTe thin film for 60 Min. deposition time.

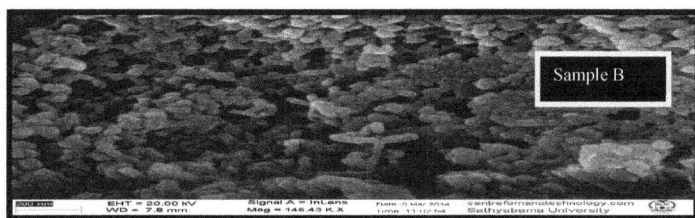

Figure 21 [B]: SEM image of CdTe thin filmfor120 Min. deposition time.

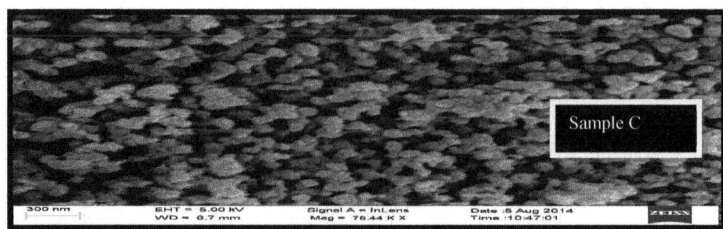

Figure 21[C]: SEM image of CdTe thin film for 180 Min. deposition time.

Figure 21 [A], [B], [C] shows SEM images of CdTe thin films for 60, 120 and 180 min deposition time respectively. It can be seen that the films are composed of a large number of nanoparticles. The CdTe films formed are highly agglomerated. SEM micrograph is showing topography of the film surface. The morphology of the particle is roughly spherical in shape. The film compactness is high, the surface's uniformity is good, the particle size is quite fine, and the particle size distribution is also narrow. The micrograph shows that the substrate is well covered with a large number of densely packed well oriented particles. The SEM pictures show the uniform coverage and the low porosity on the surface of the film. The particle forms a dense structure on the film which shows the growth in all direction. The particle overlaps one another. The SEM micrograph indicates that there are no major differences in the morphology of each particle, which shows the nanocrstalline structure of the films. Each granular structure was closely grown on the surface of film with approximately same particle size. The grain size is observed to increases with increase in deposition time. As deposition time increases, amount of solute (i.e. cadmium chloride and potassium telluride) reaching on the surface of the substrate increases and therefore the electrostatic interaction between solute particle become large increasing the grain size. These results are in good agreement with results, reported by few earlier workers [29, 30]

b)   SEM study of nanocrystalline CdTe thin film for deposition temperature variation

Figure 22 [A], [B] and [C] shows SEM micrograph and topography of the film surface at temperature of precursor solution 45°C, 60°C and 90°C respectively. The morphology of the particle is roughly spherical in shape. The film compactness is high, the surface's uniformity is good, the particles are spherical and size is quite fine, and the particle size distribution is also narrow.

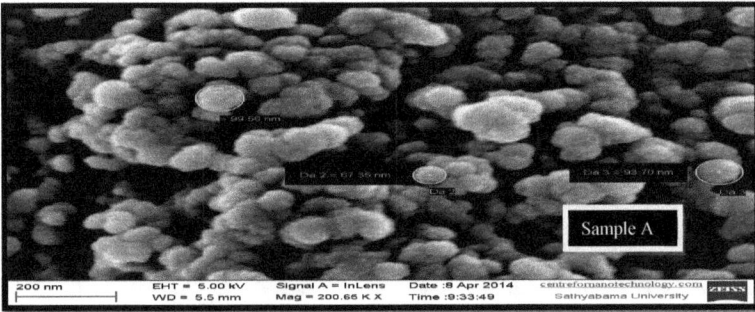

Figure 22 [A]: SEM image of CdTe thin film deposited at 45°C temperature.

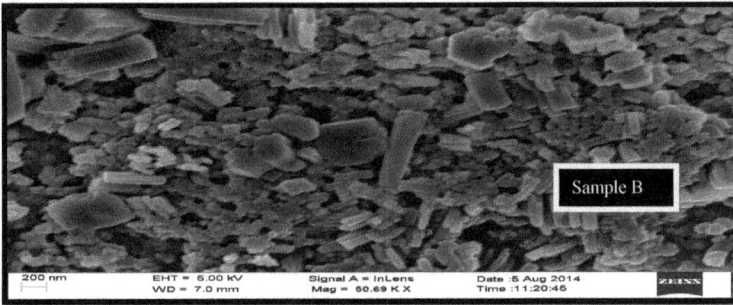

Figure 22[B]: SEM image of CdTe thin film deposited at 60°C temperature.

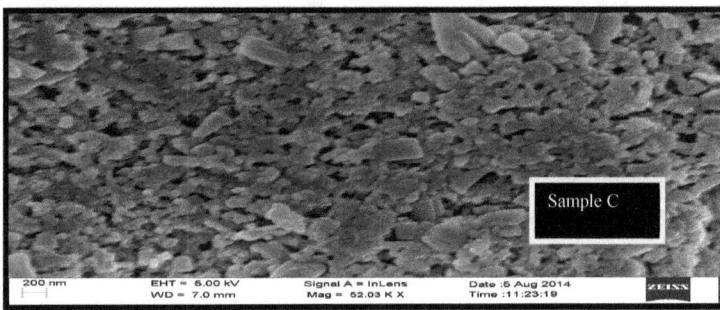

Figure 22 [C]: SEM image of CdTe thin film deposited at 90°C temperature.

d)   SEM study of nanocrystalline CdTe thin film for precursor concentration variation

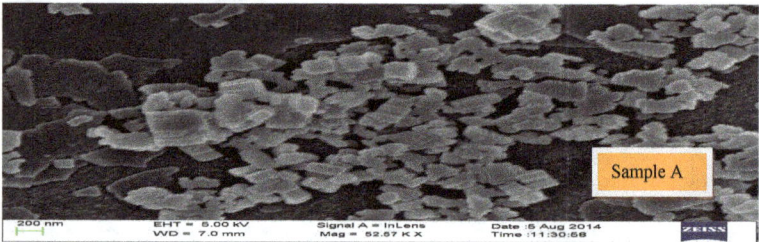

Figure 23 [A]: SEM image of CdTe thin film deposited from half concentration A.

Figure 23 [B]: SEM image of CdTe thin film deposited from normal concentration B.

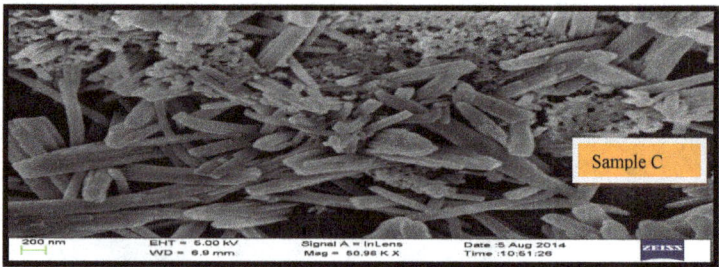

Figure 23 [C]: SEM images of CdTe thin film deposited from Double concentration C.

The micrograph shows that the substrate is well covered with a large number of densely packed well oriented spherical particles. Figure show that as temperature of precursor solution increases the crystallite size decreases. These results are in good agreement with results, reported by few earlier workers [31, 32].

SEM micrographs in Figure 23 [A], [B], [C] show different structural nanocrystalline nature of deposited films from different concentration of precursor solution viz- half concentration (named A), normal concentration (named B) and double concentration (named C). The nanocrystalline structure develops on the films as spherical or cubic particles. For concentration C the elongated structure is seen forming chips. The compact particles overlap one another. The particle forms a dense structure on the film which shows the growth in all direction. The CdTe thin films formed are highly agglomerated. As concentration of precursor solution increases the shape of the nanoparticles changes. At large magnifications, the nanoparticle clearly identified and understand their growth on entire surface of film. Figure show that as concentration of precursor solution increases the crystallite size increases. These results are in good agreement with results, reported by few earlier workers. [33, 34]

4.3.2 Atomic force microscopy

Atomic force microscopy or AFM is a method to see a surface in its full, three-dimensional glory, down to the nanometre scale. The AFM offers visualization in three dimensions. Resolution in the vertical, or Z, axis is limited by the vibration environment of the instrument, whereas resolution in the horizontal, or X-Y, axis is limited by the diameter of tip utilized for scanning.

AFM is a scanning technique to produce very high resolution, 3-D images of sample surfaces. For Atomic Force Microscopy (AFM) studies of deposited CdTe film on glass plate, AFM (NTMDT) instrument is used. (Instrument name is given by Center of Nanoscience & Nanotechecnology, Sathyabama University Chennai). The atomic force microscope (AFM) is ideally suited for characterizing nanoparticles.

a) AFM study of CdTe nanocrystalline thin film with deposition time variation

Surface morphologies of films are measured using AFM. (Figure 24 A, B, C) shows 2D atomic force microscopic image (5x5) nm of CdTe thin film for 60, 120 and 180 min deposition time and 45°C temperature.

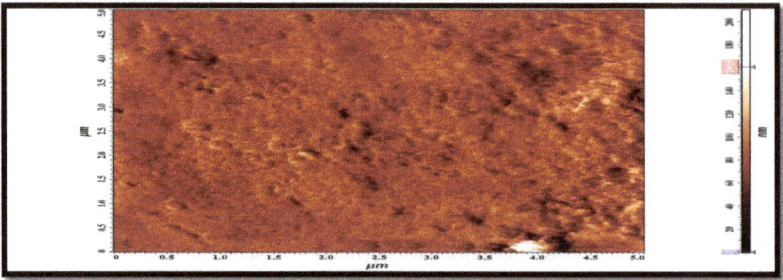

Figure 24 [A]: AFM 2D image of CdTe thin film at 60 Min deposition time.

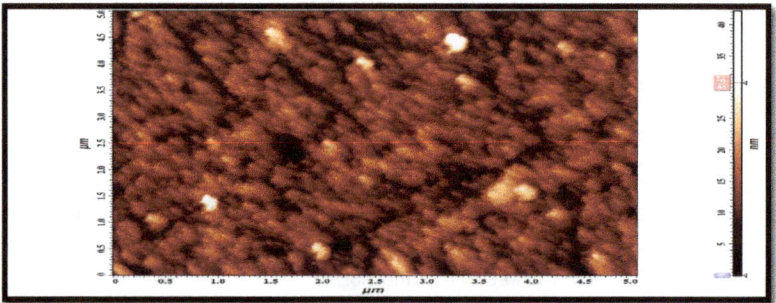

Figure 24 [B]: AFM 2D image of CdTe thin film for 120 Min deposition time.

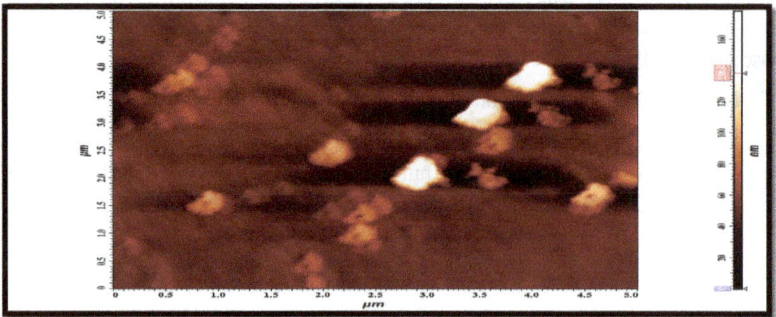

Figure 24[C]: AFM 2D image of CdTe thin film for 180 Min deposition time.

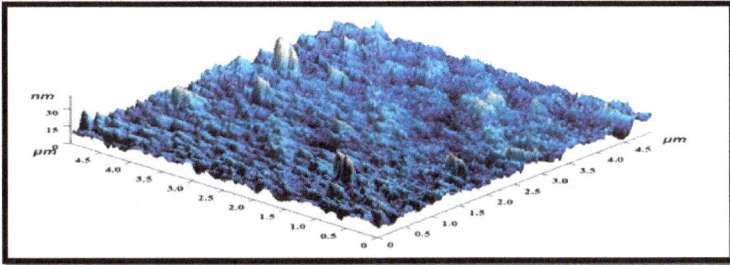

Figure 24 [D]: AFM 3D of CdTe thin film for 60 Min deposition time.

Figure 24[E]: AFM 3D image of CdTe thin film for 120 Min deposition time.

Figure 24 [F]: AFM 3D of CdTe thin film for 180 Min deposition time.

Figure 24 [D], [E], [F] shows 3D atomic force microscopic image (5x5nm) of CdTe thin film for 60, 120 and 180 min deposition time respectively. It appears that the particle size, film size and the surface roughness increases with increasing of the deposition time. We can expect this result due to the increase of the film thickness and the changes in nucleation and growth rate with the extension of the reaction time. These results are in good agreement with results, reported by few earlier workers [35].

b)   AFM study of CdTe nanocrystalline thin film with deposition temperature variation

Surface morphologies of films were measured using AFM. Figure 25 A, B and C shows 2D atomic force microscopic image (**5x5**) nm of **CdTe** thin film at **45°C, 60°C, 90°C** temperature of precursor solution.

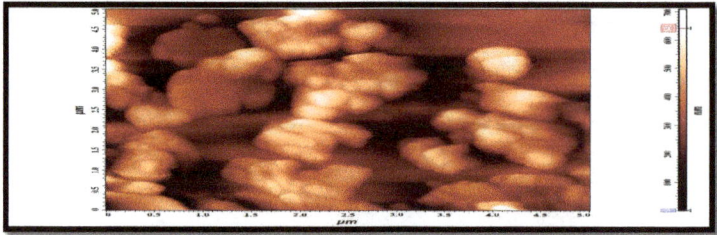

Figure 25 (A): AFM 2D image of CdTe thin film deposited at 45°C temperature.

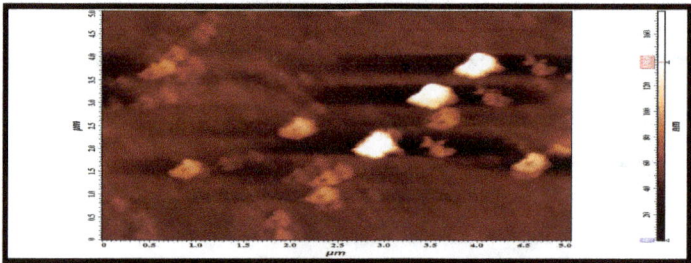

Figure 25 (B): AFM 2D image of CdTe thin film deposited at 60°C temperature.

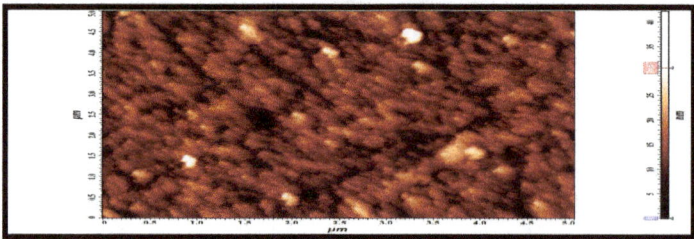

Figure 25 (C): AFM 2D image of CdTe thin film deposited at 90°C temperature.

Figure 25 [D], [E], [F]) shows 3D atomic force microscopic image (5x5nm) of **CdTe** thin film deposited at 45°C, 60°C, 90°C temperature of precursor solution. It appears that the particle size, film size and the surface roughness decreased with increasing of the temperature of precursor solution. We can expect this result due to the increase of the temperature of precursor solution and the changes in nucleation and growth rate with the increase in the temperature. These results are in good agreement with results, reported by few earlier workers [36].

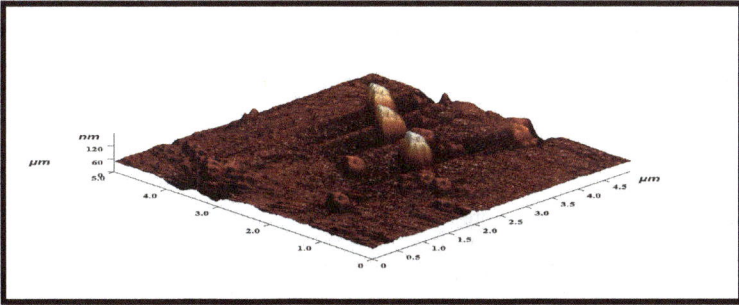

Figure 25 (D): AFM 3D image of CdTe thin film deposited at 45°C temperature.

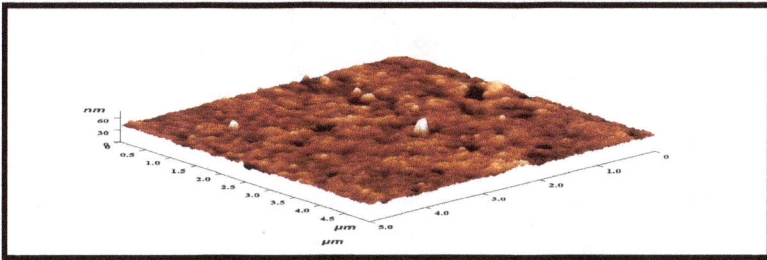

Figure 25 (E): AFM 3D image of CdTe thin film deposited at 60°C temperature.

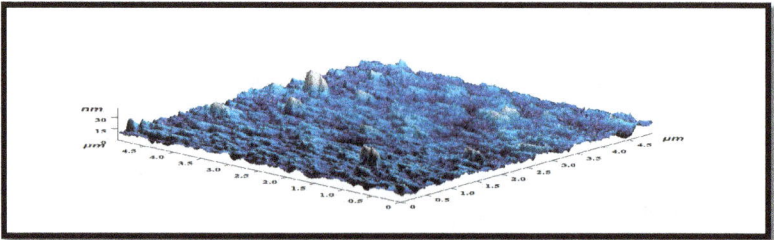

Figure 25 (F): *AFM 3D image of CdTe thin film deposited at 90°C temperature.*

c) **AFM study of CdTe nanocrystalline thin film with precursor concentration variation**

Surface morphologies of films were measured using AFM. Figure 26 [A], [B], [C] shows 2D atomic force microscopic image (**5x5**) nm of **CdTe** thin film for different concentration of precursor solution.

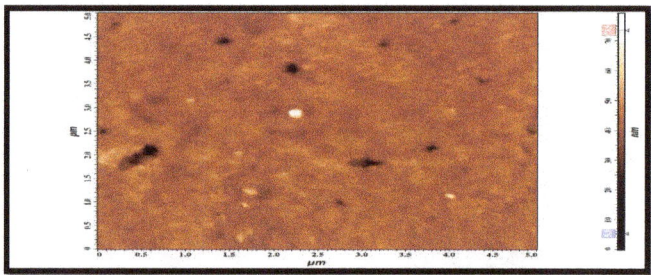

Figure 26 [A]: *AFM 2D image of CdTe thin film deposited from half concentration A.*

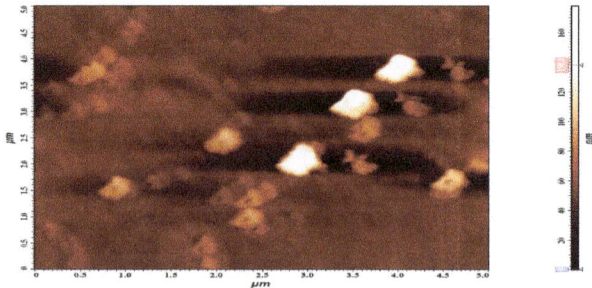

Figure 26 [B]: *AFM 2D image of CdTe thin film deposited from normal concentration B.*

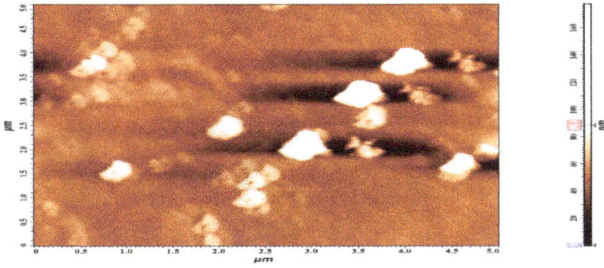

Figure 26 [C]: *AFM 2D image of CdTe thin film deposited from double concentration C.*

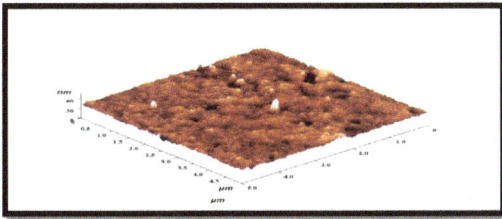

Figure 26 [D] AFM: *3D image of CdTe thin film deposited from half concentration A.*

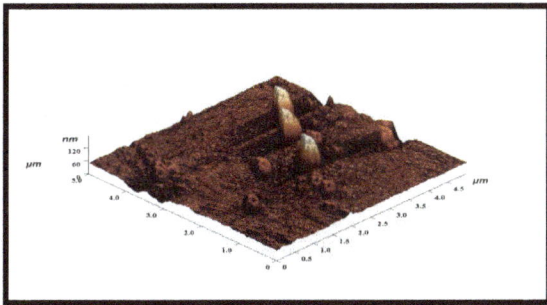

Figure 26 [E]: *AFM 3D image of CdTe thin film deposited from normal concentration B.*

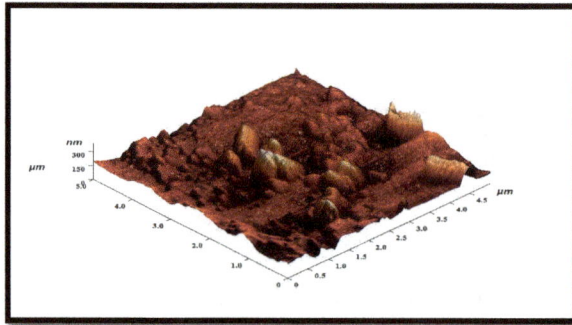

Figure 26 [F]: *AFM 3D image of CdTe thin film deposited from double concentration C.*

Figure 26 [D], [E], [F]) shows 3D atomic force microscopic image (5x5 nm) of **CdTe** thin film for different concentration viz half concentration (named-A), normal concentration (named-B) and double concentration (named-C) of precursor solution. It appears that the particle size, film size and the surface roughness increased with increasing of the concentration of precursor solution. We can expect this result due to the increase of the concentration of precursor solution and the changes in nucleation and growth rate with the extension of the temperature. These results are in good agreement with results, reported by few earlier workers [37, 38].

From 2D AFM images it is clear that the film is uniform and the substrate surface is well covered by fine spherical or elliptical particles. The spherical or elliptical particles could be attributed to faster growth. There was agglomeration of particles in most of the cases as evident from the 3D images. This observation reveals that the films are crystalline in nature. This characterization showed the CdTe films to be pinhole and crack free. The reported range of grain size, or grain agglomerate size, was determined by measuring the smallest and largest points in the AFM images.

4.4 Optical properties using UV-spectroscopy

A spectrophotometer is used for measuring optical absorption of the thin films. In fact the "absorbance" reading (i.e. photometric value) is a measure of the amount of light absorbed by the sample under specified conditions.

a) Absorption spectroscopy of nanocrystalline CdTe thin film with variation of deposition time

The spectral characteristics of the cadmium telluride thin film photo-electrodes prepared for different deposition time 60 min, 120 min, 180 min, show that the absorption edge is located at wavelength 305 nm, 313 nm, 318 nm respectively. Figure 27 [A], [B], [C] shows the optical absorption spectra of the thin films of the CdTe for 60 min, 120 min and 180 min deposition time respectively. The effective band gap is found to be greater than the band gap of bulk CdTe as 1.54 eV, which may be the consequence of quantum confinement. The absorption spectrum of the cadmium telluride thin film clearly showed that the absorption edge shift from 305nm to 318nm as the deposition time increases. The energy band gap E_{nano} is calculated from formula given below:-

$$E_{nano} = hc/\lambda$$

Where h=plank constant = 6.6261×10^{-27} erg/sec.

C=velocity of light = 2.9979×10^{10} cm/sec

λ-is wavelength at absorption edge.

The value of E_{nano} is found 4.06557eV for deposited cadmium telluride thin film photo-electrode prepared for 60 Min deposition time. Similarly the value of E_{nano} is obtained 3.9616eV and 3.8993eV for deposited cadmium telluride thin film photo-electrode prepared for 120 Min and 180 Min deposition time respectively. The increase in energy band gap E_g is calculated by-

$$\Delta E_g = E_{nano} - E_{bulk} \quad [E_{Bulk}=1.54eV] \quad [39,40,41]$$

The value of ΔE_g i.e. shift in energy band gap is obtained 2.5255 eV, 2.4216 eV and 2.3593 eV respectively for CdTe thin film photo-electrode prepared for deposition time 60 Min, 120 Min, 180 Min.

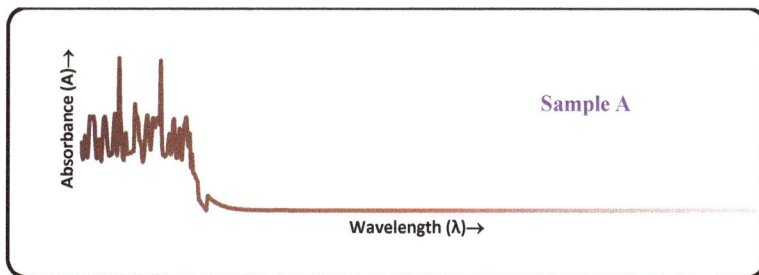

Figure 27 (A): *Optical absorption spectra of thin films of the CdTe for 60 min deposition time.*

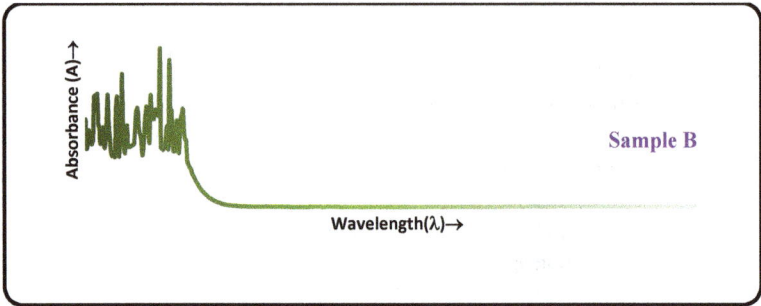

Figure 27 (B): *Optical absorption spectra of thin films of the CdTe for 120 min deposition time.*

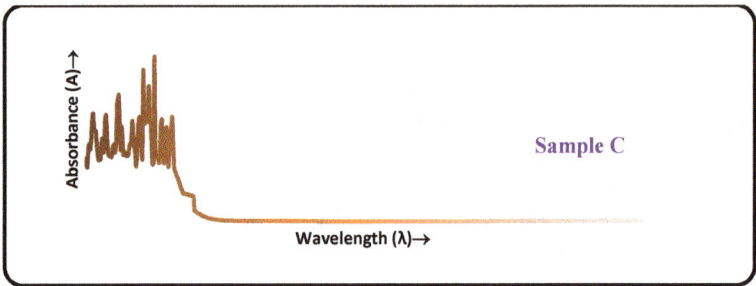

Figure 27 (C): *Optical absorption spectra of thin films of the CdTe for180 min deposition time.*

From the increase in the band gap, the particle size of CdTe has been determined from effective mass approximation using the formula:

$$r^2 = h^2\pi / \Delta E \left[1/m^*_e + 1/m^*_h \right]$$

Taking effective mass of electron of CdTe [m^*_e =0.047] [42]

Taking effective mass of holes of CdTe [m^*_h =0.879].

Table 18- Effect of deposition time on bandgap and particle size.

Deposition Time	Absorption Edge (nm)	E_{Nano} (eV)	ΔE shift in Band width (eV)	Particle size (nm)
60 Min	305	4.06557	2.5255	24.07
120 Min	313	3.9616	2.4216	24.59
180 Min	318	3.8993	2.3593	24.91

The gradual shift in absorption edge from (305-318 nm) to the higher wavelength side indicates decreased band gap with increase in particle size. The shift in the optical bandgap with increase in the deposition time may be due to the increases in particle size. The optical absorption measurements indicate that the absorption occurs due to direct transition. The results are given in Table 18. From table it is observed that the calculated value of particle size (r) for deposited CdTe thin film slightly increases with increase in deposition time, since ΔE (i.e. shift in band width) is reduced.

b) Absorption spectroscopy of nanocrystalline CdTe thin film with variation of deposition temperature

The spectral characteristics of the cadmium telluride thin film photo-electrodes prepared at different temperature of precursor solution **45°C, 60°C and 90°C** show that the absorption edge is located at wavelength **313nm, 305nm and 300nm** respectively. Figure 28 [A], [B], [C] shows the optical absorption spectra of the thin films of the CdTe deposited at different temperature of precursor solution **45°C, 60°C and 90°C** respectively. The effective band gap is found to be greater than the band gap of bulk CdTe as 1.54 eV, which may be the consequence of quantum confinement. The absorption spectrum of the cadmium telluride thin film clearly showed that the absorption edge shift from 313nm to 300 nm as the temperature of precursor solution increases.

The value of E_{nano} is found **3.9616 eV** for deposited cadmium telluride thin film photo-electrode prepared at 45°C temperature of precursor solution. Similarly the value of E_{nano} is obtained **4.06557eV** and **4.13eV** for cadmium telluride thin film photo-electrode prepared at 60°C and 90°C respectively.

The value of ΔE_g i.e. increase in energy band gap is obtained **2.4216eV, 2.5255eV** and **2.593eV** respectively for deposited CdTe thin film photo-electrode prepared at 45°C, 60°C and 90°C temperature respectively.

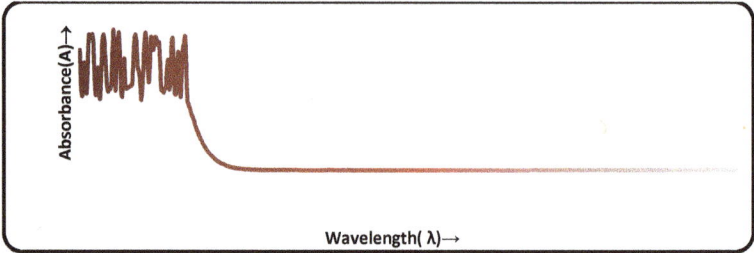

Figure 28 (A): *Optical absorption spectra of thin films of the CdTe deposited at 45°C.*

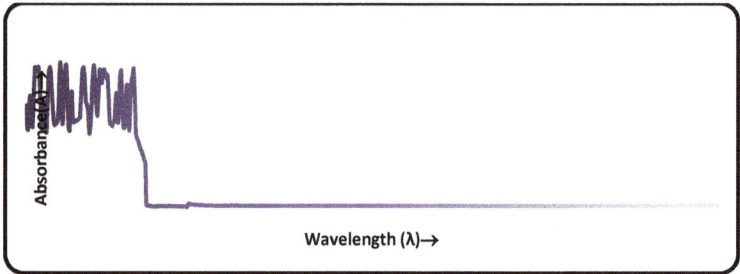

Figure 28 (B): *Optical absorption spectra of thin films of the CdTe deposited at 60°C.*

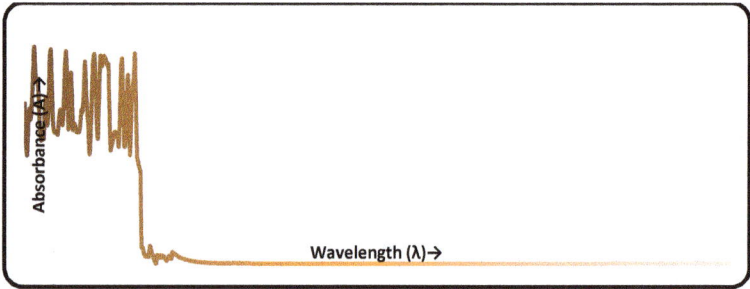

Figure 28 (C): *Optical absorption spectra of thin films of the CdTe deposited at 90°C.*

Table 19- Effect of temperature of precursor solution on bandwidth and particle size.

Temperature	Absorption Edge	E_{Nano}	ΔE increase in Band width	Particle size
45^0C	313	3.9616	2.4216	24.5902
60^0C	305	4.06557	2.5255	24.0792
90^0C	300	4.133	2.593	23.7639

The slight shift in absorption edge (313-300 nm) to the lower wavelength side indicates increased band gap and hence decrease in particle size. The shift in the optical bandgap with increase in the temperature of precursor solution may be due to the decreases in particle size. The optical absorption measurements indicate that the absorption occurs due to direct transition. The results are given in Table 19. From table it is observed that the calculated value of particle size (r) for deposited CdTe thin film slightly decreases with increase in temperature of precursor solution.

c)   **Absorption spectroscopy of nanocrystalline CdTe thin film with variation of precursor concentration**

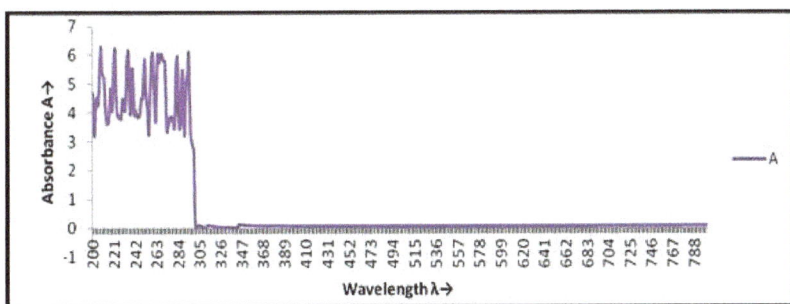

Figure 29 [A]: Optical absorption spectra of thin films of the CdTe deposited from half concentration A.

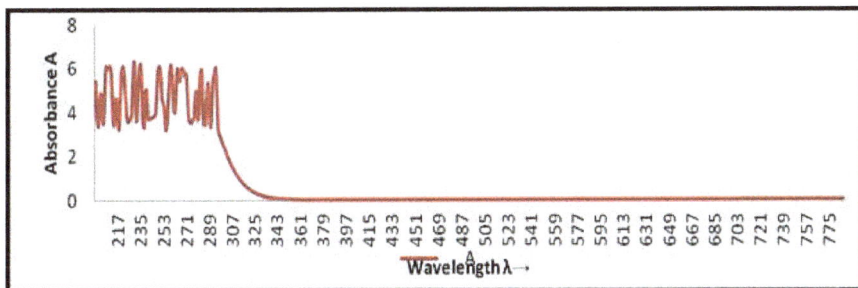

Figure 29 [B]: Optical absorption spectra of thin films of the CdTe deposited from normal concentration B.

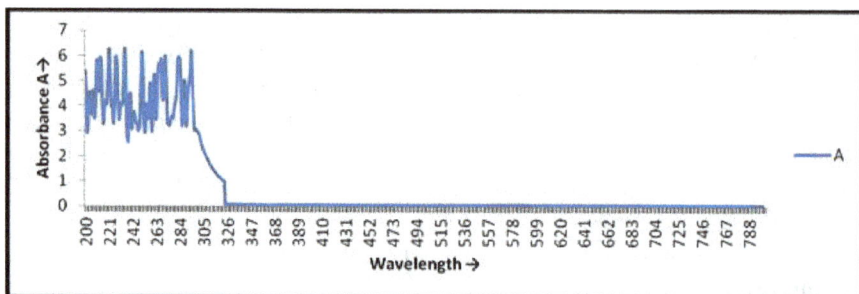

Figure 29[C]: *Optical absorption spectra of thin films of the CdTe deposited from double concentration C.*

Figure 29 [A], [B], [C] shows the optical absorption spectra of the thin films of the CdTe for different concentration of precursor solution half concentration (named-A), normal concentration (named-B) and double concentration (named-C), respectively. The effective band gap is found to be greater than the band gap of bulk CdTe as 1.54 eV, which may be the consequence of quantum confinement.

The spectral characteristics of the cadmium telluride thin film photo electrodes prepared from different concentration of precursor solution viz half (A), normal (B) and double (C), show that the absorption edge is located at wavelength **300nm, 313nm, 324nm** respectively. The absorption spectrum of the cadmium telluride thin film clearly showed that the absorption edge shift from 300 nm to 324 nm as the concentration of precursor solution increases.

The value of E_{nano} is found **4.133eV** for deposited cadmium telluride thin film photo-electrode prepared for concentration of precursor solution half A. Similarly the value of E_{nano} is obtained **3.9616eV** and **3.8271eV** for deposited cadmium telluride thin film photo-electrode prepared for concentration of precursor solution normal B and double C respectively.

The value of **ΔEg** i.e. increase in energy band gap is obtained **2.593 eV, 2.4216 eV** and **2.2871 eV** respectively for deposited CdTe thin film photo-electrode prepared from different concentration of precursor solution are half (A), normal (B) and double (C).

Table 20- Effect of concentration of precursor solution on bandwidth and particle size.

Concentration	Absorption Edge (nm)	E_{Nano} (eV)	ΔE increase in Band width	Particle size (nm)
Half A	300	4.133	2.593	23.7639
Normal B	313	3.962	2.422	24.5902
Double C	324	3.827	2.287	25.3030

The gradual shift in absorption edge (300-324 nm) slightly to the higher wavelength side indicates decreased band gap with increase in particle size. The shift in the optical bandgap with increase in the concentration of precursor solution may be due to the increases in particle size. The optical absorption measurements indicate that the absorption occurs due to direct transition. The results are given in Table 20. From table it is observed that the calculated value of particle size (r) for deposited CdTe thin film slightly increases with increase in concentration of precursor solution, since ΔE (i.e. shift in band width) is reduced.

5.   Conclusions

The investigation of photoeletrochemical cell studies and characterization of CdTe thin films synthesized for different deposition time, at different temperatures and from different concentrations of precursor solution have indicated following important conclusions:-

1. A simple and less investigated modified chemical bath deposition method may be used for depositing good quality thin films of CdTe on ITO coated glass and plane glass substrates.

2. The polarity of the dark voltage is positive toward the CdTe photoelectrode and negative toward the graphite electrode for all samples showing **p-type** semiconducting behavior.

3. Observing I-V curve it can be seen that with decrease in resistance, the terminal potential of solar cell decreases due to internal resistance of cell and the current increases.

4. The R_S and R_{SH} value increased which means that carrier mobility is reduced and leakage is reduced as deposition time and concentration of precursor solution increases. The R_S and R_{SH} value is decreased as temperature of precursor solution increases indicating higher carrier mobility.

5. With variation in the deposition time and concentration of precursor solution the photovoltaic response is maximum for film deposited for deposition time is 120 min

and from double concentration of precursor solution whereas with increase in temperature the response deteriorates.

6. The thickness increases with increase in deposition time and concentration of precursor solution increases, whereas thickness decreases with increase in temperature of precursor solution.

7. The XRD revels the structure of the CdTe thin films was **Tetragonal** (a=b=**4.0440**, c=**2.6190** and α= β = γ =**90⁰**) with a preferential orientation of **(111)** .plane **[JCPDS card number 9007154]**.

8. In XRD, particle size is calculated from XRD pattern using **Scherrer's formula**.

9. XRD results show that particle size increases with deposition time and concentration of precursor solution increases whereas particle size decreases with increase in temperature of precursor solution.

10. Morphological studies (SEM) revealed an increase in the grain size with the increase in deposition time and concentration of precursor solution increases, whereas the grain size decreases with increase in temperature of precursor solution.

11. The SEM studies show that at the optimized concentration, well covered, uniform, rounded, densely packed grains on the surface are observed.

12. AFM results show that roughness increases with increase in deposition time and concentration of precursor solution increases, whereas roughness decreases with increase in temperature of precursor solution.

13. Optical absorption spectroscopy shows that as deposition time and concentration of precursor solution increases the crystallite size slightly increases, whereas the crystallite size slightly decreases with increase in temperature of precursor solution.

14. Finally, it can be said that the CdTe thin films are potential candidates for solar cell applications.

References

[1] Internet: http://www.eia.gov/oiaf/ieo/world.html, U.S. Energy Information Administration 2010 International Energy Outlook —World Energy Demand and Economic Outlook‖ Jul. 27, 2010 [Jun. 29, 2011].

[2] Internet:http://en.wikipedia.org/wiki/Wind_power Wikipedia.WindPower‖, Jul. 3, 2011 [Jul. 10, 2011].

[3]   "Internet:http://www.ifpaenergyconference.com/Solar-Energy.html Atoms for
        Peace Energy Conference. Energy Resources —SolarEnergy, Oct. 22, 2003 [Jun.
        29, 2011].

[4]   Academicae.unavarra.es/bitstream/handle/2454/5767/577802.pdf1by S. Zazpe
        Delgado 2012

[5]   C.S. Ferekides, —Thin Films and Solar Cells of Cadmium Telluride and Cadmium
        Zinc Telluride‖, Ph.D. Dissertation, University of South Florida, 1991.

[6]   Peter Capper (1994).*Properties of Narrow Gap Cadmium-Based Compounds*. IET.
        pp. 39–.ISBN 978-0-85296-880-2. Retrieved 1 June 2012.

[7]   C.D. Lokhande, Mater. Chem. Phys., 28 (1991) 1. https://doi.org/10.1016/0254-
        0584(91)90158-Q

[8]   R.S. Mane and C. D. Lokhande, Mater. Chem. Phys., 65 (2000) 1.
        https://doi.org/10.1016/S0254-0584(00)00217-0

[9]   G. Hodes, —Chemical Solution Deposition of Semiconductor Films‖. Marcel
        Dekker., Inc.,New York, (2003) p1.

[10]   Preeti Pathak, Kavita Gour, M.Ramrakhiani, P. Mor Cosmic International Journals
        of electronics science and Technology (IJEST) Paper Titled-" Influence of
        concentration of precursor solution on the performance of nanocrystalline CdTe
        thin films used as photoelectrode in photoelectrochemical solar cell IJECT Vol. 6,
        Issue 2, April - June 2015 ISSN : 2230-7109 (Online) | ISSN : 2230-9543 (Print).

[11]   R. H. Bari, S. B .Patil, A.R. BariI, G. E. Patil, J. Aambekar, *Sensors &
        Transducers Journal* 140 (2012) 124-132.

[12]   Preeti Pathak, M.Ramrakhiani, P. Mor (, (IJSER) International journal of
        Scientific and Engineering Research, "Influence of Temperature of Precursor
        solution on structural and optical properties of Nanocrystalline CdTe thin films
        deposited chemical bath deposition technique. Vol. 6 – Issue 8 – p.no. 1023 –
        1026, ISSN – 229 - 5518) (August 2015)

[13]   S. M. Patil, P. H. Pawar* International Letters of Chemistry, Physics and
        Astronomy 17(1) (2014) 21-36 ISSN 2299-3843

[14]   Kavita Gour, Preeti Pathak, M.Ramrakhiani,P. Mor (International conference on
        Resent trends in Science and Engineering, held at Durg, C.G., from 7/01/2016 to
        9/01/2016)" Effect of different concentration of CuSe thin film as photo anode on
        its morphology and photovoltaic response.(January 2016)

[15]   Mohammed Tak International Journal of Application or Innovation in Engineering & Management (IJAIEM) Volume 2, Issue 5, May 2013 ISSN2319-4847

[16]   J. Merten, J.M. Asensi, C. Voz, A.V. Shah, R. Platzand J. Andreu, IEEE T.Electron Dev.45, 423 (1998). https://doi.org/10.1109/16.658676

[17]   Deshmukh L P, Hankare P P & Sawant V S, Solar Cell,31 (1991) 549. https://doi.org/10.1016/0379-6787(91)90097-9

[18]   Kavita Gour, Preeti Pathak, M.Ramrakhiani,P. Mor Cosmic International Journals of computer science and Technology (IJCST) Paper Titled-"Nanocrystalline CuSe thin Film used as P type Photoelectrod in photoelectrochemical solar cell." ISSN - 0976-8491(online) ISSN -2229-4333 (print) Vol.-5,Issue-2,Version-3 April-June 2014.

[19]   Kavita Gour, Preeti Pathak, M.Ramrakhiani,P. Mor International Journal of Scientific Progress & Research (IJSPR) "Effect of concentration in photo electrochemical cell of CuSe thin film as photoanode on its morphology and photovoltaic response." ISSN 2349-4689 Issue July 2014.

[20]   H.S. Hilal, Rania M.A. Ismail, A. El-Hamouz, A. Zyoud, I. Saadeddin Electrochimica Acta,54(2009) 3433. https://doi.org/10.1016/j.electacta.2008.12.062

[21]   V. B. Patil, G. S. Shahane, L. P. Deshmukh, Materials Chemistry and Physics 80 (2003) 625. https://doi.org/10.1016/S0254-0584(03)00086-5

[22]   Hanan R. A. Ali, International Letters of Chemistry, Physics and Astronomy 8 (2014) 47-55.

[23]   Raghad Y. Mohammed, S. Abduol, Ali M. Mousa, International Letters of Chemistry, Physics and Astronomy 10 (2014) 91-104 .

[24]   Raghad Y. Mohammed, S. Abduol, Ali M. Mousa, International Letters of Chemistry, Physics and Astronomy 11(2) 427.

[25]   Preeti Pathak, Kavita Gour, M.Ramrakhiani, P. Mor The International Journal of Science & Technoledge www.theijst.com "Influence of Deposition Time on Performance of Nanocrystalline CdTe Thin Films Used as Photoelectrode in Photoelectrochemical Solar Cell." (ISSN 2321 – 919X) 243-249 Vol 3 Issue 3 March, 2015

[26]   Shinde S S, Shinde P S, Bhosale C H, et al. Optoelectronic properties of sprayed transparent and conducting indium doped zinc oxide thin films. J Phys D: Appl Phys, 2008, 41: 105109. https://doi.org/10.1088/0022-3727/41/10/105109

[27] (Prof.Dr. Kadhim A. Hubeatir University of Technology Laser Engineering &
      Optoelectronic Department Glass: 3rd year Optoelectronic Engineering Subject:
      Solid state physics & material scienceAsst. Chapter 1 page -3)

[28] (Prof. Dr. Kadhim A. Hubeatir University of Technology Laser Engineering &
      Optoelectronic Department Glass: 3rd year Optoelectronic Engineering Subject:
      Solid state physics & material science Asst. Chapter 1 page -3)

[29] S. M. Patil, P. H. Pawar* International Letters of Chemistry, Physics and
      Astronomy 1

[30] Ramesh S. Kapadnis (A, B), Sanjay B. Bansode (A), Sampat S. Kale (C, *), Habib
      M. Pathan (A, Applied Science Innovations Pvt. Ltd., IndiaCarbon – Sci. Tech. 5/1
      (2013) 211- 217 ISSN 0974 – 0546.(1) (2014) 21-36 ISSN 2299-3843.

[31] S.M. Patil, P.H. Pawar* International Letters of Chemistry, Physics and
      Astronomy 17(1) (2014) 21-36 ISSN 2299-3843

[32] A.V. Kokatea,, M.R. Asabeb, P.P. HankarebB.K. Chougulea Journal of Physics
      and Chemistry of SolidsVolume 68, Issue 1, January 2007, Pages 53–58)

[33] S.M. Patil, P. H. Pawar* International Letters of Chemistry, Physics and
      Astronomy 17(1) (2014) 21-36 ISSN 2299-3843

[34] Ramesh S. Kapadnis (A, B), Sanjay B. Bansode (A), Sampat S. Kale (C, *), Habib
      M. Pathan (A, Applied Science Innovations Pvt. Ltd., IndiaCarbon – Sci. Tech. 5/1
      (2013) 211- 217 ISSN 0974– 0546

[35] Douglas L. Schulz,* Martin Pehnt, Doug H. Rose, Ed Urgiles, Andrew F. Cahill,
      David W. Niles, Kim M. Jones, Randy J. Ellingson, Calvin J. Curtis, and David S.
      Ginley CdTe Thin Films from Nanoparticle Precursors by SprayDeposition *Chem.
      Mater.* 1997, *9,* 889-900 889

[36] Deshmukh L P, Hankare P P & Sawant V S, Solar Cell,31 (1991) 549.
      https://doi.org/10.1016/0379-6787(91)90097-9

[37] Douglas L. Schulz,* Martin Pehnt, Doug H. Rose, Ed Urgiles, Andrew F. Cahill,
      David W. Niles, Kim M. Jones, Randy J. Ellingson, Calvin J. Curtis, and David S.
      Ginley CdTe Thin Films from Nanoparticle Precursors by SprayDeposition *Chem.
      Mater.* 1997, *9,* 889-900 889.

[38] Kavita Gour, Preeti Pathak,M.Ramrakhiani, P. Mor "International Journal of
      Electrical, Electronics & Computing Technology (IJEEC) (volume 8 (I) Issue 7
      March-May 2013)"ISSN 2229-3027.

[39]   *(International Nano Letters* 2013, 3:56 doi:10.1186/2228-5326-3-56 Structural, optical, and electrical properties of thioglycolic acid-capped CdTe quantum dots thin films. https://doi.org/10.1186/2228-5326-3-56

[40]   Rostam Moradian ,Mohammad Elahi, Ahmad Hadizadeh, Mahmoud Roshani, Atefeh Taghizadeh and Reza Sahraei)

[41]   Sharma S N, Kohli S, Rastogi A C. Quantum confinement ef-fects of CdTe nanocrystals sequestered in TiO2matrix: effect ofoxygen incorporation. Physica E, 2005, 25: 554. https://doi.org/10.1016/j.physe.2004.08.110

[42]   Ramos L. E. Ramos, L. K. Teles, L. M. R. Scolfaro, J. L. P. Castineira, A. L. Rosa, and J.R. Leite, Phys. Rev. B 63,165210 (2001). https://doi.org/10.1103/PhysRevB.63.165210

Chapter 7

Studies on the photovoltaic effect of CdSe based nanocrystalline multilayered photoelectrodes in photoelectrochemical solar cells

Hemraj Waxar, M. Ramrakhiani and P. Singh

Department of Post Graduate Studies and Research in Physics and Electronics, Rani Durgawati Vishwavidyalaya, Jabalpur. (MP), India

hem.raj@rediffmail.com, mramrakhiani@hotmail.com

Abstract

The use of nanostructures in solar cells may enhance the efficiency and performance. In PEC cells, the junction formation is quite easy and polycrystalline films also work very well. In the present work single layered, double layered and triple layered CdSe nanocrystalline films of different crystal size have been prepared by the pulse electrodeposition method and their performance in PEC cells with sulphide/polysulphide electrolyte have been studied.

The nanocrystaline CdSe films have been deposited on titanium substrate from 0.5 M $CdSO_4$ and 0.1 M SeO_2 electrolyte solution using the pulse electro-deposition method. ON-OFF machine based on IC-555 in astable mode have been used for various duty cycles. Seven samples of single layered, six samples of double layered and seven samples of triple layered CdSe nanocrystalline thin film with different duty cycle have been prepared. These were used as photoelectrode in 0.1 M NaOH, 1M Na_2S, & 1M S poly-sulphide electrolyte and graphite as counter electrode.

Illuminating the cell with different light intensities, I-V characteristics have been obtained and cell parameters V_{OC}, I_{SC}, F.F., & $\eta\%$ were determined. It is observed that best performance of the solar cell is obtained for the sample with duty cycle D_4 of single layered, $D_{2/3}$ of double layered and $D_{2/4/6}$ of triple layered. The best results are obtained by the PEC solar cell with photo-anode of CdSe film prepared with duty cycle $D_{2/3}$. The study reveals that nanocrystalline CdSe based photoelectrodes give better performance in PEC cells. This may be because of the larger surface area. But very small crystallites increase the surface traps and performance becomes poor. Double layered photoelecrodes give better results.

Keywords

CdSe Nanocrystals, Multilayers, Photovoltaic Effect, Photoelectrochemical Cells

Contents

1. Introduction 243

1.1 Photoelectrochemical Cells 243

1.2 Nanostructure Based Solar Cells 243

1.3 Multijunction Solar Cells...244

2. Preparation of Nanocrystalline CdSe Thin Film Photo Electrodes..244

2.1 Experimental Arrangement for Thin Films...244

2.2 Principle and Working of the ON-OFF Machine...............................245

2.3 Preparation of the Solution for the Photo-anode 248

2.4 Mechanisms of Nanocrystalline Film Formation 249

3. Photovoltaic Study of Nanocrystalline CsSe Films 250

3.1 Experimental Setup for Photoelectrochemical (PEC) Solar Cell 250

3.2 Preparation of a PEC Solar Cell 252

3.3 Circuitry and Measurements of Solar Cell Parameters 255

4. Single Layered CdSe Nanocrystalline Photoelectrodes (P_s)..............256

5. Double Layered CdSe Nanocrystalline Photoelectrodes (P_d)...........259

6. Triple Layered CdSe Nanocrystalline Photoelectrodes (P_t)..............262

6. Summary and Conclusion...267

References 267

1. Introduction

The photovoltaic effect is defined as creation of an electromotive force by the absorption of light in an inhomogeneous solid.

1.1 Photoelectrochemical Cells

In the last decade the method of converting solar energy with the aid of semiconductor-electrolyte photoelectrochemical (PEC) cells has been advanced as an alternative to well known method of energy conversion involving the use of solid state semiconductor solar cells [1-4]. This alternative was searched because the presently available solar cells are manufactured from highly pure and perfectly crystalline materials and the p-n junctions are obtained by using very sophisticated technologies. For this reason, these cells are very costly. In PEC cell, use is made of the interface which forms on mere dipping the SC into the electrolyte solution and liquid junction potential barrier can be easily established. Besides, polycrystalline semiconductor films can be used without any drastic decrease in efficiency. Thus PEC cells provide an economic chemical route for trapping solar energy [5-8]. It consists of a photosensitive p- or n-type semiconductor electrode and a counter electrode dipped in suitable electrolyte. The counter electrode can also be a photosensitive p-type semiconductor electrode. Photons are absorbed by SC electrode producing electron-hole pairs. Because of a potential barrier created in the SC after the SC/electrolyte junction formation, electrons and holes are separated and their recombination inhibited. The majority carriers (electron) flow to counter electrode through an external circuit and the minority carrier, reach the surface and effect an electrochemical reaction. In liquid junction solar cells (LJSC) a redox electrolyte system is employed where oxidation reaction at the anode is counter balanced by reduction reaction at the cathode and hence there is no net change in the chemical composition of redox electrolyte system and light is converted to DC. electricity.

1.2 Nanostructure Based Solar Cells

Nanostructure based solar cells have been previously proposed due to their potential to provide high conversion efficiency [9, 10]. As nano-sizes are comparable to carrier scattering, they significantly reduce the scattering rate and increase the carrier collection efficiency. Nanostructures have strong absorption coefficient due to increased density of states. In addition by varying the size of the nanostructures, the band gap can be tuned to absorb in a particular photon energy range.

1.3   Multijunction Solar Cells

Multijunction solar cells can extract higher solar energy compared to a single junction solar cell. Multijunction solar cells with different band gaps are attractive for effectively collecting sunlight, which has a wide-range spectrum from ultraviolet to infrared. The operation of these cells involves two or more single junction photovoltaic cells [11, 12]. The cells are arranged so that sunlight if first incident on the semiconductor with the largest energy gap. Photons of higher energy are absorbed by it and electrical energy is produced. The lower energy photons pass to the next semiconductor with next largest energy gap and so on. Thus, the whole spectrum of the incident light may be absorbed by a different semiconductor and light energy is converted into electricity. This results in high overall efficiencies in a multijunction solar cell design.

The present paper describes the photovoltaic effect in photoelectrochemical cells in which a number of layers of cadmium selenide nanocrystals of different sizes have been used as photoelectrodes.

2.   Preparation of Nanocrystalline CdSe Thin Film Photo Electrodes

Recently there have been many efforts to produce nano size materials because the electrical and optical properties can be varied via chemical control over size, stoichiometery and interparticle separation. These materials have been synthesized by various techniques including pyrolysis of organometellic compounds, sol-gel synthesis, chemical bath deposition, electrodeposition, etc.

2.1   Experimental Arrangement for Thin Films

In the present work, CdSe films were deposited by pulse electrodeposion. Figure 1 shows the experimental arrangement for pulse electrodeposition method.

According to K. R. Murali and D.C. Trivedy [13] nanocrystalline thin films were deposited by pulse electrodeposition. TEM studies indicated a crystals size in the range of 10nm -50nm depending upon the duty cycle. The size of nanoparticle could be controlled with the pulse parameter.

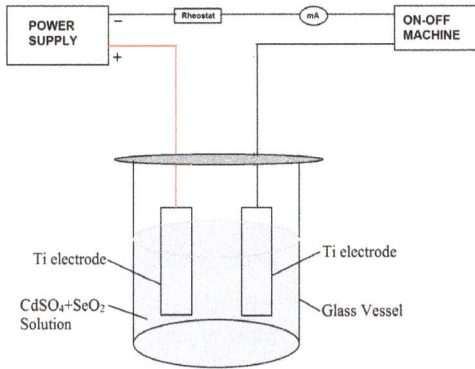

Figure 1: *Experimental arrangement for pulse electro-deposition.*

2.2   Principle and Working of the ON-OFF Machine

The ON-OFF machine was designed for adjustable pulsed output for pulse electrodeposition. The basic component of the ON-Off machine is a IC timer 555 used in astable mode. An astable circuit produces a square wave, this is a digital waveform with sharp transition between low (0V) and high (+Vs). The duration of low and high states may be different. The circuit is called astable because it is not stable in any state: the output is continuously changing between low and high [14].

The time period of the square wave is given by:

$T = 0.7 \times (R1 + 2 R2) \times C1$ and

$f = 1.4/ (R1 + 2 R2) \times C1$

The time period can be split into two parts:

$T = Tm + Ts$

Mark time (output high): $Tm = 0.7 \times (R1 + R2) \times C1$ Space time (output low): $Ts = 0.7 \times R2 \times C1$

Many circuits require Tm and Ts to be almost equal; this is achieved if R2 is much larger than R1. For a standard astable circuit Tm cannot be less than Ts, but this is not too restricting because the output can both sink the source current. Figure 2(a) shows the 555 astable output, a square wave and Figure 2(b) shows the IC555, astable circuit.

Astable operation:

With the output high (+Vs) the capacitor C1 is charged by current flowing through R1 and R2. The threshold and trigger inputs monitor the capacitor voltage and when it reaches $^2/_3$Vs (threshold voltage) the output becomes low and the discharge pin is connected to 0V.

The capacitor now discharges with current flowing through R2 into the discharge pin. When the voltage falls to $^1/_3$Vs (trigger voltage) the output becomes high again and the discharge pin is disconnected, allowing the capacitor to start charging again.

This cycle repeats continuously unless the reset input is connected to 0V which forces the output low while reset is 0V. An astable circuit can be used to provide the clock signal for circuits such as counters. Figure 2(c) shows the clock signal for circuits such as counters.

555 in astable mode:

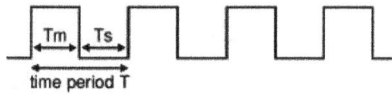

Figure 2(a): 555 astable output, a square wave (Tm and Ts may be different).

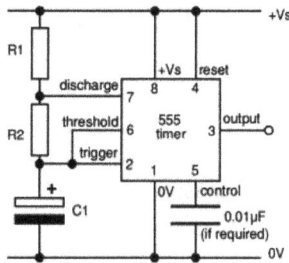

Figure 2(b): IC555 astable circuit.

246

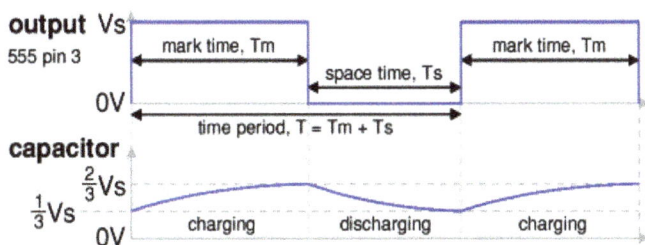

Figure 2(c): Clock signal for circuits such as counters.

A low frequency astable (< 10Hz) can be used to flash an LED on and off, higher frequency flashes are too fast to be seen clearly. Driving a loudspeaker or piezo transducer with a low frequency of less than 20Hz will produce a series of 'clicks' (one for each low/high transition) and this can be used to make a simple metronome

Duty cycle:

The duty cycle of an astable circuit is the proportion of the complete cycle for which the output is high (the mark time). It is usually given as a percentage.

For a standard 555/556 astable circuit the mark time (Tm) must be greater than the space time (Ts), so the duty cycle must be at least 50%:

$$\text{Duty cycle} = \frac{T_m}{T_m + T_S} = \frac{R1 + R2}{R1 + 2R2}$$

The circuit diagram of the on-off machine is shown below in Figure 2(d) and Figure 2(e) shows the duty cycle by changing value of Tm & Ts. A relay is connected at the output of the 555, the input to relay is –16 V DC power supply and the output is also -16 V DC but its duration is controlled by 555 and -relay. To achieve a duty cycle of less than 50% a diode can be added in parallel with R2 as shown in the diagram. This bypasses R2 during the charging (mark) part of the cycle so that Tm depends only on R1 and C1:

Tm = 0.7×R1×C1 (ignoring 0.7V across diode)

Ts = 0.7 × R2 × C1 (unchanged)

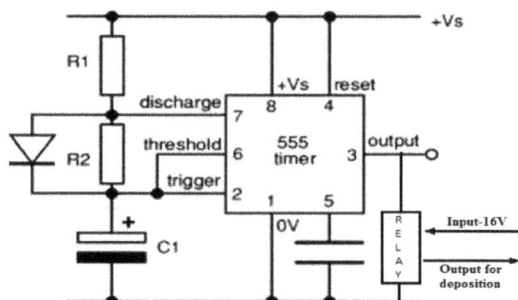

Figure 2(d): *ON –OFF machine using 555 astable circuit.*

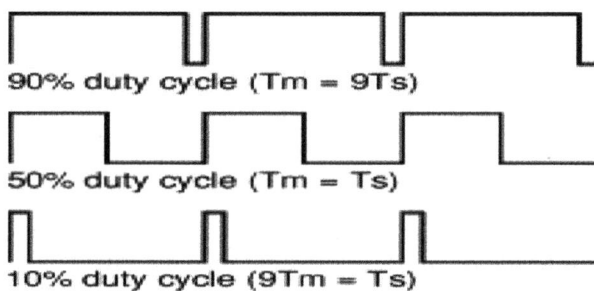

90% duty cycle (Tm = 9Ts)

50% duty cycle (Tm = Ts)

10% duty cycle (9Tm = Ts)

Figure 2(e): *Duty cycle by changing value of Tm & Ts.*

Duty cycle with diode $= \dfrac{T_m}{T_m + T_S} = \dfrac{R1}{R1 + R2}$

Thus we may change the duty cycle by changing the values of Tm and Ts and the ON-OFF machine provides desired output to one of the electrodes used for pulse electrodeposition. In the present investigation the duty cycle is taken as 1:1, i.e on time Tm=1 second and off time Ts=1 second [14].

2.3 Preparation of the Solution for the Photo-anode

The nanocrystalline thin film cadmium selenide (CdSe) were prepared by pulse electro deposition onto titanium substrate as cathode using an aqueous acidic solution of cadmium sulphate ($CdSO_4$) and selenium dioxide (SeO_2). We took 0.5 mole ($CdSO_4$) solution, this was then added 0.1 mole SeO_2 solution [13].

The solution of $0.1M/40ml$ SeO_2 is prepared as follows:

Molecular weight of SeO_2 =110.96 g/mol

Molarity of solution = weight of solute /molecular weight of solute × volume of solution in a liter

Weight of SeO_2 = 0.1M x 110.96 × 40/1000 = 444 mg

Similarly, the weight of $CdSO_4$ in the solution of 0.5 M/ 40ml

Molecular weight of $CdSO_4$ =176.407

weight of $CdSO_4$ = 4170 mg

Hence 444 mg of SeO_2 was dissolved in 40 ml of distilled water to give a 0.1 M SeO_2 solution. Similarity calculations were done for the weight of $CdSO_4$ for 0.5M/40ml= 4170 mg hence 4170 mg of $CdSO_4$ was dissolved in 40 ml of distilled water. The two solutions are then mixed to form a electrolyte. The film of CdSe was then deposited at an electrolysis current density of $7mA/cm^2$ for a time of different duty cycle by the ON/OFF machine at 80° C.

2.4 Mechanisms of Nanocrystalline Film Formation

Peniker et al. [15] have carried out a detailed electrochemical study for formation of CdTe from aqueous solution of $CdSO_4$ and TeO_2. The solution used for CdSe was $CdSO_4$ and SeO_2. Hence drawing analogy with Peniker et al's work, the formation of CdSe film may be assumed to take place in the following steps.

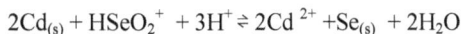

$2Cd_{(s)} \rightleftharpoons 2 Cd^{2+} + 4e^-$

$Se + 2H_2O \rightleftharpoons HSeO_2^+ + 3H^+ + 4e^-$

$2Cd_{(s)} + HSeO_2^+ + 3H^+ \rightleftharpoons 2Cd^{2+} + Se_{(s)} + 2H_2O$

$Cd_{(s)} + Se_{(s)} \rightleftharpoons CdSe$

giving

$^aCdSe/^aCd \, ^aSe = \exp (\, ^GCdSe/RT)$

or for $^aCd \, ^aSe=1$

$^aCdSe = \exp (\, ^GCdSe/RT)$

where (s) denotes the species in solution ^aCd, ^aSe, ^aCdSe denote activities of Cd and Se and CdSe respectively. Cd and Se control the stoichimetric composition of compound CdSe.

In order to deposit Cd and Se in almost equal quantities, it is necessary to use an electrolyte with high concentration of less noble component, Cd, and a low concentration of the noble component, Se. In our experiments, typical concentrations were:

$CdSO_4$ = 4170 mg/40 ml

SeO_2 = 444mg/40 ml

Thus CdSe thin films were deposited on titanium substrate by pulsed-electrodeposition technique with different duty cycles. As reported by Murali and Trivedy [13], smaller nanoparticles are formed by increasing the OFF time. Seven different single layered photoelectrode (P_S) of nanocrystalline CdSe films were prepared with duty cycles as $D_0(1:0)$, $D_1(1:1)$, D_2 (1:2), D_3 (1:3), D_4 (1:4), $D_5(1:5)$,& $D_6(1:6)$, six different double layered (P_D) of nanocrystalline CdSe films were prepared with duty cycles as $D_{0/1}$, $D_{1/2}$, $D_{2/3}$, $D_{3/4}$, $D_{4/5}$, & $D_{5/6}$ and five different triple layered/(P_T) with duty cycles as $D_{0/1/2}$, $D_{1/2/3}$, $D_{2/3/4}$, $D_{3/4/5}$, & $D_{4/5/6}$. Thus photoelectrodes have different size of CdSe nanocrystals were prepared for PEC solar cells.

3. Photovoltaic Study of Nanocrystalline CsSe Films

3.1 Experimental Setup for Photoelectrochemical (PEC) Solar Cell

Although photosynthesis is the primary energy source for life on earth, only 3% of the light energy striking a plant is actually stored as chemical fuel. Of Course, plants had certain constraints on their raw materials and growth environment, and were therefore not optimized by nature to function solely as energy conversion machines. The photovoltaic cells utilize energy much more efficiently from sunlight [16].

The basic processes that must occur in such a system to achieve efficient solar energy conversion are now well-understood scientifically. The semiconductor electrode must have a band gap so that it efficiently absorbs light from the solar spectrum. Absorption of light by the semi-conducting solid produces and excited electronic state, in this excited state, the electron and the electron vacancy (the "hole") are both more energetic than they were in their respectively ground states. The photoexited electrons and holes are generally not tightly bound to and individual atom or set of atoms in the solid.

"Basically the Photoelectrochemical solar cells produce electric energy/chemical by sunlight or visible light".

Photoelectrochemical cell or Wet solar cell or semiconductor/Liquid junction solar cell is closely related to a battery or fuel cell, in that it is composed of two electrodes and an electrolyte, in which the energy is initially stored in the reactants and is then released through the electrical discharge circuit during the formation of chemical reaction products. PECs utilize the input optical energy to drive electrochemical reactions [16]. The schematic diagram is shown in Figure 3(a) and photograph of experimental arrangement for V_{OC} & I_{SC} is given in Figure 3(b).

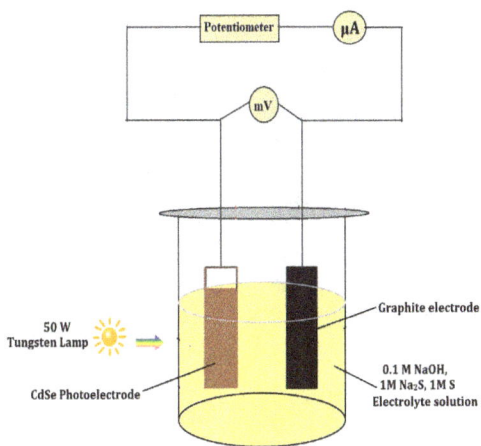

Figure 3(a): *Experimental set up of a PEC solar cell.*

In photoelectrochemical cell, one of metal electrodes of a conventional electrochemical cell is replaced by a semiconductor electrode. Light absorbed by the semiconductor electrode produces a current through the cell. Semiconducting electrode absorbs energy from sunlight and produces a current though the electrochemical cell, the current produces electrical and or chemical energy in the system. If the chemical reaction at on electrode is exactly the opposite of the reaction at the other electrode, no Net chemical change will occur in the electrolyte. The overall process is then conversion of light into electricity. In others word a photoelectrochemical cell can be used to produce electricity by choosing redox systems such that the reaction taking place at one electrode is completely reversed at the counter electrode. In this situation no net chemical reaction occurs and $\Delta G = 0$. This mode of operation is called photovoltaic of regenerative solar cell mode [6], [17], [18]. The Photoelectochemical cell operation resembles that of a photovoltaic device, and is designated as a regenerative and fuel forming.

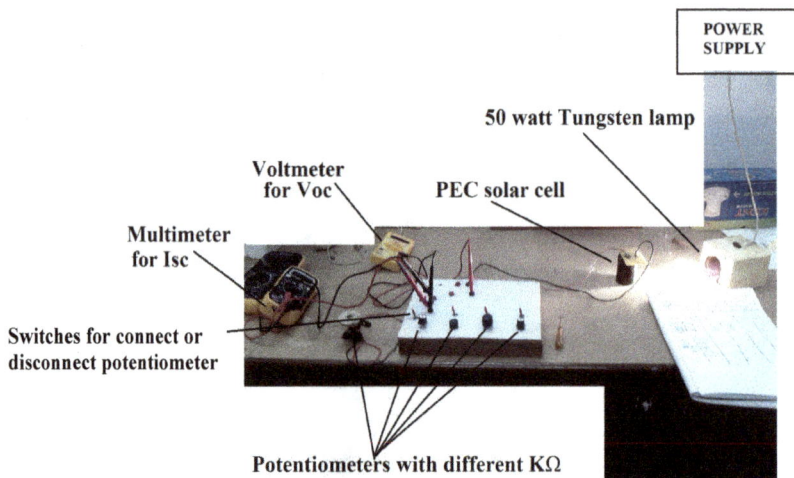

POWER SUPPLY

50 watt Tungsten lamp

Voltmeter for Voc

PEC solar cell

Multimeter for Isc

Switches for connect or disconnect potentiometer

Potentiometers with different KΩ

Figure 3(b): *Experimental set up of a PEC solar cell.*

3.2   Preparation of a PEC Solar Cell

A PEC solar cell was fabricated in a cylindrical Pyrex glass vessel by dipping CdSe film in an electrolyte containing 0.1M NaOH, 1M Na_2S, & 1M S. A standard two electrode configuration was used with CdSe photoanode and graphide as the counter electrode. CdSe photoelectrode does not take part in the chemical reaction with polysulphide electrolyte. The concentration of the CdSe film remains unchanged. This is explained as follows:

When the CdSe nanocrystalline thin film is dipped in sulphide-polysulphide electrolyte, considerable exchange of selenium by sulphure takes place. There are in effect two paths which the PEC solar cell was fabricated in a cylindrical Pyrex glass vessel by dipping CdSe film in an electrolyte containing 0.1M NaOH, 1M Na_2S, & 1M S. A standard two electrode configuration was used with CdSe photoanode and graphite as the counter electrode.

The CdSe photoelectrode does not take part in any chemical reaction with polysulphide electrolyte. The concentration of CdSe film remains unchanged. This is explained with the photogenerated hole (h^+) that can follow:

(I) Regenerative charge transfer reaction:

$$2S^{2-}+2h \rightleftharpoons S_2^-$$

This reaction stabilizes the semiconductor against photocorrosion.

(II) photocorrosion reaction:

$$CdSe + S^2 \rightleftharpoons Cd^{2+} + SeS^{2-} +2e$$

$$Cd^{2+} + S \rightleftharpoons CdS$$

The photo-corrosion product SeS^{2-} goes into the solution while CdS formed as a result of Se $\rightleftharpoons$ S exchange reaction is deposited on CdSe surface. Owing to slowness of regenerative slowness of regenerative redox reaction the degree of stabilization of CdSe is limited [19].

Counter electrodes:

The counter electrode should have a very good electronic conductivity (for current pick up) and should be chemically inert to electrolyte. Usually, large area electrode (compared to SC electrode area) is employed to avoid kinetic factors (charge transfer) influencing the PEC cell parameters. However we have used graphite as the counter electrode in PEC solar cell for it is of low cost as compared to platinum.

Electrolyte:

The electrolyte selected in this investigation consisted of a solution of 0.1M NaOH, 1M Na_2S, 1M S. Thus the electrolyte contained sulphide polysuphide redox couple. The selection of S^{2-}/S_2 as redox couple is known to facilitate the stabilization of cadmium chalcogenide electrodes.

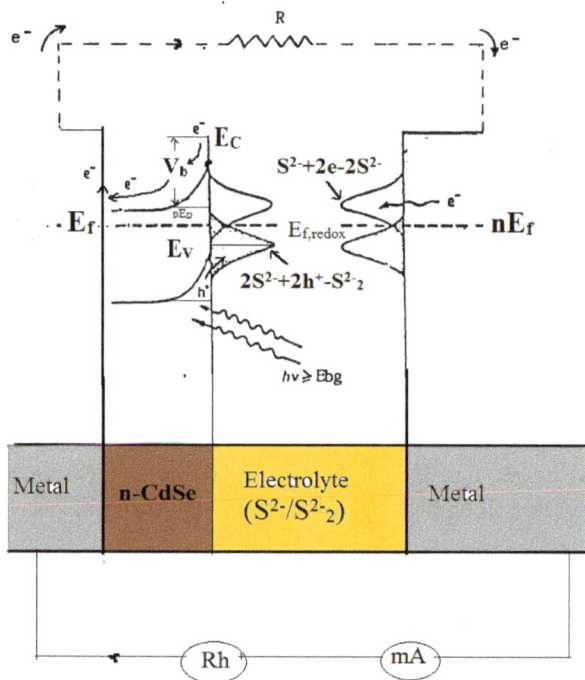

Figure 4: *Relative energy position in metal/ CdSe film/electrolyte/metal photoelectrochemical solar cell and current flow under illumination.*

Figure 4 shows the relative positions of conduction band edge (Ec), Valance band (Ev) decomposition potential (PE$_D$), Fermi level (nE$_F$) of the semiconductor along with the Fermi level of redox species (E$_f$, redox) and the metal counter electrode (mE$_f$) for specific PEC solar cell using CdSe film. The broadening of energy levels of reduced species (S^{2-}) and oxidized species (S^{2-}) is also shown by the distribution curve. The flow of photo generated holes and electron is represented by the arrows.

The electrostatic potential in space charge layer is such that the hole moves towards the semiconductor, electrolyte interface where they participate in oxidation reaction with the sulphite ions. The remaining photo generated electron moves in the semiconductor bulk. Thus the current flow results in the external circuit at the metal counter electrode.

Reduction of polysulphide by electrons take place completing the overall reaction which is expressed as –

(i) at the semiconductor electrode

$2S^{2-} + 2h \rightarrow S^{2--}$

The overlapping of the valance band edge with energy distribution curve of the sulphide ion facilitate such a charge transfer reaction

(ii) at the metal counter electrode

$S_2^{--} + 2e \rightarrow 2S^{--}$

The redox potential of reaction is -0.48V with respect to normal hydrogen electrode. In electrolyte containing 0.1M NaOH, 1M Na_2S and 1M S, Sulphur behaves as charge carriers.

3.3   Circuitry and Measurements of Solar Cell Parameters

The nanocrystalline thin film of CdSe was deposited on Ti substrate with different duty cycle at 80°C. After deposition, the film was kept for a day in air before measuring the cell parameters. The following measurements were carried out.

The PEC solar cell fabricated by above described method was then placed at a fixed distance from the light source. The light source used for illumination was a 50 watt tungsten lamp. The intensity of light was measured by a Luxmeter. The two electrodes of the solar cell were then connected to a potentiometer whose resistance varies from 0 to 447 K. Current –voltage measurements were carried out by varying the external resistance R2. Digital multimeter (model) was used to record the current and voltage of the cell.

a)   Open circuit voltage V_{oc} and short circuit current I_{sc}

The photoelectrode was illuminated by a 50 watt tungsten lamp with various intensities varying the distance varying the distance between lamp and photoelectrode under above condition open circuit voltage (V_{oc}) and short circuit current (I_{sc}) was measured by a multimeter.

b)   V-I characteristics

The intensity of incident radiation was kept fixed by putting the light source at a fixed distance from the photo anode. A potentiometer was connected in series with the PEC solar cell. By varying the resistance of the potentiometer the output voltage and current was varied and measured for different films.

c) Thickness measurement

The thickness of the film is the most significant parameter that affects the properties of the thin film. It may be measured either by in situ monitoring of the rate of the deposition or after the film is taken out from deposition chamber. The technique of the first type often referred to as monitor method generally allow both monitoring and controlling of deposition rate and film thickness. Any known physical quantity related to film thickness can be used to measure the thickness. The method chosen should be convenient, reliable and simple.

One of the most convenient and earliest methods for determining film thickness is the gravimetric method. In this method, area and weight of the film are measured. The thickness was obtained using the formula [20].

$$t = \frac{M}{\rho \times A}$$

$$M = m_1 - m_2$$

Where, t is the film thickness, M is mass of the film material (in gm), A is area of the film (in cm^2), m_1 is mass of the substrate with film, m_2 is mass of the substrate without film and ρ is density of the film material (gm/cm^3). The value pertaining to the bulk material is usually taken for 'ρ' even when the actual density of the material in the thin film form is lower. However, this method is not very accurate and actual density is different than the film material.

4. Single Layered CdSe Nanocrystalline Photoelectrodes (P_s)

The single layered photoelectrodes prepared with different duty cycles as D_0 (1:0), D_1 (1:1), D_2 (1:2), D_3 (1:3), D_4 (1:4), D_5 (1:5), D_6 (1:6) at fixed temperature of 80° C for all single layered photoelectrodes.

I.   First single layered photoelectrode with duty cycle D_0 (1:0) means 15 minute continuously electrodeposition of 1sec. ON & 1sec. OFF, total deposition time-15 minute.

II. Second single layered photoelectrode with duty cycle D_1 (1:1) means 1 sec. ON & 1 sec. OFF, total deposition time-30 minute.

III. Third single layered photoelectrode with duty cycle D_2 (1:2) means 1sec. ON & 2sec. OFF, total deposition time-45 minute.

IV.   Fourth single layered photoelectrode with duty cycle D_3 (1:3) means 1sec. ON & 3sec. OFF, total deposition time-60 minute.

V.   Fifth single layered photoelectrode with duty cycle D_4 (1:4) means 1sec. ON & 4sec. OFF, total deposition time-75 minute (or 1hour 15 min.).

VI.   Sixth single layered photoelectrode with duty cycle D_5 (1:5) means 1sec. ON & 5sec. OFF, total deposition time-90 minute (or 1hour 30 min.).

VII.   Seventh single layered photoelectrode with duty cycle D_6 (1:6) means 1sec. ON & 6sec. OFF, total deposition time-105 minute (or 1hour 45 min.).

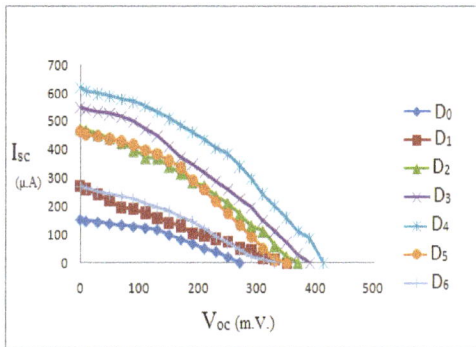

Figure 5(a): *I-V characteristics of single layered photo-anodes with different duty cycles.*

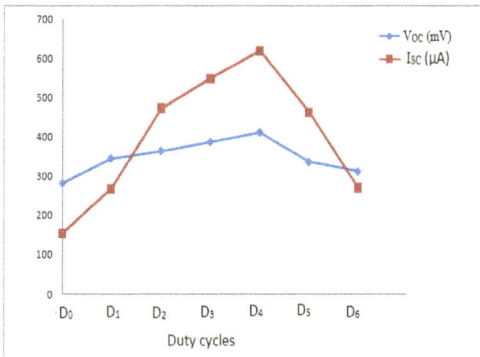

Figure 5(b): *Effect of the different duty cycles on V_{OC} & I_{SC} of single layered photo-anodes.*

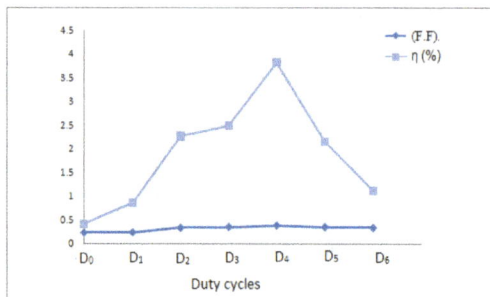

Figure 5(c): *The graph between F.F., η (%) with different duty cycles of single layered photo-anodes.*

Table 1. Solar cell Parameters for CdSe nanocrystalline Single layered photoelectrodes (P_s):

Duty cycles	Open circuit voltage V_{OC} (mV)	Short circuit current I_{SC} (μA)	Fill-factor (F.F).	efficiency η (%)
D_0	282	154	0.227	0.408
D_1	345	268	0.235	0.860
D_2	364	473	0.332	2.26
D_3	387	548	0.350	2.94
D_4	411	619	0.379	3.81
D_5	336	463	0.349	2.14
D_6	312	271	0.332	1.11

The variation of open circuit voltage (V_{OC}), short circuit current (I_{SC}), fill-factor (F.F), and efficiency (η%) with different duty cycles have been studied. These solar cell parameters with different samples are given in table 1. In the present investigation the best single layered CdSe nanocrystalline thin film, for photoelectrode are obtained for duty cycle D_4 by the pulse electrodeposited CdSe film.

The I-V curves are displaced linearly along the current axis, as a function of intensity of incident light [Figure 5(a)] hence intensity dependence of short circuit current is linear and the open circuit voltage increase as logarithmic function of light intensity. Figure 5(b) shows the effect of the different duty cycles on V_{OC} & I_{SC}, and Figure 5(c) shows the variation of F.F. and η (%) with different duty cycles.

The PEC cell parameters are obtained for all composition of CdSe nanocrystal and listed in Table 1. It is seen that the cell performance parameters increase with duty cycles 1:0 to 1:4 or (D_0 to D_4), but further increases the off time of duty cycle these 1:5 & 1:6 (D_5 & D_6) duty cycle are shows poor performance. The best result are obtained for single layered CdSe nanocrystalline photoanodes with 1:4 (D_4) duty cycle at an intensity of 6200 Lux using the 0.1M NaOH, 1M Na_2S, & 1M S, polysulfide as the redox electrolyte are V_{OC} of 411 mV, I_{SC} of 619 (μA), F.F. of 0.379 and η (%) of 3.81%. The studies show that comparatively large nanocrystals of CdSe give better result in a PEC solar cell. As the size of nanocrystal further decreases performance is very poor. Such observation may be explained as follows: Improved performance for 1:4 (D_4) duty cycle can be attributed to larger surface area available for absorption of photons in case of nanoparticles. This causes increase in photo generated electron hole pairs, which separation gives photovoltaic effect. For duty cycles 1:5 & 1:6 (D_5 & D_6) perhaps the surface states become dominate and so the photogenerated charge carriers are captured by them reducing the photovoltaic effect.

For duty cycles 1:0 to 1:6, still smaller CdSe crystals will be deposited or titanium electrode where quantum size effect is considerable. This increases the oscillator strength and hence the absorption co-efficient, but more energy is required to produce photogenerated charge carriers.

5. Double Layered CdSe Nanocrystalline Photoelectrodes (P_d)

The six samples of double layered photoelectrodes prepared with different duty cycles as $D_{0/1}$, $D_{1/2}$, $D_{2/3}$, $D_{3/4}$, $D_{4/5}$, & $D_{5/6}$, at fixed temperature of 80°C for all double layered photoelectrodes.

I.   First double layered photoelectrode with duty cycle ($D_{0/1}$), 1:0 onto 1:1, means first layered 15 minute continuously electrodeposition or 1sec. ON & 1sec. OFF, total deposition time-15 minute and second layered means 1sec. ON & 1sec. OFF, total deposition time-30 minute.

II. Second double layered photoelectrode with duty cycle ($D_{1/2}$), 1:1 onto 1:2, means first layered 1sec. ON & 1sec. OFF, total deposition time-30 minute and second layered means means 1sec. ON & 2sec. OFF, total deposition time-45 minute.

259

III. Third double layered photoelectrode with duty cycle ($D_{2/3}$), 1:2 onto 1:3 means first layered 1sec. ON & 2sec. OFF, total deposition time-45 minute and second layered means 1sec. ON & 3sec. OFF, total deposition time-60 minute.

IV. Fourth double layered photoelectrode with duty cycle ($D_{3/4}$), 1:3 onto 1:4 means first layered 1sec. ON & 3sec. OFF,total deposition time-60 minute and second layered means 1sec. ON & 4sec. OFF, total deposition time-75 minute (or 1hour 15 min.).

V. Fifth double layered photoelectrode with duty cycle ($D_{4/5}$), 1:4 onto 1:5 means first layered 1sec. ON & 4 sec. OFF, total deposition time-75 minute (or 1hour 15 min.) second layered 1sec. ON & 5sec. OFF, total deposition time-90 minute (or 1hour 30 min.).

VI. Sixth double layered photoelectrode with duty cycle ($D_{5/6}$), 1:5 onto 1:6 means first layered 1sec. ON & 5sec. OFF, total deposition time-90 minute (or 1hour 30 min.) second layered 1sec. ON & 6sec. OFF, total deposition time-105 minute (or 1hour 45 min.).

The variation of open circuit voltage (V_{OC}), short circuit current (I_{SC}), fill-factor (F.F), and efficiency ($\eta\%$) with different duty cycles have been studied. In the present investigation the best double layered CdSe nanocrystalline thin films, for photo electrodes are obtained for duty cycle $D_{3/4}$. The I-V curves [Figure 6(a)] are displaced linearly along the current axis, as a function of intensity of incident light hence intensity dependence of short circuit current is linear and the open circuit voltage increase as logarithmic function of light intensity. Figure 6(b) shows the variation of V_{OC} & I_{SC} with duty cycle and Figure 6(c) shows the graph between F.F., η (%) with different duty cycles.

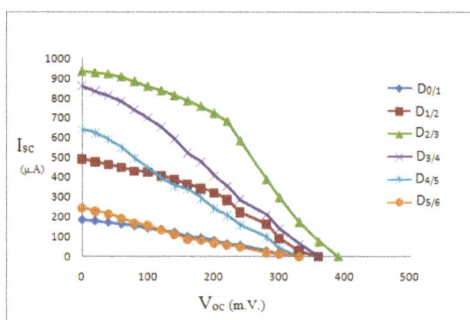

Figure 6(a): I-V characteristics of double layered photo-anodes with different duty cycles.

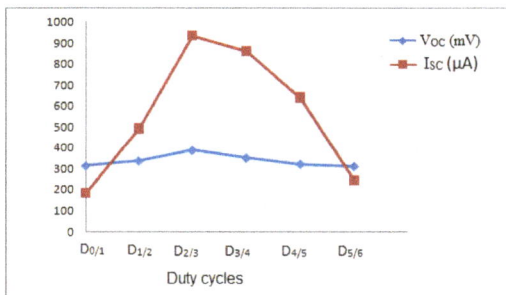

Figure 6(b): *Effect of the different duty cycles on V_{OC} & I_{SC} of double layered photo-anodes.*

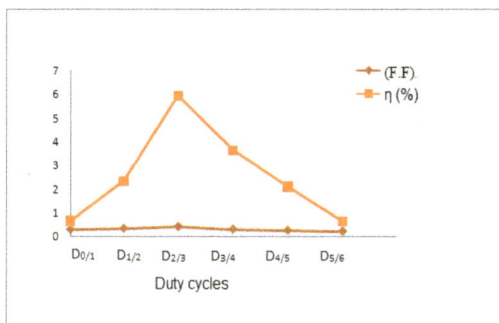

Figure 6(c): *Variation of F.F. and η (%) with different duty cycles of double layered photo-anodes.*

Table 2. *Solar cell Parameters for CdSe nanocrystalline double layered photoelectrodes (P_D):*

Duty cycles	Open circuit voltage V_{OC} (mV)	Short circuit current I_{SC} (µA)	Fill-factor(F.F.)	Efficiency η (%)
$D_{0/1}$	314	182	0.291	0.66
$D_{1/2}$	337	491	0.355	2.32
$D_{2/3}$	387	934	0.415	5.93
$D_{3/4}$	351	859	0.305	3.64
$D_{4/5}$	322	639	0.257	2.09
$D_{5/6}$	311	242	0.215	0.64

The PEC cell parameters are obtained for all composition of CdSe nanocrystal and listed in Table 2. It is seen that the cell performance improves parameters increases with duty cycles $D_{0/1}$ to $D_{2/3}$, but further increases the off time of duty cycle these $D_{3/4}$, $D_{4/5}$, & $D_{5/6}$ duty cycle are shows poor performance. The best result obtained for double layered CdSe nanocrystalline photoanodes with $D_{2/3}$ duty cycle at an intensity of 6200 Lux using the 0.1M NaOH, 1M Na_2S, & 1M S, polysulfide as the redox electrolyte are V_{OC} of 387 mV, I_{SC} of 934 (μA), F.F. of 0.415 and η (%) of 5.93%. The studies show that comparatively large nanocrystals of CdSe give better result in the PEC solar cell. As the size of nanocrystal further decreases performance is very poor.

6.   Triple Layered CdSe Nanocrystalline Photoelectrodes (P_t)

The five samples of triple layered photoelectrodes prepared with different duty cycles as $D_{0/1/2}$, $D_{1/2/3}$, $D_{2/3/4}$, $D_{3/4/5}$, & $D_{4/5/6}$, at fixed temperature of 80°C for all triple layered photoelectrodes.

I.   First triple layered photoanode with duty cycle $D_{0/1/2}$ (1:0 onto 1:1 onto 1:2) means first layered 15 minute continuously electrodeposition or 1sec. ON total deposition time-15 minute, second layered means 1sec. ON & 1sec. OFF, total deposition time-30 minute and third layered means 1sec. ON & 2sec. OFF, total deposition time-45 minute.

II.   Second triple layered photoelectrode with duty cycle $D_{1/2/3}$ (1:1 onto 1:2 onto 1:3) means first layered 1sec. ON & 1sec. OFF, total deposition time-30 minute, second layered means means 1sec. ON & 2sec. OFF, total deposition time-45 minute and third layered means 1sec. ON & 3sec. OFF, total deposition time-60 minute.

III.   Third triple layered photoelectrode with duty cycle$D_{2/3/4}$ (1:2 onto 1:3 onto 1:4) means first layered 1sec. ON & 2 sec. OFF, total deposition time-45 minute, second layered means 1sec. ON & 3 sec. OFF, total deposition time-60 minute and third layered means 1sec. ON & 4 sec. OFF, total deposition time-75 minute (or 1hour 15 min.).

IV.   Fourth triple layered photoelectrode with duty cycle $D_{3/4/5}$ (1:3 onto 1:4 onto 1:5) means first layered 1sec. ON & 3sec. OFF, total deposition time-60 minute, second layered means 1sec. ON & 4sec. OFF, total deposition time-75 minute (or 1hour 15 min.), third layered means 1sec. ON & 5sec. OFF, total deposition time-90 minute (or 1hour 30 min.).

V. Fifth triple layered photoelectrode with duty cycle $D_{4/5/6}$ (1:4 onto 1:5 onto 1:6) means first layered 1sec. ON & 4sec. OFF, total deposition time-75 minute (or 1hour 15 min.), second layered means 1sec. ON & 5sec. OFF, total deposition time-90 minute (or 1hour 30 min.) third layered means 1sec. ON & 6sec. OFF, total deposition time-105 minute (or 1hour 45 min.).

Table 3. Solar cell Parameters for CdSe nanocrystalline Triple layered photoelectrodes (P_T).

Duty cycles	Open circuit voltage V_{OC} (mV)	Short circuit current I_{SC} (µA)	Fill-factor (F.F.)	Efficiency η (%)
$D_{0/1/2}$	236	191	0.261	0.47
$D_{1/2/3}$	242	213	0.300	0.61
$D_{2/3/4}$	251	226	0.344	0.77
$D_{3/4/5}$	268	376	0.360	1.41
$D_{4/5/6}$	226	123	0.255	0.28

Table 4. Solar cell Parameters for triplelayered CdSe nanocrystalline photoelectrods (P_T) with change in duty cycle.

Duty cycles	Open circuit voltage V_{OC} (mV)	Short circuit current I_{SC} (µA)	Fill-factor (F.F.)	Efficiency η (%)
$D_{1/3/5}$	227	436	0.287	1.12
$D_{2/4/6}$	231	491	0.336	1.51

Table 5. Solar cell Parameters for CdSe nanocrystalline triple layered photo- electrode ($P_{2/4/6}$) with Annealed (P_{TA}) at 100°C.

Duty cycles	Annealed (in minute)	Open circuit voltage V_{OC} (mV)	Short circuit current I_{SC} (µA)	Fill-factor (F.F.)	Efficiency η (%)
$P_{2/4/6}$	0	231	491	0.336	1.51
$P_{2/4/6}$	15	237	514	0.342	1.65
$P_{2/4/6}$	30	241	527	0.346	1.74
$P_{2/4/6}$	45	243	561	0.349	1.83
$P_{2/4/6}$	60	241	627	0.356	2.12
$P_{2/4/6}$	75	228	412	0.281	1.04

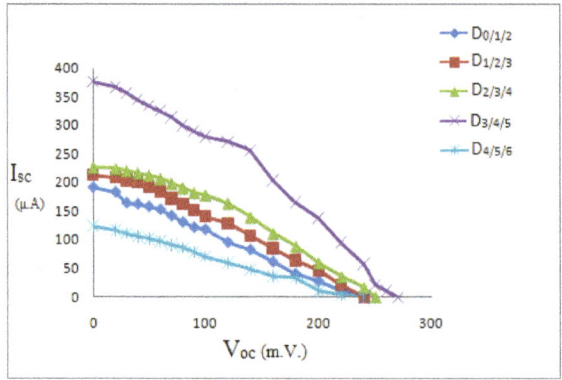

Figure 7(a): *I-V characteristics of triple layered photo-anodes with different duty cycles.*

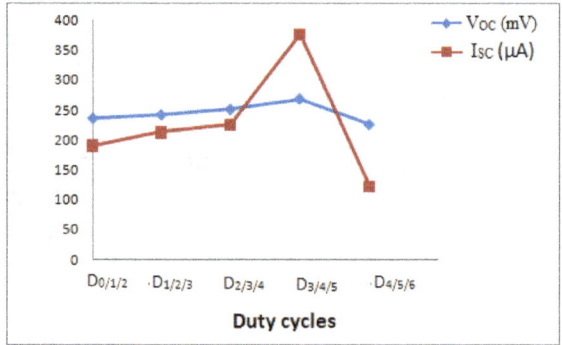

Figure 7(b): *Effect of the different duty cycles on V_{OC} & I_{SC} of triple layered photo-anodes.*

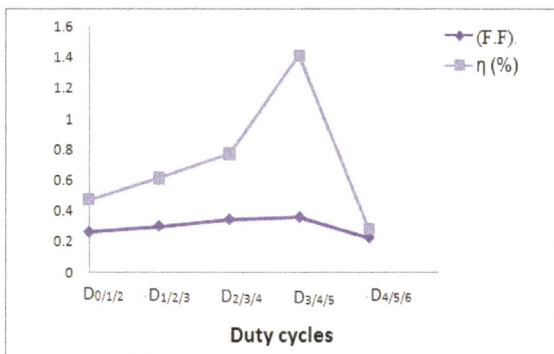

Figure 7(c): *Variation of F.F. and η (%) with different duty cycles of triple layered photo-anodes.*

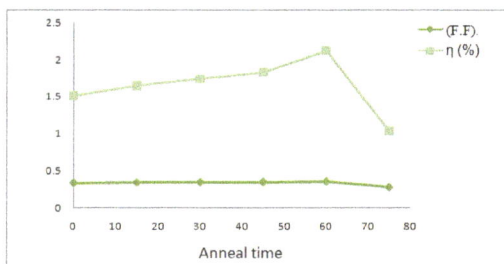

Figure 7(d): *Effect of annealing on F.F. & η (%) of triple layer photoelectrode.*

In the present investigation the best triple layered CdSe nanocrystalline thin films, for photo electrodes are obtained for duty cycle $D_{3/4/5}$. The I-V curves [Figure 7(a)] are displaced linearly along the current axis, as a function of intensity of incident light hence intensity dependence of short circuit current is linear and the open circuit voltage increase as logarithmic function of light intensity and Figure 7(b) shows the effect of the different duty cycles on V_{OC} & I_{SC} and Figure 7(c) shows the variation of F.F. and η (%) with different duty cycles.

The PEC cell parameters are obtained for all composition of triple layered CdSe nanocrystal and listed in Table 3. It is seen that the cell performance improves parameters increases with duty cycles $D_{0/1/2}$ to $D_{3/4/5}$, but further increases the off time of duty cycle $D_{4/5/6}$ shows poor performance. The best result obtained for triple layered CdSe nanocrystalline photoanodes with $D_{3/4/5}$ duty cycle at an intensity of 6200 Lux using the 0.1M NaOH, 1M Na_2S, & 1M S, polysulfide as the redox electrolyte are V_{OC} of 268 mV, I_{SC} of 376 μA, F.F. of 0.360 and η (%) of 1.41%.

Investigations are done for a few other triple layered photoelectrodes by changing the duty cycles as given in Table 4. It is observed that I_{SC} of 491 μA and η (%) of 1.51% are obtained for triple layered photoanode with the duty cycles $D_{2/4/6}$. It is also found that annealing of photoanode improves the cell performance. The best triple layered film was annealed at 100°C for various time durations ranging from 15 to 75 minutes. The solar cell parameters from annealed films are given in Table 5 and shown in Fig. (7d). It is seen that 60 minutes annealing gives the best results. By further increase in the anneal time, all parameters of PEC solar cell decrease.

The studies show that comparatively large nanocrystals of CdSe give better result in the PEC solar cell. As the size of nanocrystal further decreases performance becomes very poor.

The film thickness of CdSe nanocrystalline film deposited with different duty cycles was measured by the gravimetric method. It is seen that the film thickness decreases slightly with increasing duty cycle as given in Table 6 and shown in Figure 8.

Table 6. Thickness of Film in case of Different duty Cycles

Duty Cycle	D_0	D_1	D_2	D_3	D_4	D_5	D_6
Film Thickness (μm)	34.5	34.0	33.8	33.64	33.56	33.42	33.28

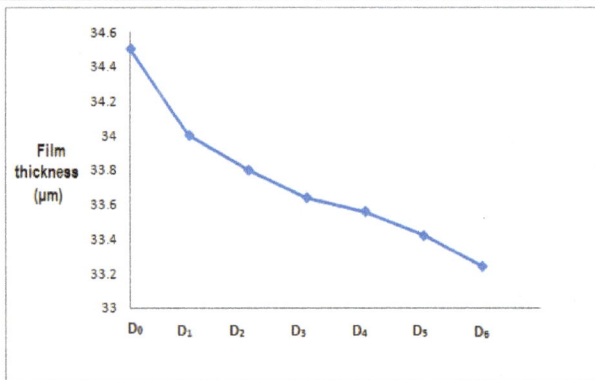

Figure 8: *Variation of film thickness with duty cycles.*

6. Summary and Conclusion

Reports are available that the use of nanostructures in solar cells may enhance the efficiency and performance. In PEC cells the junction formation is quite easy and polycrystalline films also work very well. Present attempt has been made with crystalline size in nanometer range which may be advantageous because of large surface area and high oscillator strength. At the same time quantum size effect may increase the effective band gap and hence reduce the absorption of large wave length portion of solar spectrum.

In the present work single layered, double layered and triple layered CdSe nanocrystalline films of different crystal sizes have been prepared by pulse electro-deposition method and their performance in PEC cells with sulphide/polysulphide electrolyte has been studied. The development of different type of nanocrystalline CdSe thin films solar cell and their structural and optoelectronic characteristic have been described. The growth of nanocrystalline CdSe film with different duty cycle is presented.

Electrolyte solution for PEC solar cell was 0.1 M NaOH, 1M Na_2S, & 1M S. A graphite rod was used as a counter electrode and CdSe films were used as a photoelectrodes. I-V characteristics were plotted and solar cell parameters V_{OC}, I_{SC}, F.F. and $\eta\%$ were determined. A solar cell conversion efficiency of up to 3.81% of D_4 duty cycle for single layered photoelectrode , 5.93% of $D_{2/3}$ duty cycle for double layered photoelectroode and 1.51% of $D_{2/4/6}$ duty cycle for triple layered photoelectrode was obtained.

The best results are obtained by the PEC solar cell with photo-anode of CdSe film prepared with duty cycle $D_{2/3}$. The study reveals that nanocrystalline CdSe based photoelectrodes give better results in PEC cells due to larger surface area and greater oscillator strength. Use of double layer further improves the performance.

References

[1]   Matthew C. Beard, Joseph M. Luther and Arthur J. Nozik "The promise and challenge of nanostructured solar cells" Nature nanotechnology, vol 9,p-951-952,2014

[2]   Oh, J., Yuan, H. C. & Branz, H. M. Nature Nanotech. 7, 743–748, 2012. https://doi.org/10.1038/nnano.2012.166

[3]   Kelzenberg, M. D. et al. Nature Mater. 9, 239–244,2010. https://doi.org/10.1038/nmat2727

[4]   Garnett, E. & Yang, P. Nano Lett. 10, 1082–1087,2010. https://doi.org/10.1021/nl100161z

[5]   Gerischer,H.; J.Electranalyt. Chem. and Interfacial Electrochem,58, 263, 1975.
       https://doi.org/10.1016/S0022-0728(75)80359-7

[6]   Chandra S., and Pandey R.K., Phy.Stat.Sol.(a) 59,787, 1980.
       https://doi.org/10.1002/pssa.2210590246

[7]   Di Wei and Gehan Amaratunga, "Photoelectrochemical Cell and Its Applications
       in Optoelectronics" Int. J. Electrochem. Sci., 2,897 – 912, 2007.

[8]   Fujishima, A., Inoue, T., Wantanabe, T. and Honda, K.; Chem. Lett.,357-360,
       1978. https://doi.org/10.1246/cl.1978.357

[9]   B.P. Chandra, Renewable energy source, 2008.

[10]   E-C. Cho, S. Park, X. Hao, D. Song, G. Conibeer, S-C. Park, M.A. Green "Silicon
       quantum dot/crystalline silicon solar cells", Nanotechnology 19, (2008)].
       https://doi.org/10.1088/0957-4484/19/24/245201

[11]   J. Hou and X. Guo ,W. C. H. Choy (ed.), Organic Solar Cells, Green Energy and
       Technology,DOI: 10.1007/978-1-4471-4823-4_2, _ Springer-Verlag London,
       2013. https://doi.org/10.1007/978-1-4471-4823-4_2

[12]   Butler M.A., and Ginley D.S., Review principle of photo electro chemical solar
       energy conversion (Review Pri. Of PEC, SE, Com.,1980.

[13]   K.R. Murali, K. Trivedi, ICLA, 1994.

[14]   www.Kpsec.Frecuk.com the electronic club.

[15]   Penicker M.P.R., Kanster and Kroger F.A., J. Electrochm. Soc. 125, 566, 1978.
       https://doi.org/10.1149/1.2131499

[16]   www.nature physics.com and physics@! nature.com.

[17]   Fujishima,A., Kohayawa,K. and Honda, K.; J. Electrochem. Soc., 122,1437, 1975.
       https://doi.org/10.1149/1.2134048

[18]   I- Ellise, A.B.,Kaiser, B.W. and Wrighton M.S.; J.Am. Chem. Soc., 98, 635,
       1976a. II-Ellise A.B., Kaiser B.W., and Wrighton M.S. J.Phys. Chem., 80, 635,
       1976b)

[19]   Pandey R.K., Rooz A.J.N. and Gore R.B.; Semicond. Sci. Techno., 3, 733, 1988.
       https://doi.org/10.1088/0268-1242/3/8/002

[20]   H.S. Hilal, Rania M.A. Ismail, A. El-Hamouz, A. Zyoud, I. Saadeddin,
       Electrochimica Acta, 54, 3433 2009.
       https://doi.org/10.1016/j.electacta.2008.12.062

Chapter 8

Status and potential of organic solar cells

Rashmi Swami and Sanjay Tiwari

Photonics Research Laboratory, School of Studies in Electronics & Photonics, Pt. Ravishankar Shukla University, Raipur (C.G.) 492010 India

rashmi.swami3@gmail.com

Abstract

Solar energy is clean and renewable energy which is generated from natural source sun. Solar cells are devices which convert solar energy into electricity, either directly via the photovoltaic effect, or indirectly by first converting the solar energy to heat or chemical energy. Inorganic and organic both types of solar cells are available. Organic solar cell research has developed during the past 30 years, but especially in the last decade it has attracted scientific and economic interest triggered by a rapid increase in power conversion efficiencies. This was achieved by the introduction of new materials, improved materials engineering, and more sophisticated device structures. Though efficiency of organic devices have not yet reached those of their inorganic counterparts ($\approx$ 10–24%); the perspective of low cost, low temperature and energy processing, low material requirement, can be used on flexible substrate, can be shaped to suit architectural application, are some advantages of organic solar cell that drives the development of organic photovoltaic devices further in a dynamic way. This paper gives an overview of organic solar cells. The field of organic solar cells profited well from the development of light-emitting diodes based on similar technologies, which have entered the market recently. We review here the current status of the field of organic solar cells and discuss different production technologies as well as study the important parameters to improve their performance.

Keywords

Solar Cells, Organic Electronics, Photovoltaic Devices, Semiconductors, Polymeric Materials

Contents

1.   Introduction...271

3.   Materials...271

3.1   Donor materials 272

3.2   Acceptor materials..273

4.   Working principle...273

4.1   Absorption of photons 274

4.2   Exciton diffusion 275

4.3   Charge separation...275

4.4   Charge transport 276

4.5   Charge collection 276

5.   Modes of operation 276

5.1   Short circuit condition...276

5.2   Flat-band condition 276

5.3   Open circuit condition...277

6.   Device architectures..277

6.1   Single layer structure 277

6.2   Blend structure..278

6.3   Bilayer structure 279

6.4   Laminated structure..279

7.   Parameters..281

7.1   Open circuit voltage...282

7.2   Short circuit current..282

7.3   Fill factor 282

7.4   Efficiency 282

8. Equivalent circuit...**282**

10. Future aspects **286**

11. Conclusion **287**

References..**287**

1. Introduction

Solar cell produces DC electricity when sunlight shines on it. The DC power is converted to AC power with an inverter and can be used to power local loads or fed back to the utility. An organic solar cell is a photovoltaic cell that uses organic electronics-a branch of electronics that deals with conductive organic polymers or small organic molecules for light absorption and charge transport. Organic solar cells (OSC) and polymer solar cells are built from thin films (typically 100 nm) of organic semiconductors as shown in Figure1.

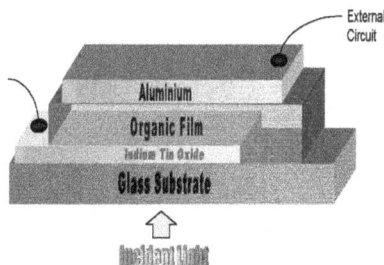

Figure 1: *Device structure of an organic solar cell. The organic film may comprise one or more semiconducting layers, a blend or a combination of these.*

The optical absorption coefficient of organic molecules is high, so a large amount of light can be absorbed with a small amount of materials. The main disadvantages associated with organic photovoltaic cells are low efficiency, low stability and low strength compared to inorganic photovoltaic cells.

3. Materials

Organic solar cells mainly consist of two organic materials [11-12] one of which is a donor (gives electron) and the other is an acceptor (accepts electrons) some of them are shown in Figure 2.

Table 2.   *History of organic solar cell.*

1839	Becquerel [1] observed the photoelectrochemical process.
1959	Kallman and Pope [2] reported the first OPV cell that used a 5 μm anthracene crystal sandwiched between two NaCl solutions. The salt solutions functioned as transparent electrodes, which were further connected with silver electrodes. With these cells, the investigators were able to achieve an efficiency of only 2×10^{-6} %.
1975	C. W. Tang *et* al [3] has reported OSC of 0.001% efficiency.
1986	The first OSC with power conversion efficiency of about 1 % under simulated AM2 illumination was reported once again by C. W. Tang [4]. In this thin-film, two-layer organic photovoltaic cell has been fabricated from copper phthalocyanine and a perylene tetracarboxylic derivative.
1993	Sariciftci [5] made the first polymer/C60 heterojunction device.
1995	Yu / Hall [6] made the first bulk polymer/polymer heterojunction PV.
2000	5% efficiency was reported by Schon et al [7].
2001	Schmidt-Mende [8] made a self-organised liquid crystalline solar cell of hexabenzocoronene and perylene.
2010	Forbes.com reported that Solarmer broke the 8% wall and reported an organic solar cell with 8.13% efficiency [9].
2011	The highest NREL (National Renewable Energy Laboratory) certified efficiency has reached 8.3% for the Konarka Power Plastic [10].

3.1   Donor materials

Polythiophene and its derivatives, Conjugated polymers like polymers with 2,1,3-Benzothiadiazole, Pyrrolo[3,4-c]pyrrole-1,4-dione (DPP) derivatives, Benzo[1,2-b;4,5-b']dithiophene based polymers, MDMO-PPV (poly[2-methoxy-5- (3,7-dimethyloctyloxy)]-1,4-phenylenevinylene), PFB (poly(9,9'- dioctylfluorene-co-bis-*N,N*'-(4-butylphenyl)-bis-*N,N*'- phenyl 1,4-phenylenediamine) act as donors .The absorption band of poly(3-alkylthiophene)s (P3AT)(example.- P3HT) is still not broad enough to get good harvest of the sunlight. Two dimensional conjugated polythiophenes (2D-PTs) provided a feasible way to broaden absorption band and also exhibit better hole mobilities than P3ATs. Polymers with 2,1,3-Benzothiadiazole (like PCDTBT,PCPDTBT etc) are low band gap polymers with strong and broad absorption band extending to near infrared region. DPP and its derivatives (like PTDPP, PFDPP) have strong absorption band in the visible range. Thiophene based DPP derivatives have well confined conjugated structures and exhibit good charge carrier mobilities for both holes and electrons. Benzo[1,2-b;4,5-b']dithiophene based polymers BDT has a symmetric and planar conjugated structure and hence tight and regular stacking can be expected for the BDT based conjugated polymers.

3.2   Acceptor materials

Fullerenes, perylene and its derivatives like F8TB (poly(9,9'- dioctylfluoreneco-benzothiadiazole) and PCBM (1-(3 methoxycarbonyl) propyl-1-phenyl[6,6]C61) act as acceptors. Fullerene C60 has well symmetric structure and exhibits good electron mobility. In comparison with PC60BM, PC70BM possesses stronger absorption in visible range. C70 is much expensive than that of C60 due to its tedious purification process which limits its application.

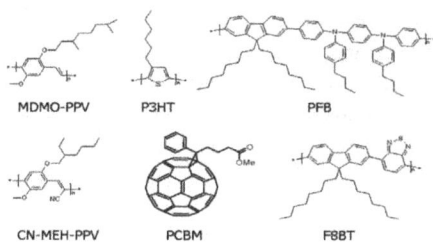

Figure 2: Poly(2-methoxy-5-(3',7'-dimethyloctyloxy)-1,4-phenylene-vinylene) (MDMO-PPV); poly-(3- hexylthiophene) (P3HT); poly(9, 9'-dioctylfluorene-co-bis-N,N'-(4-butylphenyl)-bis-N,N'-phenyl-1,4-phenylenediamine)   (PFB);   poly(2-methoxy-5-ethylhexyloxy-1,4-phenylenecyanovinylene) (CN-MEH-PPV); [6,6]-phenyl-C61 butyric acid methyl ester(PCBM); and poly(9,9'-dioctylfluorene-co-benzothiadiazole) (F8BT).

4.   Working principle

In organic semiconductors, absorption of photons leads to the creation of bound electron hole pairs (excitons) rather than free charges. These excitons carrying energy but no net charge may diffuse to dissociation sites where their charges can be separated. The separated charges then need to travel to the respective device electrodes to provide voltage and be available for injection into an external circuit. The conversion steps with regard to the special situation in organic solar cells, as shown in, Figure 3 are as follows:

4.1 Absorption of photons

Photons with energy equal to the bandgap energy are absorbed to create free electrons. Photons with less energy than the bandgap energy pass through the material. In most organic device only a small portion of the incident light is absorbed for the following reasons:

- The semiconductor bandgap is too high. A bandgap of 1.1eV (1100 nm) is required to absorb 77% of the solar radiation on earth whereas the majority of semiconducting polymers have bandgaps higher than 2.0eV (600nm) limiting the possible absorption to about 30%.
- The organic layer is too thin. The typically low charge carrier and exciton mobilities require layer thickness in the order of 100 nm. Fortunately the absorption coefficient of organic materials is generally much higher than in e.g. Si so that only about 100 nm are necessary to absorb between 60 and 90% if a reflective back contact is used.
- Reflection: Reflection losses are probably significant but little investigated in these materials. Systematic measurements of photovoltaic materials are desired to provide knowledge of their impact on absorption losses. Anti-reflection coatings, as used in inorganic devices, may then prove useful once other losses such as recombination become less dominant.

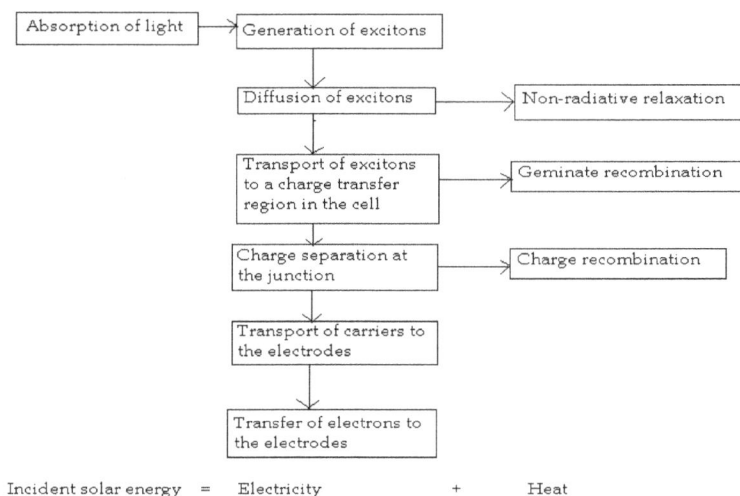

Figure 3: *Conversion steps and loss mechanism of light power into electric power. Light which is not converted to electricity is converted to heat and eventually contributes to damage.*

4.2 Exciton diffusion

Ideally, all photoexcited excitons should reach a dissociation site. Since such a site may be at the other end of the semiconductor, their diffusion length should be at least equal the required layer thickness otherwise they recombine and photons are wasted. Exciton diffusion ranges in polymers and pigments are typically around 10 nm. However, some pigments like perylenes are believed to have exciton diffusion lengths of several 100 nm.

4.3 Charge separation

Charge separation is known to occur at organic semiconductor/metal interfaces, impurities (e.g. oxygen) or between materials with sufficiently different electron affinities (EA) and ionisation potentials (IA). In the latter one material can than act as electron acceptor (A) while the other keeps the positive charge and is referred to as electron donor (D) - since it did actually *donate* the electron to A. If the difference in IA and EA is not sufficient, the exciton may just hop onto the material with the lower bandgap without splitting up its charges. Eventually it will recombine without contributing charges to the photocurrent.

4.4   Charge transport

The transport of charges is affected by recombination during the journey to the electrodes - particularly if the same material serves as transport medium for both electrons and holes. Also, interaction with atoms or other charges may slow down the travel speed and thereby limit the current.

4.5   Charge collection

In order to enter an electrode material with a relatively low work function (e.g. Al, Ca) the charges often have to overcome the potential barrier of a thin oxide layer. In addition, the metal may have formed a blocking contact with the semiconductor so that they can not immediately reach the metal.

5.   Modes of operation

Three different modes of operation are- short circuit condition, flat band condition, open circuit condition.

5.1   Short circuit condition

Generally, the work-function of the anode is considerably higher as compared to work function of cathode. The short circuit condition as shown in Figure 4(a) requires that the Fermi levels of the two electrodes must align to ensure zero potential across the device. This generates electrical field, which leads to a drift current that aids the diffusion current. Hence, total current is maximum in the device under this condition and is called short circuit current (I_{sc}). This is the upper limit on the amount of current that a device can deliver under solar cell operation. This depends on the intensity of the incident light.

5.2   Flat-band condition

As shown in Figure 4(b) this situation arises when applied voltage is equal to the work function difference of the cathode and anode. In this condition drift current is zero because there is no electric field within the device. But because of light generated carriers at the organic-organic interface there is carrier gradient with leads to diffusion current. Hence, total current is non-zero in this condition.

5.3 Open circuit condition

This condition is shown in Figure 4(c). Now if light is made to fall on the device in this condition, it will ultimately lead to creation of charges, which, due to diffusion will move towards the electrodes. Hence, there is a clear tendency for diffusion current to flow. But due to open circuit between two electrodes, a steady state current cannot flow. Therefore, this diffusion current has to be balanced by an equal and opposite drift current. There has to be a finite field inside the bulk to make this drift current flow. So there has to be some external voltage applied between the two electrodes. This voltage is called open circuit voltage (V_{OC}), which is an important parameter in determining the efficiency of the device.

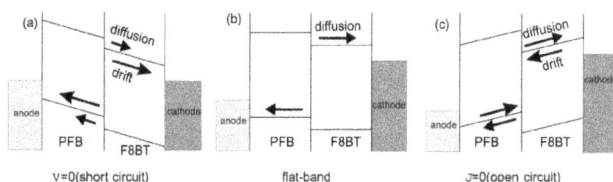

Figure 4: *Band diagram of a photo-voltaic device is shown (a) at short circuit (b) under flat band (c) at open circuit*

6. Device architectures

Four different architectures [13] are as follows-

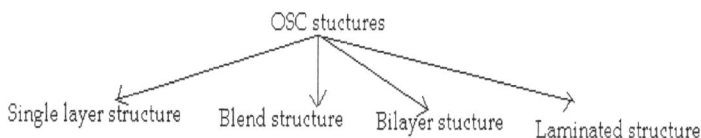

6.1 Single layer structure

These cells are made by sandwiching a layer of organic electronic materials between two metallic conductors, typically a layer of indium tin oxide (ITO) with high work function and a layer of low work function metal such as Al, Mg or Ca as shown in Figure 5.

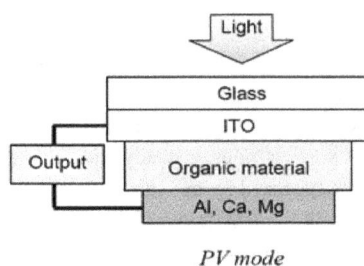

PV mode

Figure 5: *Single layer OSC structure*

The structure is simple but absorption covering the entire visible range is rare using a single type of molecule. The difference of work function between the two conductors sets up an electric field in the organic layer. In practice, single layer organic photovoltaic cells of this type do not work well. They have low quantum efficiencies (<1%) and low power conversion efficiencies (<0.1%). A major problem with them is that the electric field resulting from the difference between the two conductive electrodes is seldom sufficient to break up the photogenerated excitons. The photoactive region is often very thin and since both positive and negative photoexcited charges travel through the same material, recombination losses are generally high. Often the electrons recombine with the holes rather than reach the electrode. To deal with this problem, the multilayer organic photovoltaic cells were developed.

6.2 Blend structure

In this type of photovoltaic cell, the electron donor and acceptor are mixed together, forming a polymer blend as shown in Figure 6. The strong point of this type is the large interface area.

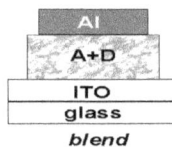

blend

Figure 6: *Blend OSC structure*

6.3 Bilayer structure

As shown in Figure 7 the active region of an organic device consists of two materials, one which acts as an electron donor and the other as an acceptor. These two layers of materials have differences in electron affinity and ionization energy, therefore electrostatic forces are generated at the interface between the two layers. The materials are chosen properly to make the differences large enough, so these local electric fields are strong, which may break up the excitons much more efficiently than the single layer photovoltaic cells do. The layer with higher electron affinity and ionization potential is the electron acceptor, and the other layer is the electron donor. This structure is also called a planar donor-acceptor heterojunction.

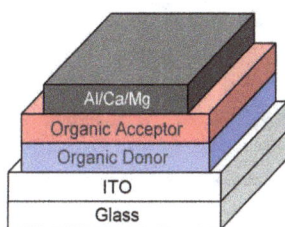

Figure 7: *Bilayer structure of OSC.*

This structure benefits from the separated charge transport layers that ensure connectivity with the correct electrode and give the separated charge carriers only little chance to recombine with its counterpart. The diffusion length of excitons in organic electronic materials is typically on the order of 10 nm. In order for most excitons to diffuse to the interface of layers and break up into carriers, the layer thickness should also be in the same range as the diffusion length. However, typically a polymer layer needs a thickness of at least 100 nm to absorb enough light. At such a large thickness, only a small fraction of the excitons can reach the heterojunction interface.

6.4 Laminated structure

In this type of photovoltaic cell shown in Figure 8(a), the electron donor and acceptor are mixed together, but in such as way that the gradient is gradual and is sandwiched between donor and acceptor layer. This architecture combines the short electron travel distance in the dispersed heterojunction with the advantage of the charge gradient of the bilayer technology. The drawback is that certain mechanical properties of the organic semiconductors are required (low glass transition temperature) to form the intermixed layer.

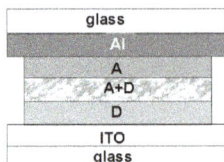

Figure 8(a): Laminated structure of OSC.

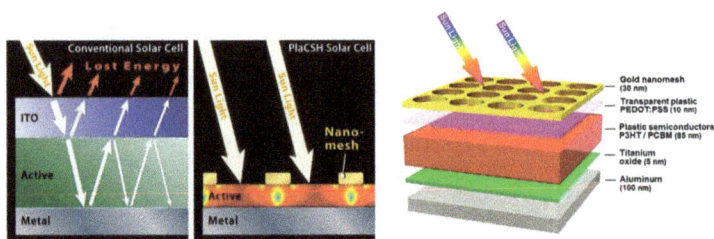

Figure 8(b): Nanostructured Sandwich Organic Solar Cells.

A research group of Princeton University has reported a new device structure addressing the two main causes limiting efficiency – first: light being reflected by the cell surface and second: the lack of an ability to fully capture the light that does enter the cell. The new device shown in Figure 8(b), is essentially a 'sandwich' of nanostructured metal and plastic that is able to trap light, increasing solar cell efficiency by 175 percent. The sandwich — called a subwavelength plasmonic cavity — has an extraordinary ability to dampen reflection and trap light. The new technique will create a solar cell that only reflects about 4 percent of light and absorbs as much as 96 percent. It demonstrates 52 percent higher efficiency in converting light to electrical energy than a conventional solar cell. The nanostructured metal film is also promising for silicon solar panels that now dominate the market. Because the PlaCSH sandwich captures light independent of what electricity-generating material is used as the middle layer, it should boost efficiency of silicon panels as well. It also can reduce the thickness of the silicon used in traditional silicon solar panels by a thousand-fold, which could substantially decrease manufacturing costs and allow the panels to become more flexible. The device efficiency increases more for indirect light, as occurs on cloudy days. By "capturing angled rays, the new structure boosts efficiency by an additional 81 percent, leading to the 175 percent total increase. The device works like a 'black hole' for light, completely trapping it.

Organic solar cells, which use a polymer-based photoactive layer that is normally placed on top of an inorganic substrate layer, have achieved efficiencies of more than 10% in the lab. Recently the researchers from Georgia Tech and Purdue University have designed a new solar cell structure using an organic, cellulose substrate layer to obtain 2.7% efficiency. Although significantly less efficient, the cellulose substrate has several potential benefits, including lower cost, minimal use of toxic chemicals and recyclability. The recent research suggests that if an organic solar cell can reach 5% efficiency and a lifetime of 5 years, it would be commercially viable. The researchers created the organic substrate by breaking down softwood pulp in sulfuric acid, then cleaning, drying, and purifying the material until they had a thin, transparent layer of cellulose nanocrystals. Once the cell is no longer usable, both the substrate and photoactive layers can be dissolved away, recycled, or reused.

7. Parameters

Open circuit voltage, short circuit current, fill factor and efficiency are four important parameters of OSC. V-I characteristic showing parameters is shown in Figure 9.

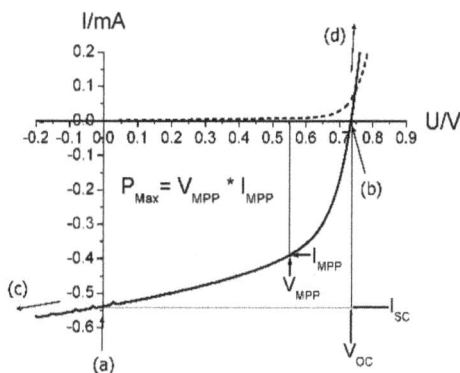

Figure 9: *Current-voltage (IV) curves of an organic solar cell (dark- dashed; illuminated - full line).*

7.1 Open circuit voltage

When the cell is operated at open circuit, $I = 0$ and the voltage across the output terminals is defined as the *open-circuit voltage* V_{OC}.

7.2 Short circuit current

When the cell is operated at short circuit, $V = 0$ and the current I through the terminals is defined as the *short-circuit current* I_{SC}.

7.3 Fill factor

The fill factor is defined as the ratio of the actual maximum obtainable power to the product of the open circuit voltage and short circuit current.

$$FF = V_{max} \, I_{max} / V_{OC} \, I_{SC}$$

7.4 Efficiency

$$\eta = P_{OUT} / P_{IN} = V_{OC} * I_{SC} * FF / P_{IN}$$

8. Equivalent circuit

Equivalent circuit is shown in Figure 10. The solar cell can be seen as a current generator which generates the current (density) J_{sc}. The dark current flows in the opposite direction and is caused by a potential between the + and - terminals. In addition, you would have two resistances; one in series (R_S) and one in parallel (R_{SH}). The series resistance is caused by the fact that a solar cell is not a perfect conductor. The parallel resistance is caused by leakage of current from one terminal to the other due to poor insulation, for example on the edges of the cell. In an ideal solar cell, you would have $R_s = 0$ and $R_{SH} = \infty$.

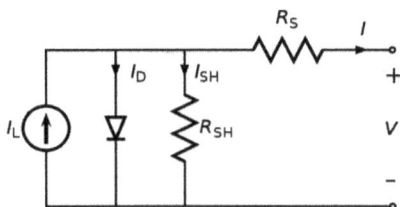

Figure 10: *Equivalent circuit.*

9.   Organic photovoltaics in a nutshell

Shrotriya et al (2006) [14] has reported about the methods that can accurately measure the current–voltage characteristics of organic solar cells under standard reporting conditions. To calculate spectral-mismatch factors for different test-cell/reference-cell combinations, four types of organic test cells and two types of silicon reference cells are used.

Belhocine-Nemmar et al (2010) [15] has reported the influence of temperature on the different parameters of organic solar cells. The short circuit current I_{sc} increases so monotonous with temperature and then saturates to a maximum value before decreasing at high temperatures. The open circuit voltage V_{oc} decreases linearly with temperature. The fill factor FF and efficiency, which are directly related with I_{sc} and V_{oc}, follow the variations of the later. The phenomena are explained by the behavior of the mobility which is a temperature activated process.

Yuan et al (2011) [16] has reported that tandem structures can boost the efficiency of organic solar cell to more than 15%, compared to the 10% limit of single layer bulk heterojunction devices. This article shows the main experimental progresses of tandem organic solar cells, and focus on the intermediate layers (charge recombination layers) in both thermal evaporated and solution processed organic tandem solar cell devices.

Monestier et al (2006) [17] has reported that the short-circuit current density of organic solar cells is based on the thickness of poly (3-hexylthiophene)(P3HT)/6,6-phenyl C61-butyric acid methyl ester (PCBM) blend. This paper shows that even if short-circuit current densities increase with higher blend thicknesses, power conversion efficiency is usually limited by low fill factor.

Boudia and Cheknane (2010) [18] has reported the results on the electrical and optical modeling of a mono or multi-layer organic photovoltaic device, in which the incident light of sun is absorbed in the active layer. The influence of the optical parameters and thicknesses of different layers had been taken into account to improve the device performance. A composite of poly (2-methoxy-5- (39,79-dim ethyloctyloxy)-1,4-phenylene-vinylene) (MDMO-PPV) and (6,6)-phenyl-C61-butyric acid methylester (PCBM) is used as photo-active material, sandwiched between a transparent Indium Tin Oxide (ITO)- electrode and an Al backside contact. The essential result of this work is the minimization of losses in the generation rate of charge carriers for an organic solar cell, produced by the insertion of a protective layer.

Chao Zhu et al (2010) [19] has reported the effect of the cathode work function, the absorption coefficient, the carrier mobility, the temperature and the thickness of the organic active layer on the short-circuit current density of single layer organic solar cells with Schottkey contacts.

Xin Yan et al (2011) [20] has reported that modeling (optical and electrical modeling) is a powerful tool to explore the device physics and speeds up research process by predicting the optimum conditions, e.g., structures, mechanisms and parameters, for improving device performances. Optical modeling is used to calculate the distribution of absorbed light energy within the device, and usually used to obtain the optimum thickness of film layers used in the device. Electrical modeling is used to provide key insights into the internal mechanism, which derives the influence of factors limiting the electrical transport processes. Numerical modeling can describe the kinetic process in OSCs intuitively.

Cheknane et al (2007) [21] has reported a study of the current–voltage characteristics of organic solar cell based on a composite of poly (2-methoxy-5-(2"- ethylhexyloxy)-1, 4-phenylenevinylene (MEH-PPV) with [6, 6]-phenyl C60 butyric acid methyl ester (PCBM) as a function of illumination intensity. The variation of the photocurrent, the fill factor, the open-circuit voltage, the saturation current and the maximum output power under different light intensity values (100 mW/cm2, 60 mW/cm2 and 24 mW/cm2), were investigated experimentally.

Barker et al (2003) [22] has reported a numerical model to predict the current-voltage curves of bilayer conjugated polymerphotovoltaic devices. The model accounts for charge photogeneration, injection, drift, diffusion, and recombination and includes the effect of space charge on the electric field within the device. They find that the short-circuit quantum efficiency is determined by the competition between polaron pair dissociation and recombination. The model shows a logarithmic dependence of the open-circuit voltage on the incident intensity, as seen experimentally. This additional intensity-dependent voltage arises from the field required to produce a drift current that balances the current due to diffusion of carriers away from the interface.

Cheknane et al (2007) [23] has reported the characteristics of organic bulk heterojunction solar cells. A composite of poly(2-methoxy-5-(39,79-dimethyloctyloxy)-1,4-phenylenevinylene)(MDMO-PPV) and (6,6)-phenyl-C61- butyric acid methyl ester (PCBM) is used as photo-activematerial, sandwiched between a transparent Indium Tin Oxide (ITO)-electrode and an Al back side contact. This study aims to show the electrical effect of an extra interfacial layer of poly (3, 4ethylenedioxythiophene) (poly(styrenesulfonate) (PEDOT/PSS) on top of the ITO-electrode. Current voltage characteristics of devices with and without this additional interfacial layer are compared and modeled by an equivalent electrical diagram. The simulation results clearly demonstrated that the current-voltage characteristics of the bulk heterojunction solar cell are affected by the presence of the PEDOT/PSS layer.

Thompson and Frechet (2008) [24] has reported that the use of a TiOx layer inserted between aP3HT/PCBM composite layer and the Al electrode has been used as an optical spacer to increase the absorption of light in the active layer and has been shown to give considerable enhancement in the photocurrent generated across the visible spectrum (maximum EQE value ca. 90%).

Hwang et al (2009) [25] has reported a time-dependent drift-diffusion model incorporating electron trapping and field-dependent charge separation to explore the device physics of organic bulk-heterojunction solar cells based on blends of poly(3-hexylthiophene) (P3HT) with a red polyfluorene copolymer. The model is used to reproduce experimental photocurrent transients measured in response to a step-function excitation of light of varied intensity.

Vervisch et al (2011) [26] has reported Organic Solar Cells (OSCs) simulation using finite element method. Optical modeling is performed via Finite Difference Time Domain method whereas the continuity and Poisson"s equations are solved to obtain electrical characteristics of the OSC. In this work, simulation results point out the influence of physical parameters such as the exciton diffusion coefficient or the exciton lifetime on OSC performances. They demonstrated the influence of the excitonic parameters on the J_{sc} values. No influence was observed on the V_{oc} which is normal as the V_{oc} is mainly influenced by the energy levels of materials. The comparison of modeling results and experimental measurement allows the exciton recombination, dissociation rate and lifetime to be determined.

Chuan et al (2011) [27] has described the characteristics of organic solar cell. Several characteristics impact on the conversion efficiency of organic solar cells, including the series resistance (R_s), the shunt resistance (R_{sh}) and so on. A circuit model of the solar cell provides a quantitative estimate for losses in the solar cell to interpret the characteristics of the solar cell. The conventional circuit model, which is suitable for inorganic solar cells, has emerged. However, compared with inorganic solar cells, organic solar cells lack a three-dimensional crystal lattice, different intramolecular and intermolecular interactions, local structural disorders, amorphous and crystalline regions, and chemical impurities. Therefore, an appropriate circuit model is represented in the paper.

Ray et al (2011) [28] has reported to achieve high efficiency/reliability for OPV, a systematic theoretical approach is required to optimize the underlying device fabrication process. They use an anneal-time dependent process-device co-simulation framework (the phase-field model for phase separation coupled with the self-consistent drift-diffusion transport for free carriers) to explore the effects of the process conditions (e.g., annealing temperature, anneal duration) on the performance of organic solar cells. They have developed a conceptual and computational framework to connect process conditions to the ultimate device performance for organic solar cells.

Sievers et al (2006) [29] has reported the device characteristics of polymer based bulk-heterojunction photovoltaic cells incorporating Poly[2-methoxy-5-(2'-ethyl-hexyloxy)-1,4-phenylene vinylene] and methanofullerene ([6,6]-phenyl C61-butyric acid methyl ester) as the active materials and examined as a function of active layer thickness. The effects of polymer layer thickness on device operation parameters such as short circuit current, open-circuit voltage, fill factor, and series resistivity are measured. Considering the variation of above mentioned parameters, an optimized power conversion efficiency, as high as 1.8% under simulated air mass 1.5 global conditions, was achieved for a device with a polymer layer thickness of 55 nm.

Galagan et al (2011) [30] has reported the presence of a transparent conductive electrode such as indium tin oxide (ITO) limits the reliability and cost price of organic photovoltaic devices as it is brittle and expensive.

Poh et al (2011) [31] has reported that the optical enhancement is demonstrated in a bilayer P3HT-C60 solar cell by embedding gold nanoparticles directly into the P3HT layer of the photovoltaic device. FDTD simulations are used to model the observed performance gain.

10. Future aspects

By studying the electrical and photoelectrical properties of organic molecules it is possible to influence V_{OC}, I_{SC}, FF and efficiency. To enhance cell performance and stability we can use interfacial layers between the active layer and electrode.

- Improving the nanoscale morphology together with the development of novel low bandgap materials is expected to lead to power conversion efficiencies approaching 10%. By introducing low band gap materials the solar photon harvesting can be increased which can increase the efficiency.

- Tandem structures can boost the efficiency of OSC to more than 15%, compared to the 10% limit of single layer bulk heterojunction devices.

- Improvements can be made by using nanotubes if the nanotubes are well dispersed in the medium and preferably oriented for straightforward charge transport capability to the electrodes. Due to their abundance and stability, CNT can improve crystallinity and ordering of the BHJ active layer if processed properly towards enhancing cell performance for viability of organic photovoltaic technology.

- Light trapping mechanisms on thin active layers of polymeric materials with simple patterning technique can improve efficiency.

- Organic materials are very sensitive to oxygen and moisture hence OSC degrade very fast. Water is the primary species to degrade the solar cell. So the encapsulation of the cell should be such that the oxygen and moisture should not come in contact directly with organic material. The research in the encapsulation of solar cell and the actual reason of degradation needs to be studied in detail. An ideal solution is to search for stable materials that are less sensitive to oxygen and moisture.

- We know that there are so many different fabrication techniques are available. The techniques that have been suited for fabrication in small sunstrates are spin coating, doctor blading and casting. The problem is that in terms of industrial production these processes are not cost effective. So, we should work to find a way to improve both the economic and technical aspects of energy production. Roll to roll fabrication, which is low cost and doesn't compromise efficiency, may become a possible way of industrial production of organic solar cell.

11. Conclusion

Organic solar cells offer unique opportunities in future due to low-cost high volume distributed production, and they are environmentally benign devices. Organic solar cells would create renewable energy, rely less on fossil fuels, be recyclable, and be cheaper than current cells. The next generation of microelectronics is aiming for applications of "electronics everywhere," and such organic semiconductors will play a major role in these future technologies. Combinations of organic solar cells with batteries, fuel cells, and so forth, will enhance their product integration. This integrability of organic solar cells into many products will be their technological advantage. If cheap, efficient renewable solar cells sound too good to be true, bear in mind that the technology is still being developed and is still years away from being commercialized.

References

[1]   Becquerel AE. Mémoire sur les effets électriques produits sous l'influence des rayons solaires. Comt. Rend. Acad. Sci. 9, 1839:561-567 .

[2]   Kallmann H, Pope M. *Photovoltaic effect in organic crystals*. J Chem Phys. 1959; 30: 585-586. https://doi.org/10.1063/1.1729992

[3]   Tang CW & Albrecht AC. *Chlorophyll-a photovoltaic cells*. *Nature*. 1975;254: 507 - 509. https://doi.org/10.1038/254507a0

[4]   Tang CW. *Two-layer organic photovoltaic cell*. Applied Physics Letters, 1986;48:183-185. https://doi.org/10.1063/1.96937

[5]   Sariciftci NS, Smilowitz L, Heeger AJ, Wudl F. Semiconducting polymers (as donors) and Buckminsterfullerene (as acceptor) — photoinduced electron-transfer and heterojunction devices. Synth Met 1993 Vol: 59:333-352.

[6]   Yu G, Gao J, Hummelen JC, Wudl F, Heeger AJ. Polymer photovoltaic cells: Enhanced efficiencies via a network of internal donor-acceptor heterojunction. Science, Vol.270, 1995:1789-1791. https://doi.org/10.1126/science.270.5243.1789

[7]   Schön JH, Kloc CH, Batlogg B. Efficient photovoltaic energy conversion in pentacene-based heterojunctions. Appl. Phys. Lett. 2000;77:2473. https://doi.org/10.1063/1.1318234

[8]   Schmidt Mende L. Self-Organized Discotic Liquid Crystals for High-Efficiency Organization Photovoltaics. Science 293 2001: 1119-1122. https://doi.org/10.1126/science.293.5532.1119

[9]   Li J, "Solarmer energy, Inc. breaks psychological barrier with 8.13% OPV efficiency", Solarmer Energy, Inc. 2010. Available: http://www.forbes.comlfeeds/businesswire/201 0107127 /businesswire142993163 .html.

[10]   pv-tech.org.

[11]   Nunzi JM. Organic photovoltaic materials and devices. C. R. Physique 3. 2002: 523–542. https://doi.org/10.1016/S1631-0705(02)01335-X

[12]   Ratier B, Nunzi JM, Aldissi M, Kraft TM, Buncel E. Organic solar cell materials and active layer designs-improvements with carbon nanotubes:a review. Polym Int. 2012;61:342-354. https://doi.org/10.1002/pi.3233

[13]   Wohrle D and Meissner D. Organic Solar Cells. *Adv.* Mater. 3.1991; No. 3:129-138.

[14]   Shrotriya V, Li G, Yao Y, Moriarty T, Emery K and Yang Y. Accurate Measurement and Characterization of Organic Solar Cells . Adv. Funct. Mater. 2006;16 : 2016–2023. https://doi.org/10.1002/adfm.200600489

[15]   Belhocine-Nemmar F, Belkaid MS, Hatem D and Boughias O. Temperature Effect on the Organic Solar Cells Parameters. World Academy of Science, Engineering and Technology. 2010; 64: 132-134.

[16]   Yuan Y, Huangand J, Li G. Intermediate Layers in Tandem Organic Solar Cells . Green. 2011;1: 65–80. https://doi.org/10.1515/green.2011.009

[17]   Monestier F, Simon JJ, Torchio P, Escoubas L, Flory F, Bailly S et al. Modeling the short-circuit current density of polymer solar cells based on P3HT:PCBM blend. Sol. Energy Mater. Sol. Cells. 2006.

[18]   Boudia MRM , Cheknane A . A modelling and simulation approach of electromagnetic field in organic photovoltaic devices. World Journal of Modelling and Simulation. 2010; 6(3): 198-204.

[19]   Chao Zhu MA, Min MW, Quan PY, Sheng WR, Hua LR, Wei XH et al. Numerical study on short-circuit current of single layer organic solar cells with Schottkey contacts. Sci China Tech Sci. 2010;53(4):1023−1027. https://doi.org/10.1007/s11431-009-0403-y

[20]   Xin Yan Z, BaoXiu M, Qiang GZ, Wei H. Recent progress in the numerical modeling for organic thin film solar cells. Sci China Phys Mech Astron. 2011;54(3): 375-387. https://doi.org/10.1007/s11433-011-4248-6

[21]   Cheknane A, Benyoucef B, Chaker A. The effects of parasitic resistances on organic solar cell's. *Sciences &* Technologie A – N°25. 2007: 63-66.

[22]   Barker JA, Ramsdale CM, and Greenham NC. Modeling the current-voltage characteristics of bilayer polymer photovoltaic devices. Physical Review B 67. 2003: 075205. https://doi.org/10.1103/PhysRevB.67.075205

[23] Cheknane A, Aernouts T, MeradBoudia M. Modelling and Simulation of organic bulk heterojunction solar cells. Revue des Energies Renouvelables ICRESD-07 Tlemcen. 2007: 83 – 90.

[24] Thompson BC and Freche JMJ. Polymer–Fullerene Composite Solar Cells. Angew. Chem. Int. Ed. 2008; 47: 58–77. https://doi.org/10.1002/anie.200702506

[25] Hwang I, McNeill CR, and Greenham NC. Drift-diffusion modeling of photocurrent transients in bulk heterojunction solar cells. J. Appl. Phys. 2009;106: 094506 . *https://doi.org/10.1063/1.3247547*

[26] Vervisch W, Biondo S, Rivière G, Duché D, Escoubas L, Torchio P et al. Optical-electrical simulation of organic solar cells: excitonic modeling parameter influence

on electrical characteristics. Appl. Phys. Lett.98. 2011: 253306.
https://doi.org/10.1063/1.3582926

[27] Chuan J, Tianze L, Luan H, Xia Z. Research on the Characteristics of Organic Solar
Cells. Journal of Physics: Conference Series 276. 2011: 012169.
https://doi.org/10.1088/1742-6596/276/1/012169

[28] Ray B , Nair PR, Alam MA. Annealing dependent performance of organic bulk-
heterojunction solar cells: A theoretical perspective. Solar Energy Materials &
Solar Cells 95. 2011: 3287–3294. https://doi.org/10.1016/j.solmat.2011.07.006

[29] Sievers DW, Shrotriya V, Yang Y. Modeling optical effects and thickness dependent
current in polymer bulk heterojunction solar cells. J. Appl. Phys. 100,
2006:114509. https://doi.org/10.1063/1.2388854

[30] Galagan Y, Rubingh JEJM, Andriessen R, Fan CC, Bloma PWM, Veenstra SC et
al. ITO-free flexible organic solar cells with printed current collecting grids. Solar
Energy Materials & Solar Cells 95 .2011: 1339–1343.

[31] Poh CH, Rosa L, Juodkazis S, Dastoor P. FDTD modeling to enhance the
performance of an organic solar cell embedded with gold nanoparticles. Optical
materials express. 2011;1(7):1326. https://doi.org/10.1364/OME.1.001326

Chapter 9

Efficiency rise in PCDTBT:PC$_{70}$BM organic solar cell using interface additive

Rashmi Swami*, Rajesh Awasthi, Sanjay Tiwari

Photonics Research Laboratory, SOS in Electronics , Pt. Ravishankar Shukla University, Raipur (C.G.) 492010, India,

Email: rashmi.swami3@gmail.com

Abstract

Low efficiency is one of the biggest problems with organic solar cell. In order to increase the efficiency of bulk hetero-junction organic solar cell we are using interface surfactant additive poly(oxyethylene tridecyl ether) (PTE) with blend photoactive layer. Here we are reporting on the enhanced photovoltaic (PV) effects by means of a polymer bulk-hetero-junction (BHJ) layer having PCDTBT as a low-band gap e' donor/HTL polymer and PC$_{70}$BM as an acceptor/ETL, doped with poly(oxyethylene tridecyl ether) (PTE) which is an interface surfactant additive. For PCDTBT:PC$_{70}$BM OSC , we recorded 0.886 V open-circuit voltage (V$_{OC}$), 11.7 mA/cm^2 short-circuit current density (J$_{SC}$), 47.3% fill factor (FF$)$ and PCE of 4.9%. For PCDTBT:PCBM70:PTE organic solar cell, we recorded V$_{OC}$ of 0.904 V, higher values of J$_{SC}$ of 13.8 mA/cm^2, FF of 48.2% and improved PCE of 6.0% for a PTE concentration of $ca.$ 0.164 wt%. Power conversion efficiency (PCE) reaches to 6.0%, by the addition of PTE to a PCDTBT:PC$_{70}$BM system which is much higher than a reference device not including the additive (4.9%). Increase in efficiency is because of the increase in lifetime of charge carrier, which is due to the existence of PTE molecules at the interfaces sandwiched between the BHJ photovoltaic active layer and the anode and cathode, in addition to the phase-separated BHJ domains interfaces.

Keywords

Organic Solar Cell, PCDTBT, PCBM, PTE, IPCE, Bulk Hetero-junction

Contents

1.   Introduction...292

2.   Materials used 293

2.1 Donor Molecule..293

2.2 Acceptor molecule..293

2.3 PTE additive...294

3.   Experimental details..294

4.   Results and discussion 296

5.   Conclusions...297

Acknowledgement 297

Refrences...297

1.   Introduction

Solar cell can be designed with photoactive layer of organic and inorganic materials. The global rising demand for low-priced electricity has triggered deep research on solar cells comprising organic semiconductors. Organic solar cell (OSC) technology has received significant attention over the past decade due to low cost, low temperature and energy processing, low material requirement, can be used on flexible substrate and can be shaped to suit architectural application. The key development of organic solar cells has been made with the pioneering concept of ''bulk hetero-junction (BHJ)'' photoactive layers [1-2]. The bulk hetero-junction (BHJ) PSC [1][3] is of particular interest, due to the efficient photo-induced generation of charge in its blended photovoltaic (PV) layer, that is consisted of interpenetrating, channel-like domains of separated fullerene and polymer. Following the annealing of the BHJ structure at elevated temperatures, PSCs with PV layers of P3HT which is poly(3-hexylthiophene) and PCBM60 which is phenyl C61-butyric acid methyl ester have shown high power conversion efficiencies (PCEs) of 3-5%. Efficiency of P3HT:PCBM organic solar cell is upto 5% because of the limitations of conventional P3HT, whose bandgap lies at around 1.9 eV, which limits absorbance to wavelengths below 650 nm [4]. To improve the efficiency of PSC we need new active materials having lower bandgap to harvest more solar photons. More recently, a PCE of 5-6% was reported for a BHJ PSC that used a blend of PCBM70 and PCDTBT having a bandgap of 1.88 eV [5, 6]. Using 'processing additives', PCE of organic solar cell can be increased [7-9]. To increase carrier lifetimes (reduce recombination loss) we modify the BHJ interfaces between the phase-separated domains of the donor-conjugated polymer and the acceptor fullerene, and added a non-ionic surfactant poly(oxyethylene tridecyl ether) (PTE) as an additive to the PV layer. In this paper we report J-V characteristic and IPCE spectra of PCDTBT:PC$_{70}$BM organic solar cell with and without PTE.

2. Materials used

Organic materials used in this structure are as follows –

2.1 Donor Molecule

Next generation HTL/donor material for organic photovoltaics is Poly[[9-(1-octylnonyl)-9H-carbazole-2.7-diyl]-2.5-thiophenediyl-2.1.3 benzothiadiazole-4.7-diyl-2.5-thiophenediyl] (PCDTBT) shown in Figure (1), which can produce better efficiencies and lifetimes. The main qualities of PCDTBT are -

- lower HOMO and LUMO levels
- narrow band gap
- Increased open circuit voltage
- Longer wavelength absorption
- Lower concentration and material usage
- Improved stability under ambient conditions
- High electron and hole generation rate and high mobility of electron and hole.

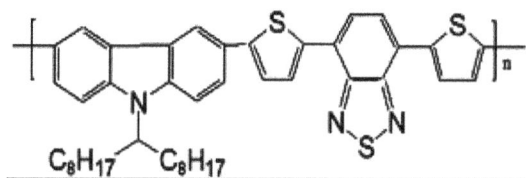

Figure 1: Molecular structure of PCDTBT

2.2 Acceptor molecule

Extremely symmetrical cage-shaped molecules of carbon atoms is Fullerenes PC$_{70}$BM which is [6,6]-phenyl C70 butyric acid methyl ester as shown in Figure (2). For the separation of photoexcited exciton into free charge carriers, blending of conjugated polymers (electron donor) with fullerenes (electron acceptors), is extremely efficient way.

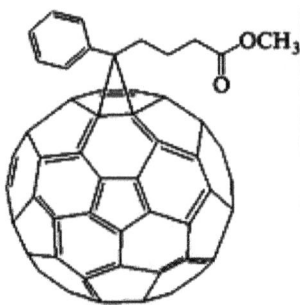

Figure 2: Molecular structure of $PC_{70}BM$

2.3   PTE additive

Poly(oxyethylene tridecyl ether) (PTE) shown in Figure (3) as an additive have low (- 8.1 eV) highest- occupied-molecular-orbital (HOMO) and high (- 2.1 eV) lowest-unoccupied-molecular-orbital (LUMO) [10–12].

PTE

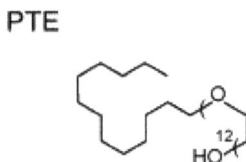

Figure 3: Molecular structure of PTE

3.   Experimental details

The sample BHJ PSCs were fabricated in a sandwich structure with an anode of indium tin oxide (ITO) and an Al:Li/Al cathode. Patterned 80-nm-thick ITO glass was cleaned by sequential ultrasonic treatment in detergent, deionized water, acetone and isopropanol, and then treated in an ultraviolet-ozone chamber for 15 min. Then, a *ca.* 40-nm-thick hole-collecting PEDOT:PSS buffer layer was spin-coated onto the ITO electrode. On the top of the PEDOT:PSS layer spin coat the blended solution of PCDTBT (0.456 wt%), PCBM70 (1.824 wt%), and PTE additive in dichlorobenzene. The PV layer was about 85 nm thick. Finally, for the cathode, a *ca.* 1-nmthick Al:Li alloy (Li: 0.1 wt%) layer and a pure Al (*ca.* 50-nm-thick) layer were created on the photovoliaic layer through thermal deposition (0.5 nm/s), at a foundation pressure below 2×10^{-4} Pa. The sample device structure studied was therefore [ITO/PEDOT:PSS/PCDTBT:PC$_{70}$BM:PTE/Al:Li/Al] as shown in Figure (4). The active area of the fabricated device was 3×3 mm^2. For comparison, a reference PSC was fabricated with the structure [ITO/PEDOT:PSS/PCDTBT:PC$_{70}$BM/Al:Li/Al] as shown in Figure (5). The performance of the PSCs was measured in 100 mW/cm^2 illumination intensity produced by an AM 1.5G light resource. With the help of a source meter (Keithley 2400) the photocurrent-versus-voltage (*J-V*) characteristics were measured. The IPCE (incident photon-to-current collection efficiency) spectra were measured for the PSCs studied using an IPCE measurement system.

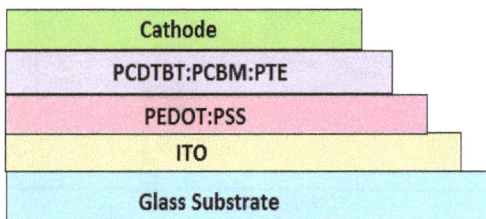

Figure 4: *ITO/PEDOT:PSS/ PCDTBT:PC$_{70}$BM:PTE /Al:Li/Al Organic Solar Cell*

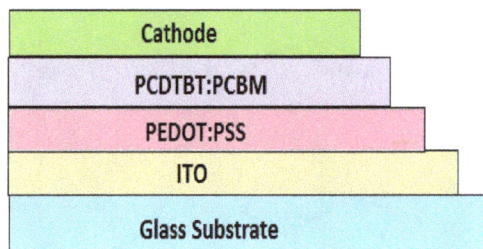

Figure 5: *ITO/PEDOT:PSS/ PCDTBT:PC$_{70}$BM /Al:Li/Al Organic Solar Cell*

4.   Results and discussion

As shown in Figure (6) for PCDTBT:PC$_{70}$BM organic solar cell, under an illumination of AM 1.5G and 100 mW/cm2, we recorded 0.886 V open-circuit voltage (V$_{OC}$), 11.7 mA/cm^2 short-circuit current density (J$_{SC}$) and 47.3% of fill factor (FF) and PCE of 4.9%; a value comparable with those reported by others [6]. For PCDTBT:PC$_{70}$BM:PTE organic solar cell, we recorded V$_{OC}$ of 0.904 V, higher values of J$_{SC}$ of 13.8 mA/cm^2, FF of 48.2% and improved PCE of 6.0% for a PTE concentration of *ca.* 0.164 wt%. These increased values resulted in an improved efficiency of 6.0%, which led to a PCE that was up to 22% higher than that of PCDTBT:PC$_{70}$BM based organic solar cell.

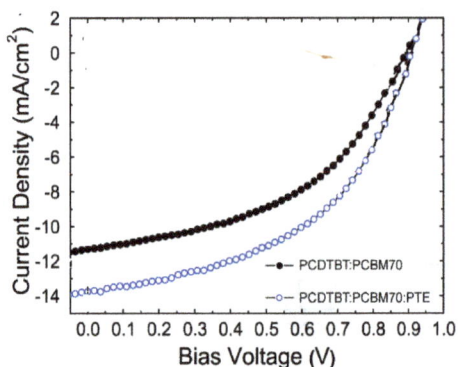

Figure 6: *The current-voltage characteristics of BHJ OSCs with and without the PTE additive*

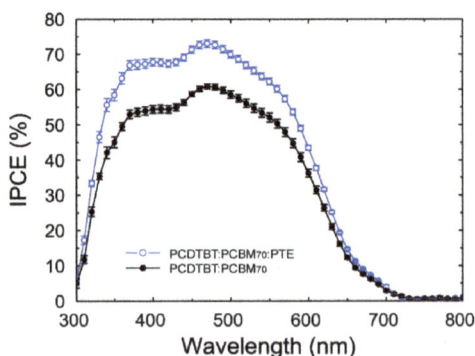

Figure 7: *IPCE spectra of PCDTBT:PC$_{70}$ BM OSCs with and without the PTE additive*

We further investigated the PV performance of the OSCs that incorporated the PTE additive by studying the IPCE spectra. Figure (7) shows the observed IPCE spectrum of the PSC devices. It can be seen that the IPCE values are consistent with the variations in J_{SC} for the OSCs with and without the PTE additive. The maximum IPCE was 73.0% at 470 nm for the sample device with the PTE additive, which corresponded to the highest J_{SC} *(13.8 mA/cm^2)*, while the IPCE value was about 60.9% for the reference device without the additive, which had the lowest J_{SC} (11.7 mA/cm^2).

5. Conclusions

In conclusion, we have reported on the use of a low-bandgap PCDTBT:PC$_{70}$BM-based PV layer that incorporates a PTE surfactant, which was used to the BHJ interfaces in OSCs. We have shown that BHJ OSCs that contain the interface PTE additive are more efficient than conventional OSCs. A high PCE (6.0%) was obtained for our PCDTBT:PC$_{70}$BM (1:4 w/w) OSC device using 0.164 wt% of the PTE additive, which yielded improvements in PCE of up to 22%. This improvement may be attributed to the increased selective flow of dissociated charge carriers, not only at the interfaces of the PV layer and the electrodes, but also at the BHJ interfaces between the PCDTBT and PC$_{70}$BM domains. Our findings show that a combination of PTE interface additives and high-performance low-band gap PV materials holds great potential for the development of a new generation of highly efficient OSCs.

Acknowledgement

I am extremely thankful to my organization Pt.Ravishankar Shukla University, Raipur (C.G.) for funding my research work.

Refrences

[1] G. Yu, J. Gao, J.C. Hummelen, F. Wudl, A.J. Heeger. Polymer Photovoltaic Cells:Enhanced Efficiencies via a Network of Internal Donor-Acceptor Heterojunctions. Science, New Series, 1995, 270(5243): 1789-1791.

[2] J.J.M. Halls, C.A. Walsh, N.C. Greenham, E.A. Marseglia, R.H. Friend, S.C. Moratti, A.B. Holmes. Efficient photodiodes from interpenetrating polymer networks. Nature, 1995, 376: 498–500. https://doi.org/10.1038/376498a0

[3] C. J. Brabec, N. S. Sariciftci, and J. C. Hummelen. Plastic solar cells. Adv. Funct. Mater. 2001, 11(1): 15–26. https://doi.org/10.1002/1616-3028(200102)11:1<15::AID-ADFM15>3.0.CO;2-A

[4]   K. M. Coakley and M. D. McGehee. Conjugated polymer photovoltaic cells. Chem. Mater., 2004, 16(23): 4533–4542. https://doi.org/10.1021/cm049654n

[5]   S. H. Park, A. Roy, S. Beaupré, S. Cho, N. Coates, J. S. Moon, D. Moses, M. Leclerc, K. Lee and A. J. Heeger. Bulk heterojunction solar cells with internal quantum efficiency approaching 100%. Nat. Photonics, 2009, 3(5): 297–302. https://doi.org/10.1038/nphoton.2009.69

[6]   J. Zhou, X. Wan, Y. Liu, F. Wang, G. Long, C. Li, and Y. Chen. Synthesis and photovoltaic properties of a poly(2,7-carbazole) derivative based on dithienosilole and benzothiadiazole. Macromol. Chem. Phys., 2011, 212(11): 1109–1114. https://doi.org/10.1002/macp.201100060

[7]   J. Peet, J. Y. Kim, N. E. Coates, W. L. Ma, D. Moses, A. J. Heeger, and G. C. Bazan. Efficiency enhancement in low-bandgap polymer solar cells by processing with alkane dithiols. Nat. Mater., 2007, 6(7): 497–500. https://doi.org/10.1038/nmat1928

[8]   G. Garcia-Belmonte and J. Bisquert. Open-circuit voltage limit caused by recombination through tail states in bulk heterojuction polymer-fullerene solar cells. Appl. Phys. Lett., 2010, 96(11): 113301. https://doi.org/10.1063/1.3358121

[9]   Y. Liang, Z. Xu, J. Xia, S.-T. Tsai, Y. Wu, G. Li, C. Ray, and L. Yu. For the bright future-bulk heterojunction polymer solar cells with power conversion efficiency of 7.4%. Adv. Mater. (Deerfield Beach Fla.), 2010, 22(20): E135–E138. https://doi.org/10.1002/adma.200903528

[10]   Y. I. Lee, M. Kim, Y. Ho Huh, J. S. Lim, S. Cheol Yoon, and B. Park. Improved photovoltaic effect of polymer solar cells with nanoscale interfacial layers. Sol. Energy Mater. Sol. Cells, 2010, 94(6): 1152–1156. https://doi.org/10.1016/j.solmat.2010.02.045

[11]   B. Park, Y. H. Huh, and M. Kim. Surfactant additives for improved photovoltaic effect of polymer solar cells. J. Mater. Chem., 2010, 20(48): 10862–10868. https://doi.org/10.1039/c0jm02091e

[12]   J. H. Park, S. S. Oh, S. W. Kim, E. H. Choi, B. H. Hong, Y. H. Seo, G. S. Cho, B. Park, J. Lim, S. C. Yoon, and C. Lee. Double interfacial layers for highly efficient organic light-emitting devices. Appl. Phys. Lett., 2007, 90(15): 153508. https://doi.org/10.1063/1.2721872

Chapter 10

Recent advances in polymer solar cells

Verma, A. K.[#], Agnihotri, P., Patel, M. , Sahu, S. and Tiwari, S.[*]

Photonics Research Laboratory, School of Studies in Electronics & Photonics Pt. Ravishankar Shukla University, Raipur(C.G.) India

[#]akverma.prsu@gmail.com [*]drsanjaytiwari@gmail.com

Abstract

Polymer solar cells belongs to promising class of next-generation photovoltaic, because they hold promise for the realization of mechanically flexible, lightweight, large-area devices that can be fabricated by room-temperature solution processing. High power conversion efficiencies of ~15% in tandem polymer solar cells based on semiconducting polymers are fabricated from solution-processing techniques and have unique prospects for achieving low-cost solar energy harvesting, owing to their material and manufacturing advantages. The potential applications of polymer solar cells are broad, ranging from flexible solar modules and semitransparent solar cells in windows, to building applications and even photon recycling in liquid-crystal displays. This review covers the scientific origins and basic properties of polymer solar cell technology, material requirements and device operation mechanisms, while also providing a synopsis of major achievements in the field over the past few years. Potential future developments and the applications of this technology are also briefly discussed.

Keywords

Polymer Solar Cells, Flexible Transparent Solar Cells

Contents

1. Introduction...300

2. Morphology in PSCS 303

3. New advances in pscs and future scopes.............................304

Conclusion...306

References 306

1. Introduction

Harnessing solar energy is one of the most promising ways to tackle today's energy issues. Although the present dominant photovoltaic (PV) technology is based on inorganic materials, high material and manufacturing costs limit its wide acceptance [1]. Intensive research has been conducted towards the development of low-cost PV technologies, of which organic photovoltaic (OPV) devices are one of the promising. OPV devices are based on organic semiconductors-carbon-based materials whose backbones are comprised mainly of alternating C–C and C=C bonds. Electron delocalization along the conjugated backbone is responsible for the semiconducting properties of OPV devices [2]. One of the major differences between organic semiconductors and inorganic semiconductors is the presence of tightly bonded excitons (electron–hole pairs) resulting from their low dielectric constant ($\varepsilon r \approx 2$–4). The binding energy of the Frenkel exciton is in the range of 0.3–1 eV [2, 3]. Such a large binding energy prevents exciton dissociation by an electrical field (a non-radiative decay channel) and can achieve a high electroluminescent efficiency in organic light-emitting devices. The weak intermolecular Van de Waals interaction enables the realization of low-cost, large-area deposition technologies such as roll-to roll printing [3]. In recent years, organic electronic devices such as organic light-emitting diodes (OLEDs), organic thin film transistors, OPVs and organic memory devices have attracted considerable attention, owing to their potential low cost and high performance characteristics. OLED displays have gained a considerable share in the portable electronics market, for use in devices such as smart phones. However, research into OPV cells continues to lag behind, despite the first patent and the first paper [4] by Tang appearing ahead of those of OLEDs, probably owing to the fact that developing alternative energy sources has been viewed, until recently, as being relatively unimportant.

OPVs are divided into two different categories according to whether their constituent molecules are either small or large (polymers). These two classes of materials are rather different in terms of their synthesis, purification and device fabrication processes. Polymer solar cells (PSCs) are processed from solution in organic solvents, whereas small-molecule solar cells are processed mainly using thermal evaporation deposition in a high-vacuum environment. Using the solution process to fabricate small-molecule solar cells has recently been gaining momentum [5], although the film quality and crystallization is expected to be an issue. PSCs are attractive owing to a number of advantageous features [6], including their thin-film architecture and low material` consumption resulting from a high absorption coefficient, their use of organic materials, which are abundant, their utilization of efficient solution processes and low manufacturing energy requirements. Other advantages include their low specific weight, mechanical flexibility, tunable material properties and high transparency. This review is primarily devoted to recent advances in PSCs.

Tang [4] introduced the donor–acceptor bilayer planar heterojunction to the OPV cell in 1979, and achieved power-conversion efficiencies (PCEs) of around 1%. The energy difference between the lowest unoccupied molecular orbital (LUMO) of the donor and highest occupied molecular orbital (HOMO) of the acceptor provides the driving force for the dissociation of Frenkel excitons. The separated holes and electrons are then collected at the anode and cathode, respectively. One of the major breakthroughs in OPV technology was the adoption of C60 fullerene and its derivatives (such as [6, 6]-phenyl-C61-butyric acid methyl ester, PCBM) to replace the n-type molecules in OPV devices. Owing to their strong electro negativity and high electron mobility, C60 derivatives have become standard n-type molecules in OPV devices. In the early 1990s, Heeger et al. and Yoshino et al. independently demonstrated electron transfer between a conjugated polymer and fullerene derivatives [7, 8]. They observed an extremely fast photo-induced electron transfer process of around 50–100 fs, which dominates over all other photo-physical processes present. These discoveries provided a solid foundation for OPV technology. In 1993, researchers made the first demonstrations of planar heterojunction PSCs [9]. The planar junction concept has certain limitations, including a small surface area between the donor–acceptor interfaces and the requirement of long carrier lifetime to ensure that the electrons and holes reach their respective electrodes. This problem can be addressed by introducing a bulk heterojunction, which involves mixing donor–acceptor materials in the bulk body of an OPV device. This concept was first demonstrated by Hiramoto et al. through the co-evaporation of donor and acceptor molecules under high-vacuum conditions [10]. The first efficient bulk heterojunction PSCs were independently realized in 1995 by the groups of Heeger and Friend in polymer–fullerene and polymer–polymer blends [11]. Polymer–fullerene systems currently dominate the field of high-efficiency PSCs. PSC efficiencies are now approaching 10%, which indicates remarkable progress towards a promising future [12].

Although the bulk heterojunction concept is powerful as a solution for addressing the issue of exciton disassociation, in 2005 researchers discovered that the morphology (the donor– acceptor phase separation) also plays a critical role in achieving proper charge transport channels for collecting the electrons and holes[13,14].

A typical PV process involves the creation of free carriers from incident photons. The physics and the energy diagram of polymer–fullerene-based PSCs are illustrated in Figure 1(a). The external quantum efficiency (EQE) as a function of wavelength (λ) is the ratio between the collected photogenerated charges and the number of incident photons, ultimately being the product of four efficiencies (η): absorption (A), exciton diffusion

(ED), charge separation (CS) and charge collection (CC), giving EQE(λ) = $\eta A(\lambda) \times \eta ED(\lambda) \times \eta CS(\lambda) \times \eta CC(\lambda)$. The photovoltage (or open-circuit voltage, V_{OC}) is directly linked to the energy difference between the LUMO level of the acceptor and the HOMO level of the donor, thereby providing the primary driving force for charge separation. Figure 1(b) compares the solar spectrum to the EQE spectrum of the representative PSCs (bandgap of 1.9 eV). The short-circuit current density (J_{SC}) is equal to the integral of the product between cell responsivity and incident solar spectral irradiance.

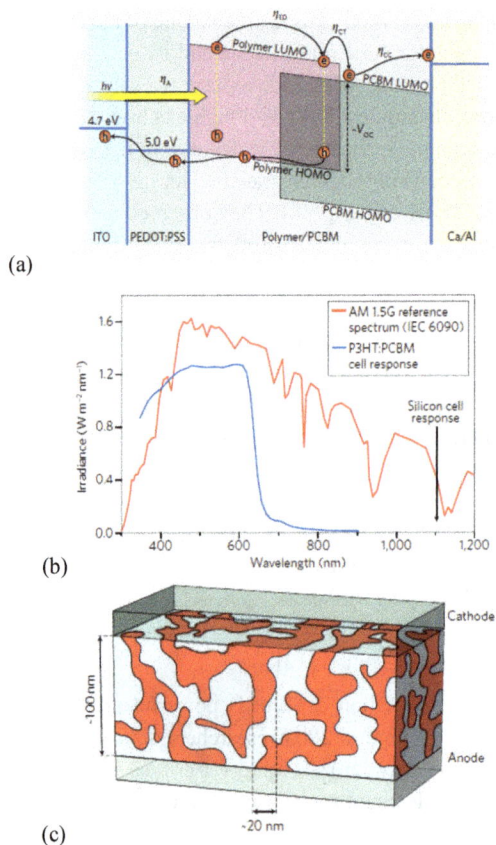

(a)

(b)

(c)

Figure 1: *Introduction to PSCs (a) The operating mechanism of a PSC. (b) Comparison between solar spectrum and the photo-response of a P3HT: PCBM solar cell. (c) Conceptual morphology model with bi-continuous interpenetration network of the polymer and the acceptor [15].*

Thus, it is necessary to utilize a broader solar spectrum and enlarge the energy level difference between the LUMO level of the donor and the HOMO level of the acceptor, resulting in high values of J_{SC} and V_{OC}, respectively. Innovations in materials science have provided efficient ways of achieving these goals. Morphology is another critical factor in bulk heterojunction PSCs. The preferred morphology of bulk heterojunction is a bicontinuous interpenetration network (Figure 1(c)) [18]. Donor and acceptor domains should be twice the size of the exciton diffusion length (around 10 nm), which allows excitons to diffuse to the donor–acceptor interface and thus achieves efficient $\eta ED(\lambda)$ and $\eta CS(\lambda)$ for charge generation. After charge separation at the donor–acceptor interface, holes and electrons must travel to the positive and negative electrodes through donor and acceptor networks, respectively. The third key factor is the organic–electrode interface, where the charges are extracted to external circuits. The charge collection efficiency $\eta CC(\lambda)$ accounts for both carrier transport in the networks and the extraction steps.

2. Morphology in PSCS

Morphology control is critical in bulk-heterojunction PSCs. Thermal annealing [16] and solvent annealing [13] are currently the most popular methods for controlling morphology. It was not until 2005 that both thermal and solvent annealing was shown to enhance PSC efficiency by a significant amount [15, 16]. Many other approaches are also effective for improving polymer–fullerene morphology, such as solvent selection [17] and solvent mixture [18] techniques, and the use of additives [19].

The characterization of polymer morphology involves a variety of different technologies. Microscopic techniques provide a direct view of polymer morphology. Atomic force microscopy in tapping mode is suitable for soft PSC films and can provide high resolution surface topographical and surface donor–acceptor distribution data on the nanoscale. Figure 2 shows a phase image of a high-crystallinity P3HT: PCBM film [20]. The polymer nanofibrillar structure is consistent with that of a pure P3HT film. The nanofibrillar width of P3HT is 20–30 nm, which is consistent with the morphology model. Another powerful imaging technique is transmission electron microscopy (TEM). Yang et al. studied a P3HT: PCBM film using bright-field TEM [21] (Figure. 2(b)). The specific density difference of P3HT and PCBM enables the polymer- and fullerene-rich regions to be mapped, and thus provides information on the dimensions of the P3HT nanostructure. Cross-section TEM provides another critical piece of morphological information. Heeger et al. observed bi-continuous interpenetrating polymer and fullerene domains in cross-sectional TEM images of a P3HT: PCBM solar cell (Figure 2) [22]. The combination of surface topological and cross-sectional imaging tools provides important information regarding the morphology of quasi-optimized P3HT: PCBM solar cells.

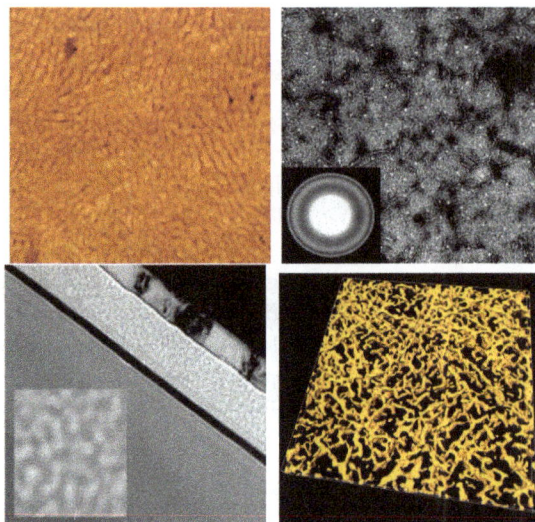

Figure 2 : *Morphology in PSCs. a, Tapping mode atomic force microscopy image of a solvent-annealed P3HT: PCBM film [20], TEM image of a thermally annealed P3HT:PCBM film[21], Defocused cross-sectional TEM image of a P3HT:PCBM film[22]. Three-dimensional electron tomography image of thermally annealed P3HT–PCBM film [23].*

3. New advances in pscs and future scopes

The maximum power conversion efficiencies of PSCs must rise above 15% in the laboratory (corresponding to a module efficiency of around 10–12%) before they can become practically useful. The tandem solar cell concept [24], which combines two or more sub-cells with different absorption ranges, is obviously a very attractive way to improve power efficiencies because it can significantly enhance photon utilization efficiencies and thus preserve V_{OC}. The key components include high-efficiency sub-cells with maximized V_{OC}, sub-cell spectra matching, and robust and transparent solution-processing interconnection layers. Over the past two years, device innovation, in combination with the advancement of photoactive materials and improvements to the interface layer, have provided a solid foundation for all solution-processed tandem PSCs. Several groups [25-27] have demonstrated the feasibility of this concept, for which the current state-of-the-art cell efficiency is 8.6%.

The unique properties of PSCs open the door for many novel applications. Stretchable PSCs enable the concept of conformal photovoltaics on curved or wrinkled surfaces, such as textiles and fabrics [28]. PSCs can also be integrated into liquid crystal displays (LCDs) to recycle wasted photons by creating a PV polarizer [29]. In this work, the researchers oriented the conjugated polymer chain by employing a simple rubbing process, which caused the material to polarize incoming light. This PV LCD panel can harvest energy from sunlight, ambient light or the backlight of LCDs. Guo et al. demonstrated a PV colour filter concept for integrated PV-LCD panels [30].

Stability is one of the major hurdles that must be tackled before PSCs can enter the market. Brabec et al. investigated the requirements that PSC technology must meet for it to become competitive [31]. They claim that a module efficiency of 7% and a lifetime of seven years is the threshold for roll-to-roll-processed PSCs. Many research groups have shown that inverted PSCs have a much longer lifetime than traditional PSCs. Mc Gehee et al. recently showcased a PCDTBT-based PSC rigid device with a lifetime of seven years [32]. In industry, Konarka's first-generation flexible PSC panel has a three-year lifetime with flexible encapsulation. All of these technological advances are encouraging [33-35]. Progress in the OLED industry has shown that realizing a long-lifetime PSC, although challenging, is not impossible. Rapid progress in the development of PSCs in recent years provides significant confidence in this promising technology [36-40].

In Table-1, we provide a summary of the best results obtained thus far in the field of organic polymer solar cells based on both small molecules and polymers. It is apparent that power conversion efficiencies are approximately 15 %, a value which is at the lower limit of interest in crystalline and thin film Si, respectively.

TABLE-1 Summary of the some best Polymer Solar cells results:

Sr. No.	Year	Device Structure	J_{sc} (mA/cm^2)	V_{oc} (V)	FF (%)	PCE (%)	Ref.
1.	1986	Bilayer	-	-	65	1.0	4
2.	2002	LiF/Au	-	0.763	54	2.3	32
3.	2003	PCBM:MDMO-PPV	7.6	0.77	51	3.0	33
4.	2005	BHJ of P3HT/PCBM	10.6	0.61	67.4	4.37	15
5.	2009	PCDTBT/PC70BM	10.6	0.88	66	6.1	34
6.	2011	PBnDT-FTAZ	11.83	0.79	72.9	7.10	35
7.	2013	PBT3IT	12.9	0.859	77.8	8.66	36
8.	2014	PBDTT-SeDPP:PC71BM	20.6	0.94	62.0	12.0	37
9.	2015	Dopant-free DOR3T-TBDT	20.7	0.97	74	14.9	38
10.	2015	PC71BM	19.4	0.708	73.4	10.1	39
11.	2016	DERDTS–TBDT	21.2	1.05	72.8	16.2	40

Conclusion

Polymer solar cells show a good candidate in the advancement of low cost photovoltaic alternatives. A controlled and organized research is necessary in the field to effectively utilize the foreseen advantages of polymer solar cells.

The review on basic physics, charge transfer dynamics, mechanism and recent approaches and advancement in the field are carried out. It is very important to understand the basic mechanism of operation in conversion of light into electricity. Since all the design, architecture and materials developments revolve around our level of understanding of the basic mechanism, it accounts chief importance in all researches going on in the field of polymer solar cells.

References

[1]   www.eia.gov/aer

[2]   Pope, Martin, and Charles E. Swenberg "Electronic processes in organic crystals and polymers." Oxford University Press on Demand, (1999).

[3]   Forrest, Stephen R. "The path to ubiquitous and low-cost organic electronic appliances on plastic." Nature 428.6986 (2004): 911-918.4. Tang, C. W. Multilayer organic photovoltaic elements. US patent 4, 164,431 (1979).

[4]   Tang, Ching W. "Two-layer organic photovoltaic cell' Applied Physics Letters 48.2 (1986): 183-185. https://doi.org/10.1063/1.96937

[5]   Liu, Y. S. et al. "Spin-coated small molecules for high performance solar cells." Adv. Energy Mater. 1, 771–775 (2011). https://doi.org/10.1002/aenm.201100230

[6]   Dennler, G., Scharber, M. C. & Brabec, "C. J. Polymer-fullerene bulk hetero junction solar cells." Adv. Mater. 21, 1323–1338 (2009). https://doi.org/10.1002/adma.200801283

[7]   Sariciftci, N. S., Smilowitz, L., Heeger, A. J. &Wudl, F. "Photoinduced electron-transfer from a conducting polymer to Buckminsterfullerene." Science 258, 1474–1476 (1992). https://doi.org/10.1126/science.258.5087.1474

[8]   Morita, S., Zakhidov, A. A. & Yoshino, K. "Doping effect of Buckminsterfullerene in conducting polymer change of absorption spectrum and quenching of luminescence." Solid State Commun. 82, 249–252 (1992). https://doi.org/10.1016/0038-1098(92)90636-N

[9]   Sariciftci, N. S., Smilowitz, L., Heeger, A. J. & Wudl, F. "Semiconducting polymers (as donors) and Buckminsterfullerene (as acceptor) photo induced

electron-transfer and heterojunction devices." Synth. Met. 59, 333–352 (1993). https://doi.org/10.1016/0379-6779(93)91166-Y

[10] Hiramoto, M., Fujiwara, H. & Yokoyama, M. "P‑I‑N like behavior in 3-layered organic solar-cells having a co-deposited interlayer of pigments." J. Appl. Phys. 72, 3781–3787 (1992). https://doi.org/10.1063/1.352274

[11] Yu, G., Gao, J., Hummelen, J. C., Wudl, F. & Heeger, A. J. "Polymer photovoltaic cells enhanced efficiencies via a network of internal donor– acceptor heterojunction." Science 270, 1789–1791 (1995). https://doi.org/10.1126/science.270.5243.1789

[12] Li, G. et al. "High-efficiency solution processable polymer photovoltaic cells by self-organization of polymer blends." Nature Mater. 4, 864–868 (2005). https://doi.org/10.1038/nmat1500

[13] Ma, W. L., Yang, C. Y., Gong, X., Lee, K. & Heeger, A. J. "Thermally stable, efficient polymer solar cells with nanoscale control of the interpenetrating network morphology." Adv. Funct. Mater. 15, 1617–1622 (2005). https://doi.org/10.1002/adfm.200500211

[14] Li, G. et al. "High-efficiency solution processable polymer photovoltaic cells by self-organization of polymer blends." Nature Mater. 4, 864–868 (2005). https://doi.org/10.1038/nmat1500

[15] Padinger, F., Rittberger, R. S. & Sariciftci, N. S. "Effects of postproduction treatment on plastic solar cells." Adv. Funct. Mater. 13, 85–88 (2003). https://doi.org/10.1002/adfm.200390011

[16] Shaheen, S. E. et al. "2.5% efficient organic plastic solar cells. Appl. Phys. Lett. 78, 841–843 (2001)." https://doi.org/10.1063/1.1345834

[17] Zhang, F. L. et al. "Influence of solvent mixing on the morphology and performance of solar cells based on polyfluorene copolymer/fullerene blends." Adv. Funct. Mater. 16, 667–674 (2006). https://doi.org/10.1002/adfm.200500339

[18] Peet, J. et al. "Efficiency enhancement in low-bandgap polymer solar cells by processing with alkane dithiols." Nature Mater. 6, 497–500 (2007). https://doi.org/10.1038/nmat1928

[19] Li, G. et al. "Solvent annealing' effect in polymer solar cells based on poly(3-hexylthiophene) and methanofullerenes." Adv. Funct. Mater. 17, 1636–1644 (2007). https://doi.org/10.1002/adfm.200600624

[20] Zhang, R. et al. "Nanostructure dependence of field-effect mobility in regioregular poly (3-hexylthiophene) thin film field effect transistors." J. Am. Chem. Soc. 128, 3480–3481 (2006). https://doi.org/10.1021/ja055192i

[21] Yang, X. N. et al. "Nanoscale morphology of high-performance polymer solar cells." Nano Lett. 5, 579–583 (2005). https://doi.org/10.1021/nl048120i

[22] Moon, J. S., Lee, J. K., Cho, S. N., Byun, J. Y. & Heeger, A. J. "Columnlike' structure of the cross-sectional morphology of bulk heterojunction materials." Nano Lett. 9, 230–234 (2009). https://doi.org/10.1021/nl802821h

[23] Van Bavel, S. S., Sourty, E., de With, G. & Loos, J. "Three-dimensional nanoscale organization of bulk heterojunction polymer solar cells." Nano Lett. 9, 507–513 (2009). https://doi.org/10.1021/nl8014022

[24] King, R. R. et al. "40% efficient metamorphic GaInP/GaInAs/Ge multijunction solar cells." Appl. Phys. Lett. 90, 183516 (2007). https://doi.org/10.1063/1.2734507

[25] Gilot, J., Wienk, M. M. & Janssen, R. A. J. "Double and triple junction polymer solar cells processed from solution." Appl. Phys. Lett. 90, 143512 (2007). https://doi.org/10.1063/1.2719668

[26] Kim, J. Y. et al. "Efficient tandem polymer solar cells fabricated by all-solution processing. Science" 317, 222–225 (2007). https://doi.org/10.1126/science.1141711

[27] Sista, S. et al." Highly efficient tandem polymer photovoltaic cells." Adv. Mater. 22, 380–383 (2010). https://doi.org/10.1002/adma.200901624

[28] Lipomi, D. J., Tee, B. C. K., Vosgueritchian, M. & Bao, Z. N. "Stretchable organic solar cells. Adv. Mater. 23, 1771–1775 (2011). https://doi.org/10.1002/adma.201004426

[29] Zhu, R., Kumar, A. & Yang, Y. "Polarizing organic photovoltaics." Adv. Mater. 23, 4193–4198 (2011). https://doi.org/10.1002/adma.201101514

[30] Park, H. J., Xu, T., Lee, J. Y., Ledbetter, A. & Guo, L. J. "Photonic color filters integrated with organic solar cells for energy harvesting." ACS Nano 5, 7055–7060 (2011). https://doi.org/10.1021/nn201767e

[31] Brabec, C. J., Shaheen, S. E., Winder, C., Sariciftci, N. S. & Denk, P. "Effect of LiF/metal electrodes on the performance of plastic solar cells." Appl. Phys. Lett. 80, 1288–1290 (2002). https://doi.org/10.1063/1.1446988

[32] Peters, C. H. et al. "High efficiency polymer solar cells with long operating lifetimes." Adv. Energy Mater. 1, 491–494 (2011). https://doi.org/10.1002/aenm.201100138

[33] Wienk, M. M. et al. "Efficient methanofullerene/MDMO-PPV bulk heterojunction photovoltaic cells." Angew. Chem. Int. Ed. 42, 3371–3375 (2003). https://doi.org/10.1002/anie.200351647

[34] Park, S. H. et al. "Bulk heterojunction solar cells with internal quantum efficiency approaching 100%." Nature Photon. 3, 297–302 (2009). https://doi.org/10.1038/nphoton.2009.69

[35] Price, S. C., Stuart, A. C., Yang, L. Q., Zhou, H. X. & You, W. "Fluorine substituted conjugated polymer of medium band gap yields 7% efficiency in polymer–fullerene solar cells." J. Am. Chem. Soc. 133, 4625–4631 (2011). https://doi.org/10.1021/ja1112595

[36] Guo, Xugang, et al. "Polymer solar cells with enhanced fill factors." Nature Photonics 7.10 (2013): 825-833. https://doi.org/10.1038/nphoton.2013.207

[37] Liu, Yongsheng, et al. "Integrated perovskite/bulk-heterojunction toward efficient solar cells." Nano letters 15.1 (2014): 662-668. https://doi.org/10.1021/nl504168q

[38] Vohra, Varun, et al. "Efficient inverted polymer solar cells employing favourable molecular orientation." Nature Photonics 9.6 (2015): 403-408. https://doi.org/10.1038/nphoton.2015.84

[39] Kim, Jeehwan, et al. "10.5% efficient polymer and amorphous silicon hybrid tandem photovoltaic cell." Nature communications 6 (2015). https://doi.org/10.1038/ncomms7391

[40] Liu, Yongsheng, et al. "Perovskite Solar Cells Employing Dopant-Free Organic Hole Transport Materials with Tunable Energy Levels." Advanced Materials 28.3 (2016): 440-446. https://doi.org/10.1002/adma.201504293

Chapter 11

Advancement in simulation and modeling of organic solar cells

Pooja Agnihotri[a*], M.Patel[a], A.Verma[a], S.Sahu[a] ,Sandeep Pathak[b] and Sanjay Tiwari[a*]

[1] Photonics Research Laboratory, S.O.S. in Electronics and Photonics, Pt. Ravishankar Shukla University, Raipur (C.G.)-492010 India

[b]Center for Energy Studies, Indian Institute of Technology, New Delhi 110016,India

drsanjaytiwari@gmail.com , pooja.agni28@gmail.com

Abstract

Organic solar cells (OSCs) has recently received a great impulse as it is safe & clean substitute for the existing fossil fuel power plants, can be manufactured by low temperature processes at low-cost with a promising energy balance. During the past years OSCs performance has improved significantly but needs further improvements. Simulations/modelling are powerful tools for optimization of OSCs, reveal new insights, and predict the behaviour, performance, limitations, stability, dependency of OSCs & maximum attainable efficiency. In this paper we review a chain of simulation models (optical/electrical) for modelling state of the art devices, corresponding development in recent years on the basis of device physics and working principle, analyzing photo-absorption, quantum efficiency, short-circuit current, open-circuit voltage and fill-factor of the device to meet photovoltaic needs.

Keywords

Organic Solar Cells, Drift Diffusion Model, Photo-active Polymer, Non-uniform Generation, Exciton

Contents

1. Introduction 311

2. Organic solar cells 312

3. Performance parameters of organic solar cells 313

4.   **Device models of OSCs**...314

4.1   **Electrical model** 314

4.2   **Optical model** 318

5.   **Modelling results**...322

6.   **Annotations on simulation/modeling**...327

References 328

1.   Introduction

The greatest threats for the future of life on earth are energy crises, global warming, emission of greenhouse gases (water vapor, carbon dioxide, methane, nitrous oxide, etc.). Among all of the renewable resources, the most promising and secure fields that can be adopted as a safe and clean substitute for their existing fossil fuel power plants is Photovoltaic (PV) energy. As the records show [1], among renewable energy sources PV has the fastest growing rate. As an ideal green and renewable energy, solar energy attracts increasing attention with the growing importance of energy sources and environmental protection. As a promising novel solar energy source, organic solar cells (OSCs) have been extensively studied because of their merits of flexibility, light weight, low cost and ease of fabrication.

In 1959 first organic photovoltaic (OPV) was discovered, with the configuration as anthracene single crystal sandwiched between two electrodes with efficiency below 0.1% [2]. This extremely low efficiency obstructed its commercial application for many years. In recent years power conversion efficiency (PCE) of OSCs has been improved rapidly. Recently OSCs with efficiency of 11.0±0.3% were achieved [3]. PV structure having the power conversion efficiency up to about 1% was reported by Tang in 1986, with the first bilayer heterojunction organization where two organic layers (electron donor and electron acceptor) collectively formed an organic heterojunction (D-A interface), so high dissociation efficiency of photo-generated excitons was achieved at the heterojunction interface. [4]. The PCE of bilayer heterojunction OSCs using C60 as an acceptor material is close to 4% [5]. For additional enhancement of efficiency, a revolutionary development with a concept of bulk heterojunction (BHJ) solar cell was proposed by Yu et al. in 1995 [6]. Thereafter bulk heterojunction solar cells have become the most potential structure of OSCs giving highest efficiencies.

Further developing new device structures and designing new functional materials for finding the dependency of devices efficiency on their structure parameters and material properties, device simulation/modeling can help predicting performances of devices and indicate optimization direction. Computer simulations are important tool in predicting the response of OSCs. Many cases of simulation/modeling of OSCs have been seen in recent years and deduction of informative and guidable conclusions. This paper gives a review of recent progress in simulation/modeling for OSCs. We illustrate how a drift-diffusion model can incarcerate the transient behaviour of electron, hole and exciton concentrations in heterogeneous devices. To emphasize the predictive capabilities of computer simulations the systematic variation of device morphologies and the effect this has on photovoltaic performance is revealed. Thus, one can associate and predict the device performance with the device's internal structure and customize the polymer morphology to meet the photovoltaic needs. To get a more profound understanding of the device behavior upon certain variation of parameters such as the doping level, the bimolecular-recombination rate or the charge-carrier mobilities, proposed model for the device is discussed. First, the mechanism of the bulk heterojunction organic solar cells is explained and then, the parameters used to characterize the performance of solar cells are presented, chain of simulation models (optical and electrical) is summarized and analyzed. Finally, this review ends with conclusions and our outlook.

2. Organic solar cells

In an OSC, as shown in Figure 1, the initially photo-generated state is an exciton (a strongly bound electron-hole pair) which is charge neutral and therefore moves by diffusion. For photovoltaic action the exciton requires to be split into its constituent charges and a familiar way of achieving this is the insertion of a second material in the environs of the absorption site. Then the exciton can be separated where one material has a larger electron affinity and ionisation potential and the electron and hole are transferred, assuming the photogeneration has occurred [7]. In photogeneration mechanism of OSCs, the processes of charge generation and separation happen simultaneously at a material heterojunction. The organic-organic interface separates the exciton into its constituent charges. In reality, after exciton dissociation the resulting state is not a free electron and free hole in the acceptor and donor respectively, instead a bound charge-transfer state known as a polaron pair, extending across the junction between the two materials. Likelihood in a process known as energy transfer, the exciton is transferred between the phases. Energy of the charge-transfer state, lower than that of an exciton on either material respectively, is the key point for the occurrence of charge transfer [8]. The charges will then diffuse in different directions as they have opposite spatial derivatives.

Polaron pairs can recombine to give back excitons and can also decay directly back to bound electrons. The direct promotion from bound electrons to Polarons was also observed [9]. A difficulty in exciton dissociation is that excitons have a finite diffusion length, after which they will probably recombine instead of giving out free carriers [10]; this set a limitation on the device thickness of organic cells.

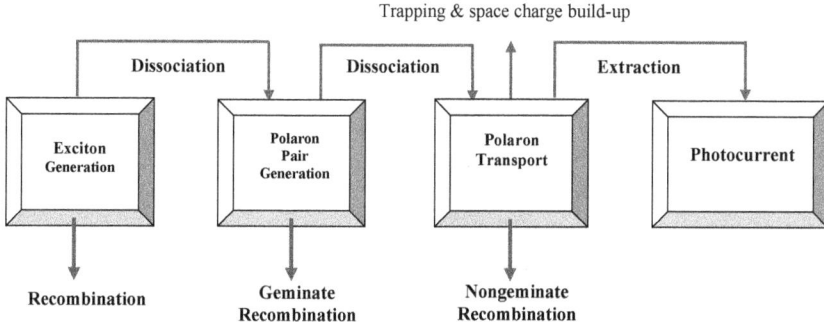

Figure 1 : *Steps of photocurrent generation in an OSC*

3. Performance parameters of organic solar cells

The device physics relevant to the study of solar cells is now considered. Of particular relevance are the four parameters:

1. Short circuit current

2. Open circuit voltage

3. Fill factor

4. Power conversion efficiency at maximum power point

J_{sc} - The short circuit current density, is defined as the current density under illumination at zero applied bias and under these conditions the potential difference between the electrodes is called the built-in bias, V_{bi}.

V_{oc} - The open circuit voltage, is the applied bias at which the photogenerated, extracted current equals the injected current and therefore the total current is zero.

η_{mpp} - The efficiency at maximum power point is the power density delivered to the external circuit divided by the incident illumination power density, P_0,

$$\eta_{mpp} = \frac{J_{mpp}V_{mpp}}{P_0} \tag{1}$$

FF = Fill factor is a measure of the 'squareness' of the current-voltage characteristic and is defined as,

$$FF = \frac{J_{mpp}V_{mpp}}{P_0} = \frac{\eta_{mp}P_0}{J_{sc}V_{oc}} \tag{2}$$

Under AM1.5 illumination conditions these performance parameters are measured where AM represents air mass and 1.5 represents the increased path length of the light through the Earth's atmosphere compared to that with the sun overhead, particularly, when the angle of incidence (θ) is such that Cosec(θ)=1.5, θ=42° [11].

The performance of OSCs is a function of several variables, such as the layer thickness, energy band-gap, absorption coefficient, carrier mobility, contact barrier, recombination coefficient and device architecture [12, 13, 14, 15, 16]. These factors impart a noteworthy influence on the performance of BHJ OSCs. For device optimization and further improvement, broad understanding of the physics of the device is necessary. For exploring the physics of the device modeling and numerical simulation are powerful tool which provides key insights into the internal mechanisms of OSCs.

4. Device models of OSCs

In a typical OSC the incident light passes through a transparent substrate and its successive layers. Excitons are formed and undergo dissociation into free electrons and holes in the BHJ layer after light absorption. The charge carriers are driven by an internal electric field and collected by the electrodes before recombination and at the metal electrode the light reflects back. The electrical model considers the generation, recombination, drift, diffusion, and collection process of the electron and hole of the photovoltaic structure and the optical model is applied to calculate the numbers of the absorbed photon in the structure.

4.1 Electrical model

Electrical transport models fall into three main categories: Equivalent circuits, Microscopic models and Continuum models.

Equivalent circuits - Based on equivalent lumped circuits few models, such as the single diode model (SDM), the double diode model (DDM) (Figure 2), the SDM-based approach by the Lambert W-function [17] and the two-diode model also known as

Mazhari's model [18] are often used to characterise the electrical characteristics owing to simplicity and flexibility; however, the features of these models do not fully take into account the physical phenomena of organic cells.

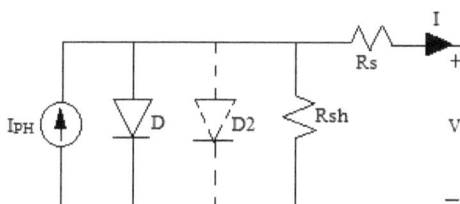

Figure 2 : *Single Diode Model and Double Diode Model (including D2).*

Microscopic Models - Microscopic (discrete) model based on kinetic Monte Carlo (kMC) simulations. The main advantage of kMC is the possibility to correctly include the effect of the real morphology on, for example, the internal quantum efficiency [19]. The kMC represents an invaluable tool to investigate these processes in time domain and suggest optimization strategies for the device fabrication. The main drawback of kMC is the extremely high computational cost, especially if the entire active layer has to be included in a 3D structure. For this reason in many cases, only a fraction of the entire device has been effectively simulated.

Continuum Models - Most continuum models consider the blend as a homogeneous medium. Good agreement has been achieved with experimental current–voltage (J–V) characteristics in OSCs for such models [20]. However, computationally efficient at the same time such models are unable to describe carrier interactions in sufficient detail.

The Effective Medium Model - The BHJ layer is a blend of donors and acceptors exhibiting a complex morphology. In the effective medium model, the BHJ layer is considered a homogeneous semiconductor. The energy difference between the LUMO of the acceptor and the HOMO of the donor functions is considered the effective band gap (Eg) of the semiconductor. The transport of electrons and holes in the semiconductor under operation condition is a shown in schematic in Figure 3.

315

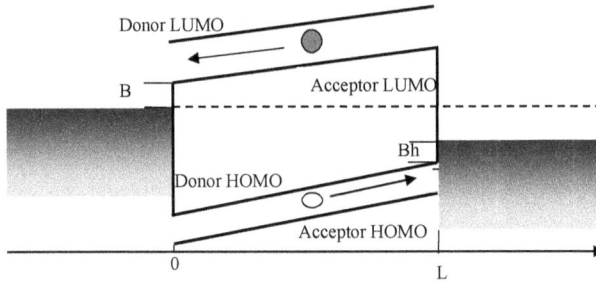

Figure 3 : *The electrical model showing energy level of materials with Be and Bh representing the energy barrier for the electron and hole respectively.*

In the BHJ layer the one-dimensional equations that describe the behavior of electrons and holes are alike to those used in inorganic semiconductor devices. The Poisson equations, where q is the elementary charge, ε is the dielectric constant, the electrical potential ψ, N_{eff} as the effective density of charge and n and p are the electron and hole densities respectively.

$$\nabla^2 \psi = \frac{-q}{\varepsilon(n - p + N_{eff})} \tag{3}$$

The photo-generated charges inside the active layer on absorption of photons move; driven by the external electric field (drift) and by their concentration gradient (diffusion). The relevant current equations are;

$$\overline{J_n} = qD_n\nabla n + qn\mu_n(-\nabla\psi) \tag{4}$$

$$\overline{J_p} = -qD_p\nabla p + qp\mu_p(-\nabla\psi) \tag{5}$$

μ_n (μ_p) is the electron(hole) mobility and Dn (Dp) is the electron(hole) diffusion coefficient given by the Einstein relation;

$$\frac{D_n}{\mu_n} = \frac{D_p}{\mu_p} = \frac{k_BT}{q} \tag{6}$$

where k_B is theBoltzmann constant and T is the temperature.

The current continuity equations are;

$$-\frac{1}{q}\nabla.\overline{J_n} = G_n - R_n \tag{7}$$

$$\frac{1}{q}\nabla.\overline{J}_p = G_p - R_p \tag{8}$$

where G_n & G_p are the optical generation rate R_n & R_p as recombination rate of electrons and holes respectively.

To obtain a unique solution, appropriate specification of the boundary conditions is necessary. At the contact interface, the semiconductor is always in thermodynamic equilibrium (assumption). Considering the energy barrier for electrons and holes as B_e and B_h (Figure 3) respectively, using Boltzmann statistics

$$n(0) = N_c\, e^{\left(-B_e/k_BT\right)} \tag{9}$$

$$p(0) = N_v\, e^{-\left(E_{gap}-B_e/k_BT\right)} \tag{10}$$

$$n(L) = N_c e^{-\left(E_{gap}-B_h/k_BT\right)} \tag{11}$$

$$p(L) = N_v e^{\left(-B_h/k_BT\right)} \tag{12}$$

where N_c (N_v) is the effective density of states of conduction (valence) band. There is no energy barrier for electrons (holes) if the contact for electron (hole) is ohmic. The boundary condition for the potential is

$$\Psi(L) - \Psi(0) = \frac{E_{gap} - B_e - B_h}{q} - V_a \tag{13}$$

where V_a is the applied voltage.

The electrostatic attraction between the charge carriers in organic semiconductors is not efficiently screened because of their low dielectric constant; thus, the charge carriers experience stronger attraction. The absorption of a photon primarily leads to the formation of a strongly bound exciton in organic semiconductors. To obtain free charge carriers, the exciton must be dissociated, which can be achieved by blending two organic semiconductors with different energy levels so that an electron can easily undergo a charge-transfer process from the bound exciton state to transform into a less tightly bound charge-transfer exciton [21] leading to efficient generation of free electrons and holes. Onsager developed an equation for the relative effect of an applied electric field on the dissociation of a weak electrolyte [22]. It is a solution to a steady-state diffusion equation which describes the dissociation and recombination kinetics of ion pairs in an applied electric field. The ions in solutions are infinitely long lived. The relative increase of the dissociation rate constant K(E) to the dissociation equilibrium constant K(0) is

$$K(E)/K(0) = J_1(2\sqrt{-2b})/\sqrt{-2b} = 1 + b + b^2/3 + b^3/18 + \ldots\ldots \tag{14}$$

Where J_1 is the Bessel function of order one, $b = \frac{q^3 E}{8\pi\varepsilon_r\varepsilon_0 k^2 T^2}$ and E is the electric field. Braun applied this result to describe the electric field assisted dissociation of charge transfer states in donor-acceptor system [23]. The dissociation rate is

$$k_{diss} = \frac{3k_R}{4\pi a^3} e^{-E_B/_{kT}} J_1 \frac{(2\sqrt{-2b})}{(\sqrt{-2b})} \tag{15}$$

Where $E_B = q^2/_{(4\pi\varepsilon_r\varepsilon_0 a)}$ is the electron-hole pair binding energy, 'a' is the pair distance and $k_R = q(\mu_n + \mu_p)/_{\varepsilon_r \varepsilon_0}$. The probability P of electron-hole pair dissociation as the bound electron-hole pair may decay to the ground state with a decay rate k_f or dissociate into free carriers with a dissociation rate k_{diss}; is given by

$$P = k_{diss}/_{(k_{diss} + k_f)}$$

Blom et al. introduced Braun's theory; the charge transfer state is considered an intermediate state, leading to modification of the free charge carrier generation described as $G = PG_0$ where G_0 is the photo-absorption induced exciton generation rate calculated by the optical model and P is the dissociation probability of the exciton and depends on the temperature, field and morphology, in some ideal cases can approach unity [24, 25].

When the carrier mean free path is smaller than the Coulomb capture radius (in low-mobility materials) Langevin recombination is expected to occur. The recombination rate is determined by the probability of charge carriers meeting in space. Langevin recombination process is described as a bimolecular process with a rate $R = \gamma (np - n_i^2)$ where the recombination coefficient γ is given by

$$\gamma = \frac{q}{\varepsilon} (\mu_n + \mu_p) \tag{16}$$

Newton iteration and Gummel iteration are commonly used to solve drift-diffusion equations. Newton iteration is a kind of coupling method that solves the three equations synchronously, which is more stable and easy to converge but more complicated. Gummel iteration [26] is a kind of uncoupling method that solves the three equations by sequence, which is more simple but relatively hard to converge.

4.2   Optical model

To elucidate the optical potential of OSCs and to have a clear understanding of the illumination direction and active layer thickness effects on the performance and transparency of the device, optical simulations are necessary [27]. The preliminary step requires the calculation of the respective number of photons absorbed in the active layer (N_{ph}) under AM1.5. The optical interference effect should not be ignored in the multilayer structure, particularly when the highly reflective metal electrode is used because the layer thicknesses in OSCs are comparable with the wavelength of sunlight. In multiple layer thin-film stacks, for optical interference the mathematical treatment has long been established [28].

The optical transfer-matrix theory introduced by Heavens [29] was applied to organic heterojunction solar cells by Pettersson et al. [30] and others [31, 32] which considered the interference effects for light wave propagating in thin films, to OSCs. From then on the optical transfer-matrix theory has been widely applied to simulate optical absorption and light intensity distribution within organic thin film solar cells. Persson et al. [33] modeled the AM1.5 standard solar irradiation as a blackbody radiator and achieved a rather good approximation:

$$I_0(\lambda) = \frac{2h\pi c^2}{\lambda^5} \frac{1}{e^{hc/\lambda k_B T} - 1}$$

(17)

Where $I_0(\lambda)$ is the monochromatic intensity of light source at wavelength λ, h is the Planck constant, k_B is the Boltzmann constant, c is the velocity of light in vacuum, and T is the temperature.

The optical transfer-matrix theory is based upon the Fresnel relations, i.e., the light wave undergoes Fresnel reflection and transmission, and employs matrix formulae to calculate optical intensity in multiple thin layers. For applying the theory some assumptions are: layers of OSCs are homogeneous and isotropic, parallel and flat interfaces, and the incident light is a plane wave normal to the substrate interface. Based on these assumptions a one-dimensional model is proper to calculate the optical electric field. As demonstrated in Figure 4, the layers ($j = 1, 2, \ldots, m$) in organic thin film solar cells are considered to be embedded between two semi-infinite layers ($j = 0, j = m + 1$). Since the glass substrate is much thicker compared to the light coherent length, light transmission through it should not be calculated by wave interference, but rather be taken into account by reflection and transmission rates. Each layer is described by its complex index of refraction, $\tilde{n}_j = n_j + i\kappa$ which is a function of wavelength (λ), and thickness of the layer (d_j).

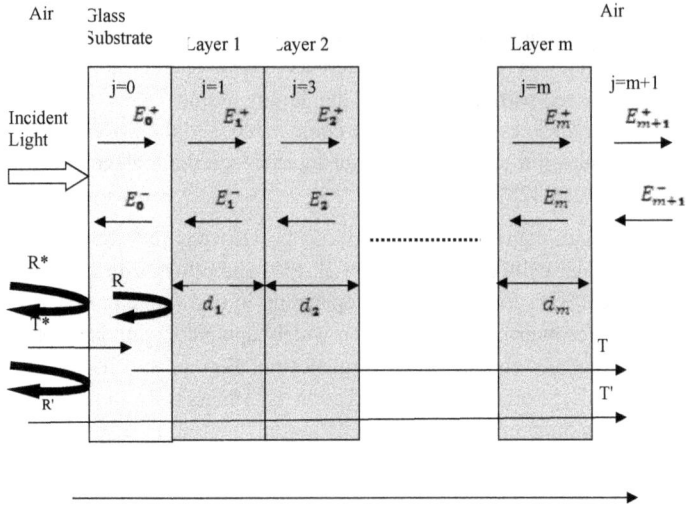

Figure 4 : *Multilayer structure of m layers with semi-infinite layer 0 and m+1*

The incident light enters from the left side along the x direction and is reflected at the interfaces inside the device, so a backward light wave is generated. The forward and backward light waves are described with a positive superscript and a negative superscript respectively, e.g., at a position x in layer j, and the forward and backward light electric fields are described by $E_j^+(x)$ and $E_j^-(x)$ respectively. The light propagation across an interface between layers j and k is described by the 2×2 matrix:

$$I_{jk} = \frac{1}{t_{jk}} \begin{bmatrix} 1 & r_{jk} \\ r_{jk} & 1 \end{bmatrix} \tag{18}$$

where $r_{jk} = (\tilde{n}_j - \tilde{n}_k)\big/(\tilde{n}_j + \tilde{n}_k)$ and $t_{jk} = 2\tilde{n}_j\big/(\tilde{n}_j + \tilde{n}_k)$ are Fresnel complex reflection and transmission coefficients respectively. A phase shift of light and absorption caused by propagating through layer j are described by the 2×2 matrix called layer matrix L_j.

$$L_j = \begin{bmatrix} e^{-i\xi_j d_j} & 0 \\ 0 & e^{i\xi_j d_j} \end{bmatrix} \tag{19}$$

Where $\xi_j = (2\pi/\lambda)\tilde{n}_j$. The electric field between the substrate (j=0) and ambient side (j=m+1) is given by the total system transfer matrix S,

$$\begin{bmatrix} E_0^+ \\ E_0^- \end{bmatrix} = S \begin{bmatrix} E_{m+1}^+ \\ E_{m+1}^- \end{bmatrix}$$

(20)

Where S is the product of all interface and layer matrices that light propagates through orderly as

$$S = \begin{bmatrix} S_{11} & S_{12} \\ S_{21} & S_{22} \end{bmatrix} = \left(\prod_{v=1}^{m} I_{(v-1)} L_v \right) . I_{m(m+1)}$$

(21)

The total transmission coefficient (t) and reflection coefficient (r) are given as,

$$t = \frac{E_{m+1}^+}{E_0^+} = \frac{1}{S_{11}}$$

(22)

$$r = \frac{E_0^-}{E_0^+} = \frac{S_{21}}{S_{11}}$$

(23)

Within layer j=0 and j=m+1 the total transmissivity and reflectivity is given as,

$$T = |t|^2 \frac{n_{m+1}}{n_0}$$

(24)

$$R = |r|^2$$

(25)

The absorption efficiency of the whole device is expressed as, $\eta_A = 1 - T' - R'$ where

$$T' = \frac{T^{\bullet} T}{1 + R^{\bullet} R}$$

(26)

$$R' = \frac{R^{\bullet} + R}{1 + R^{\bullet} R}$$

(27)

According $n_0 t|^2$ Fresnel relation for transmission and reflection $T^{\bullet} = \left| \frac{2}{1 + n_0} \right|^2$ and
$R^{\bullet} = \left| \frac{n_0 t}{1 + n_0} \right|^2$.

The total electric field at an arbitrary position $x \ (0 \leq x \leq d_j)$ inside layer j is given by

$$E_j(x) = E_j^+(x) + E_j^-(x) = \left(t_j^+ e^{i\xi_j x} + t_j^- e^{-i\xi_j x} \right) E_0^+$$

$$= t_j^+ \left[e^{i\xi_j x} + r_j'' e^{i\xi_j (2d_j - x)} \right] E_0^+$$

(28)

At a given position the number of excitons generated is directly dependent on the energy absorbed by the material. Based on the concept of Poynting vector the time-averaged absorbed-power is given as

$$Q_j(x) = \frac{4\pi c \varepsilon_0 \kappa_j n_j}{2\lambda} |E_j(x)|^2 = \frac{1}{2} c \varepsilon_0 \alpha_j n_j |E_j(x)|^2 \tag{29}$$

Where c is the speed of light, ε_0 is the permittivity of free space and $\alpha_j = 4\pi kj/\lambda$ is the absorption coefficient. The photon absorption of glass is strong below 300nm and organic materials hardly absorb photons beyond 900nm. Thus for OSCs the time averaged light energy dissipation Q(x) is integrated over this wavelength range as

$$Q(x) = \int_{300}^{900} Q(x,\lambda)d\lambda \tag{30}$$

5. Modelling results

Barker et al. in 2003 developed a numerical model to predict the current-voltage curves of bilayer conjugated polymer photovoltaic devices and expressions for the field dependence of the dissociation rate. The model showed a logarithmic dependence of the open-circuit voltage on the incident intensity, as seen experimentally. They found that the short-circuit quantum efficiency is determined by the competition between polaron pair dissociation and recombination. Increasing the dissociation rate will increase the efficiency. Decreasing the recombination rate will have a similar effect, although this may lead to a saturation of the polaron pair density at high intensities. The measured open-circuit voltage will depend not only on the difference in electrode work functions, but also on the intensity and the charge densities at the electrodes. Finally, the fill factor is optimized when polaron pair dissociation competes effectively with recombination even with a small field in the positive direction. Under these conditions the quantum efficiency rises rapidly to unity at voltages only slightly below the open-circuit voltage, leading to a rectangular current-voltage curve with a large fill factor [34]

Mihailetchi et al. in 2004 developed a numerical model that calculates the steady-state charge distributions within the active layer with Ohmic contacts. This numerical model describes the full current-voltage characteristics in the dark and under illumination, including the field dependent generation rate and showed the experimental photocurrent together with the numerical calculation for two different temperatures. The photocurrent in 20:80 blends of OC1C10-PPV:PCBM solar cells has been interpreted using a model based on Onsager's theory of geminate charge recombination. The model explains the field and temperature dependence of the photocurrent in a large voltage regime. Under short-circuit conditions at room temperature, 60% of the bound e-h pairs are separated and contribute to the photocurrent [35].

Pavel Schilinsky et al. in 2004 suggested a modified one diode model to describe the illumination dependence of current–voltage characteristics of polymeric bulk heterojunction solar cells at different light bias levels taking into account the effective reduction of the mean distance which the charge carriers cover when sweeping the electrical bias through the fourth quadrant of the solar cell. Moreover, estimated the built-in voltage V_{bi} and of the mean carrier lifetime τ or of the $\mu\tau$ product if the mobility μ of the tested material. The model does not take the statistical distribution of the mobility and lifetime into account thus small deviations were observed [36]

Florent Monestier et al. in 2006 investigated the short-circuit current density of organic solar cells based on poly (3-hexylthiophene)(P3HT)/6,6-phenyl C61-butyric acid methyl ester (PCBM) blend. Full optical modeling of the cell was performed in order to model charge collection efficiencies with respect to short circuit density in such blends and the rate of exciton generation was computed from the distribution of the electromagnetic field and was used as input in the transport equations of holes and electrons. For computing short-circuit current densities generated in the cell, charge densities at steady state were obtained as solutions. The dependence of short-circuit current densities versus the thickness of the blend was analyzed and compared with experimental data and with data extracted from the literature. It was found that the influence of the carrier recombination with Jsc variations depends on the experimental procedure. Even if short-circuit current densities increase with higher blend thicknesses, power conversion efficiency is usually limited by low fill factor [37]

Gavin A Buxton et al. in 2007 demonstrated the transient behaviour of electron, hole and exciton concentrations in heterogeneous devices by considering bilayer devices where the interface is sinusoidal, not planar and considered the systematic variation of device morphologies and the effect this has on photovoltaic performance. In these heterogeneous devices the model enabled to visualize the charge-carrier concentrations and flux, along with the concentration and flux of photo-generated excitons. At the donor acceptor interface the exciton concentration is observed to decrease as excitons are dissociated and this variation in concentration drives more excitons towards the interface. Upon dissociation of excitons, because of the internal electric field the free charge carriers both drift and diffuse away from the donor acceptor interface and towards the electrodes because of concentration gradients. This current is then accessible to an external circuit [38].

Groves et al. in 2008 used Monte Carlo model to examine the dynamic behaviors of geminate pair dissociation in polymer-polymer photovoltaic devices for the first time and found that increasing one or both carrier mobilities aids geminate separation yield particularly at low fields which leads to improved maximum power output from polymer-

polymer blend photovoltaics, even when carrier mobilities are unbalanced. In a bilayer geminate pairs become effectively free when separated by ~ 4 nm, which is far smaller than the thermal capture radius of 16 nm. As the average domain size increases from 4 to 16 nm it was found that geminate separation yield in a blend improves continuously and showed that although a small degree of separation may be available in a blend, the limited number of possible routes to further separation makes charge pairs in blends more susceptible to recombination than charge pairs in a bilayer. Substantial increases in maximum power output of 75% and 110% when either one or both carrier mobilities, respectively, are increased by a factor of 10 [39].

JonnyWilliams, in 2008, developed an organic solar cell model with a two dimensional geometry, to consider the morphology by including an optical field intensity profile that allows for interference effects and found much greater photo-generated currents than if an exponential decay in the intensity is assumed and shown that the current is sensitive to the component of the electric field normal to the interface which influences the power efficiency and fill factor, whose variation with intensity and interface length have been explored and showed that because of the increase in the open circuit voltage the fill factor decreases from 40% at low intensities to 20% at solar intensities and decreases much more rapidly at higher intensities due to the decrease in the power efficiency[40].

Junsangsri, in 2010, proposed a model that captures the physical phenomena behind the operation of OSCs and parameter extraction through a novel method. By simulation, it is showed that the I-V characteristics of the single diode model, the double diode model and the proposed model are in good agreement and close to experimental values. An extraction method is proposed and validated by fitting simulated and experimental results. Moreover, it showed that, when the output voltage is lower than V_{OC}, the I-V characteristic of OSCs using SDM, DDM and the proposed model have very close values even though minor differences are encountered for I_{max} and V_{max} [41].

C.De Falco et al. in 2012 carried out extensive numerical simulations of Two-dimensional realistic device structures with various interface morphologies and investigated the impact of the model for k_{diss} on the main device properties (short circuit current and open circuit voltage). Simulation results indicated that, if the electric field orientation relative to the interface is taken into due account, the device performance is determined not only by the total interface length but also by its shape [42].

Biswajit Ray in 2013 showed that by introducing a "fixed charge layer" at the donor–acceptor interfaces, the FF of the OPV devices made of the conventional low-mobility materials can be radically improved (>80%) and at the interface the fixed charges prevent free-carrier accumulation and, hence, minimize charge loss due to interfacial recombination. To estimate the fixed charge density required for significant performance gain in typical P3HT:PCBM cells, a detailed device simulation was used and concluded by suggesting several strategies for introducing such charged interfaces within the conventional OPV device structure [43].

Fallahpour in 2014 presented a complete opto-electronic model. The core of the simulation is constituted by drift-diffusion equations, where the photo-charge density is calculated starting from a detailed transfer matrix model. To properly describe many of the phenomena occurring in an operating cell several microscopic models are introduced and the disordered nature of the material is taken into account by hopping mobility and by a proper description of Gaussian density of states. Such a complex model has been parameterized using measured values from the literature in order to reproduce experimental data. Critical effects of the degree of energetic disorder on the cell performance was also discussed and is represented in an effective way by the spread of the Gaussian distribution of the density of states. By increasing the degree of disorder, the Langevin recombination reduces due to a lower mobility of free charges. On the other hand, the decreased effect on the mobility is actually dominant respect to the decrease of recombination, turning in a detrimental effect of the disorder on the efficiency of the cell. Finally, the correlation between the disorder parameter and the thickness of the active layer was studied [44].

Mathew Jones et al. in 2014 examined CT separation dynamics in an OPV including the effects of energetic disorder and bulk heterojunction morphology using Monte Carlo simulations and observed strongly bi-exponential decay of the recombination dynamics, similar to experimental. The slow component is due to trapping of charges within energetic disorder and the bulk heterojunction morphology, because increasing the electric field reduces the importance of this feature. An alternative kinetic framework, which includes an intermediate quasi-free state between the CT and free charge states was proposed. The model is shown to fit very well the dynamic behavior of CT state recombination and separation efficiency as a function of electric field. The CT state separation behavior was examined using an altered recombination rate in the model and showed that kinetic models can be used to successfully fit a wide variety of data describing CT state behavior, but the derived rate does not have exact correspondence with the physical processes occurring in the OPV [45]

Farrokhifar et al. in 2014 presented a comprehensive model for organic photovoltaic devices considering the optical and electrical simulation and showed that the results of optical simulation are consistent with the calculated ones. Within the structure due to interference, the peak amplitude of wave and its position change and cause the exciton generation to oscillate inside the active layer. Electrical simulation of devices was done and seen that with enhancing the thickness, the solar cell fill factor decreases due to the increased serial resistance. For the thickness of 90nm the fill factor of solar cell is 75% however by increasing the thickness to the 320nm it decreases to 42%. In OSCs the effect of active layer thickness on the short circuit current and efficiency was studied. Simulation results were compared with experimental data which confirmed that the model is well simulated. The effect of device thickness on the power conversion efficiency of OSCs was investigated and shown that by increasing the thickness of the device the efficiency reduces as non-geminate recombination rate enhances due to the low mobility of the organic materials [46].

Hossein Movla et al. in 2015 studied the electrical characteristics of BHJ OSCs based on P3HT:PCBM in the dark and under different illumination. It was shown; Voc is almost constant and is about 0.5 V in different thicknesses of BHJ solar cells. For 100 nm P3HT:PCBM blend thicknesses FF of the cell is about 0.50, Power conversion efficiency is 1.6% which relatively is a good agreement with reported data. Based upon the results of this model it is possible to design nanostructured photovoltaics and optimize them toward higher efficiencies [47].

Inche Ibrahim et al. in 2015 proposed an analytical approach to describe the current-voltage (J-V) characteristics of BHJ OSCs with the assumption of uniform bimolecular recombination rate that equal to the average of the actual bimolecular recombination rates. There are a few advantages of this analytical expression compared to numerical simulation of the drift-diffusion mode such as on applying the analytical expression to experimental J-V data enabled direct extraction and to analyze the overall recombination loss of BHJ OSC. Furthermore, it was found that when the injection barrier of a large band gap BHJ OSC is high, the numerical program fails to give the desired output and this analytical approach can be used to analyze the large band gap BHJ OSC. Therefore, when the numerical simulation fails to produce the desired results for a specific BHJ OSC, the analytical expression is useful in analyzing them [48].

Davide Bartesaghi et al. in 2015, showed that the FF in the whole range of 0.26–0.74 governed by the competition between charge recombination and extraction, can be quantified by the parameter θ (the ratio of the rates of recombination and extraction of free charges) and gave a relationship between FF and θ valid for a large number of donor:acceptor combinations (assuming Ohmic contacts). It was supported by experimental data collected from polymer:fullerene, polymer:polymer and small-molecule devices characterized with steady-state and transient extraction techniques and also evident from the drift-diffusion simulations of OSCs performing varying charge-carrier mobilities, recombination rate, light intensity, energy levels and active-layer thickness over a wide range. It provided new insights into the physical phenomena governing the FF of OSCs and explained why the significant change of FF with material [49].

Madogni et al. in 2015, determined the excitons distribution function in BHJ OSC-ITO/PEDOT:PSS/rrP3HT: PC70BM (1:0.7 weight ratio)/Yb/Al using the Laplace transforms with the residue theorem and studied the influence of the electron-hole pair separation distance on the excitons dissociation probability, the quantum efficiency of charge carriers generation and the photo-generated current density on the excitons binding energy. The simulated results indicate that the linear decreasing of the excitons generation function is attributed to the loss of free charge carriers at the electrodes/organic interfaces, causing the increase of the interfacial dissociation rate and a diffusivity of the electron-hole pair at DA interface and remarked that the total exciton generation rate does not monotonously increase with the increase of the active layer thickness, but behaves wave-like which induces the corresponding variation of Jsc. The carrier lifetimes also influence Jsc greatly. When the lifetimes of both electrons and holes are long enough, dissociation probability plays an important role in the thick active layer. The excitons dissociation probability decreases exponentially and the internal quantum efficiency charge generation monotonously increases when the electron-hole pair separation distance increases. The results were in agreement with those predicted by the literature. It was concluded that the potential improvement of the internal quantum density of charge generation depends on the exciton dissociation probability into free charge carriers and strongly depends on the photons optical absorption [50].

6. Annotations on simulation/modeling

OSCs have received a great impulse and during the past years its performance has improved significantly. For optimization of OSCs, to reveal new insights, predict the behaviour, performance, limitations, stability, dependency of OSCs and maximum attainable efficiency, simulations/modeling are powerful tools. The modeling of OSCs is divided into two parts: electrical modeling and optical modeling, often used jointly to simulate the performances of a device. Electrical modelling is used to provide key insights into the internal mechanism, which derives the influence of factors limiting the electrical transport processes generally based on the drift-diffusion model which consists of the Poison's equation and continuity equations and Optical modelling is used to calculate the distribution of absorbed light energy within the device, and to obtain the optimum thickness of film layers used in the device.

Although large active layer thickness can improve the short-circuit current density and external quantum efficiency, thicker devices require higher mobility to ensure the separated charges to reach their respective electrodes. Electrical modeling with a constant exciton generation rate will be accurate for BHJ OSCs with an active layer thinner than 250 nm, but the optical interference effects should be considered for devices with a thicker active layer. Materials used in active layer of OSCs with small energy gap will enhance the device efficiency but lowering this will decrease the open-circuit voltage. Thus optimized efficiencies of OSCs are obtained at balanced level tuning and the band gap. In OSCs recombination is the dominant loss mechanism. By reducing recombination, power conversion efficiency can be significantly improved. By reducing bimolecular recombination, finer phase separation leads to increased short-circuit current. Higher mobility of the blend leads to higher efficiencies but recently simulations predicted finite optimum charge carrier mobility for the highest power conversion efficiency. Hence Langevin-type recombination will not result in finite optimum mobility of free charge carriers, but surface recombination will. Due to the temperature dependence of carrier mobility and bimolecular recombination the short-circuit current is dependent on temperature. The local electric field influences charge dissociation and transport to the appropriate electrodes. The performance of OSCs is also dependent on the collection of charges at these electrodes. Higher anode work function results in higher energy conversion efficiency. The performance of OSCs has significantly improved but needs further improvement.

References

[1]   A. J. Waldau, "PV Status Report 2003", European Commission, EUR 20850 EN, 2003.

[2]   Kallmann H, Pope M. "Photovoltaic effect in organic crystals.", J Chem Phys, 1959, 30: 585–586 https://doi.org/10.1063/1.1729992

[3]   Hosoya M, Oooka H,Nakao H, Gotanda T, Mori S, Shida N, Hayase R,Nakano Y, Saito M. "Organic thin film photovoltaic modules", Proceedings of the 93rd Annual Meeting of the Chemical Society of Japan 21–37, (2013).

[4]   Tang C W. "2-layer organic photovoltaic cell", Appl Phys Lett, 1986, 48: 183–185 https://doi.org/10.1063/1.96937

[5]   Peumans P, Yakimov A, Forrest S R. "Small molecular weight organic thin-film photodetectors and solar cells", J Appl Phys, 93:3693–3723 (2003). https://doi.org/10.1063/1.1534621

[6]   Yu G, Gao J, Hummelen J C, et al. "Polymer photovoltaic cells-enhanced efficiencies via a network of internal donor-acceptor heterojunctions", Science, 270:1789–1791 (1995) https://doi.org/10.1126/science.270.5243.1789

[7]   A.J. Heeger. "Nobel lecture: Semiconducting and metallic polymers: The fourth generation of polymeric materials", Reviews of Modern Physics, 73:681, (2001) https://doi.org/10.1103/RevModPhys.73.681

[8]   B. O'Regan and M. Gr"atzel. "A low cost, high efficiency solar cell based on dyesensitised colloidal TiO2 films", Nature, 353:737, (1991) https://doi.org/10.1038/353737a0

[9]   H.Hoppe, N.S.Sariciftci. "Organic solar cells: An overview", J. Mater. Res, (2004) https://doi.org/10.1557/JMR.2004.0252

[10]   B. A. Gregg and M. C. Hanna. "Comparing organic to inorganic photovoltaic cells: Theory, experiment and simulation", J. Appl. Phys., 93:3605, (2002). https://doi.org/10.1063/1.1544413

[11]   J. Nelson. "The Physics of Solar Cells", Imperial College Press, London, (2003). https://doi.org/10.1142/p276

[12]   C. Shuttle, B.O'Regan, A.Ballantyne, J.Nelson, D.Bradley, J.Durrant, "Bimolecular recombination losses in poly thiophene:fullerene solar cells", Phys.Rev.B:Condens.Matter 78, 113201, (2008). https://doi.org/10.1103/PhysRevB.78.113201

[13]   J. Guo, H.Ohkita, "Charge generation and recombination dynamics in poly(3-hexylthiophene)/fullerene blend films with different region regularities and morphologies", Adv.Mater., 132, 6154–6164,(2010).

[14]   M. Lenes, M.Morana, C.J.Brabec, P.W.M.Blom, "Recombination-limited photo currents in low bandgap polymer/fullerene solar cells", Adv. Funct. Mater.19 1106–1111, (2009). https://doi.org/10.1002/adfm.200801514

[15]   K.J. Li, L.J. Li, J.C. Campbell, "Recombination lifetime of free polarons in polymer/fullerene bulk heterojunction solar cells", J. Appl. Phys. 111, 034503 (2012). https://doi.org/10.1063/1.3680879

[16]   Z.He,C.Zhong, S.Su,M.Xu,H.Wu,Y.Cao, "Enhanced power-conversion efficiency in polymer solar cells using an inverted device structure", Nat. Photonics 6:591–595, (2012). https://doi.org/10.1038/nphoton.2012.190

[17]   Jain, A., and Kapoor, A., "A new approach to study organic solar cell using Lambert W-function", Sol. Energy Mater. Sol. Cells, 86, 197–205 (2005). https://doi.org/10.1016/j.solmat.2004.07.004

[18] Mazhari, B., "An improved solar cell circuit model for organic solar cells", Sol. Energy Mater. Sol. Cells, 90,1021–1033 (2006). https://doi.org/10.1016/j.solmat.2005.05.017

[19] P. K. Watkins, A. B. Walker, and G. L. B. Verschoor, "Dynamical Monte Carlo modelling of organic solar cells: The dependence of internal quantum efficiency on morphology," Nano Lett. 5, 1814–1818 (2005). https://doi.org/10.1021/nl051098o

[20] Blom P W M, Mihailetchi V D, Koster L J A and Markov D E, Adv. Mater. 19, 1551 (2007). https://doi.org/10.1002/adma.200601093

[21] S. Singh,Z.Vardeny, "Ultrafasttransientspectroscopyofpolymer/fullerene blends for organic photovoltaic applications", Materials 6, 897–910, (2013). https://doi.org/10.3390/ma6030897

[22] L. Onsager, "Deviations from Ohm's law in weak electrolytes", J.Chem.Phys.2 599–615, (1934). https://doi.org/10.1063/1.1749541

[23] C.L.Braun, "Electric field assisted dissociation of charge transfer states as a mechanism of photo carrier production", J.Chem.Phys. 80, 4157–4161, (1984). https://doi.org/10.1063/1.447243

[24] L. Koster, E. Smits, V. Mihailetchi, P. Blom, "Device model for the operation of polymer/fullerene bulk heterojunction solar cells", Phys. Rev. B: Condens. Matter 72, 085205, (2005). https://doi.org/10.1103/PhysRevB.72.085205

[25] V. Mihailetchi, L.Koster, J.Hummelen, P.Blom, "Photocurrent generation in polymer–fullerene bulk heterojunctions", Phys. Rev. Lett. 93, 216601 (2004). https://doi.org/10.1103/PhysRevLett.93.216601

[26] Gummel H K. "A self-consistent iterative scheme for one-dimensional steady state transistor calculations", IEEE Trans Electron Device, ED-11, 1964: 455–465 https://doi.org/10.1109/T-ED.1964.15364

[27] J. P. Hugonin, P. Lalanne, "Reticolo software for grating analysis", Institut d'Optique, Orsay, France (2005).

[28] E. Hecht (Ed.), Optics, fourth ed., Pearson, San Francisco, (2002).

[29] O.S. Heavens, 'Optical Properties of Thin Solid Films", Dover, New York, (1965).

[30] L.A.A.Pettersson,L.S.Roman,O.Inganas, "Modeling photocurrent action spectra of photovoltaic devices based on organic thin films", J.Appl.Phys. 86, 487–496 (1999). https://doi.org/10.1063/1.370757

[31]F.Monestier, J.-J.Simon, P.Torchio, L.Escoubas, F.Flory, S.Bailly, R.de Bettignies, S.Guillerez, C.Defranoux, "Modeling the short-circuit current density of polymer solar cells based on P3HT:PCBM blend", Sol. Energy Mater. Sol. Cells 91, 405–410 (2007). https://doi.org/10.1016/j.solmat.2006.10.019

[32] G.F.Burkhard, E.T.Hoke, M.D.McGehee, "Accounting for interference, scattering, and electrode absorption to make accurate internal quantum efficiency measurements in organic and other thin solar cells', Adv. Mater. 22, 3293–3297 (2010). https://doi.org/10.1002/adma.201000883

[33] Persson N K, Schubert M, Inganäs O. "Optical modeling of a layered photovoltaic device with a polyfluorene derivative/fullerene as the active layer", Sol Energy Mater Sol Cells, 83: 169–186, (2004). https://doi.org/10.1016/j.solmat.2004.02.023

[34] J. A. Barker, C. M. Ramsdale, and N. C. Greenham, "Modeling the current-voltage characteristics of bilayer polymer photovoltaic devices", Physical Review, B 67, 075205 (2003). https://doi.org/10.1103/PhysRevB.67.075205

[35] V. D. Mihailetchi, L. J. A. Koster, J. C. Hummelen, and P.W.M. Blom, "Photocurrent Generation in Polymer-Fullerene Bulk Heterojunctions", Physical Review Letters 93, 21 (2004). https://doi.org/10.1103/PhysRevLett.93.216601

[36] Pavel Schilinsky, Christoph Waldauf, Jens Hauch, and Christoph J. Brabec, "Simulation of light intensity dependent current characteristics of polymer solar cells", Journal of Applied Physics, 95, 5 (2004). https://doi.org/10.1063/1.1646435

[37] Florent Monestier, Jean-Jacques Simon, Philippe Torchio, Ludovic Escoubas, Francois Flory, Sandrine Bailly, Remi de Bettignies, Stephane Guillerez, Christophe Defranoux, "Modeling the short-circuit current density of polymer solar cells based on P3HT:PCBM blend", Solar Energy Materials & Solar Cells 91, 405–410 (2007). https://doi.org/10.1016/j.solmat.2006.10.019

[38] Gavin A Buxton and Nigel Clarke, "Computer simulation of polymer solar cells", Modelling Simul. Mater. Sci. Eng., 15, 13–26 (2007). https://doi.org/10.1088/0965-0393/15/2/002

[39] C. Groves, R. A. Marsh, and N. C. Greenham, "Monte Carlo modeling of geminate recombination in polymer-polymer photovoltaic devices", The journal of Chemical Physics 129, 114903 (2008). https://doi.org/10.1063/1.2977992

[40] JonnyWilliams and Alison B Walker, "Two-dimensional simulations of bulk heterojunction solar cell characteristics", Nanotechnology 19, 424011, (2008). https://doi.org/10.1088/0957-4484/19/42/424011

[41] P. Junsangsri and F. Lombardi, "Modelling and extracting parameters of organic solar cells", Electronics Letters, 46, 21 (2010). https://doi.org/10.1049/el.2010.2232

[42] C. De Falco, M. Porro, R. Sacco, and M. Verri, "Multiscale Modelling and Simulation of Organic Solar Cells, arXiv:1206.1440v3 [math.NA] (2012).

[43] Biswajit Ray and Muhammad Ashraful Alam, "Achieving Fill Factor Above 80% in Organic Solar Cells by Charged Interface', IEEE JOURNAL OF PHOTOVOLTAICS (2013).

[44] A. H. Fallahpour, A. Gagliardi, F. Santoni, D. Gentilini, A. Zampetti, M. Auf der Maur, and A. Di Carlo, "Modeling and simulation of energetically disordered organic solar cells", Journal of Applied Physics, 116, 184502 (2014). https://doi.org/10.1063/1.4901065

[45] Matthew L. Jones, Buddhapriya Chakrabarti, and Chris Groves, "Monte Carlo Simulation of Geminate Pair Recombination Dynamics in Organic Photovoltaic Devices: Multi-Exponential, Field-Dependent Kinetics and Its Interpretation", J. Phys. Chem., 118, 85−91, (2014). https://doi.org/10.1021/jp408063f

[46] M. Farrokhifar, A. Rostami, and N. Sadoogi, "Opto-Electrical Simulation of Organic Solar Cells", IEEE, (2014). https://doi.org/10.1109/ems.2014.73

[47] Hossein Movla, Amin Mohammadalizad Rafi, Nima Mohammadalizad Rafi, "A model for studying the performance of P3HT:PCBM organic bulk heterojunction solar cells', Optik 126, 1429–1432 (2015). https://doi.org/10.1016/j.ijleo.2015.04.020

[48] M. L. Inche Ibrahim,a Zubair Ahmad, and Khaulah Sulaiman, "Analytical expression for the current-voltage characteristics of organic bulk heterojunction solar cells", AIP Advances 5, 027115 (2015). https://doi.org/10.1063/1.4908036

[49] Davide Bartesaghi, Irene del Carmen Pe´rez, Juliane Kniepert, Steffen Roland, Mathieu Turbiez, Dieter Neher & L. Jan Anton Koster, "Competition between recombination and extraction of free charges determines the fill factor of organic solar cells", Nature Communications | DOI: 10.1038/ncomms8083, (2015). https://doi.org/10.1038/ncomms8083

[50] V. I. Madogni, W. Yang, B. Kounouhéwa, M. Agbomahéna, S. A. Hounkpatin, C. N. Awanou, Dynamic, "Charge Photogeneration and Excitons Distribution Function in Organic Bulk Heterojunction Solar Cells", Open Journal of Applied Sciences,5,509-525,(2015) https://doi.org/10.4236/ojapps.2015.58050

Chapter 12

Fill factor analysis of organic solar cell

Rashmi Swami, Sanjay Tiwari

Photonics Research Laboratory, SOS in Electronics, Pt. Ravishankar Shukla University, Raipur (C.G.) 492010, India,

rashmi.swami3@gmail.com

Abstract

Solar cell is a device used to convert light into electricity. It can be made by organic and inorganic materials. Its most important parameters are open circuit voltage, short circuit current, fill factor and conversion efficiency. This paper is based on the analysis of factors that affect the fill factor of organic solar cell using MATLAB. Fill factor is calculated using conventional organic solar cell model without series and shunt resistances and constant light generated current for two different cases –first using Exponential dark characteristic and second using Polynomial dark characteristic. We get that for exponential V-I relationship, increase in ideality factor n will reduce the fill factor and for polynomial V-I relationship, increase in m will increase fill factor. A large dependence of light generated current I_{ph} on increasing operating voltage would cause a significant drop in fill factor. Increase or decrease in an additional factor $\frac{I_d(V_{oc})}{I_{sc}}$ would accordingly change fill factor. Dark current can be varied in two ways, one by varying mobility and other by varying injection barrier heights. In both the cases fill factor increases with $\frac{I_d(V_{oc})}{I_{sc}}$.

Keywords

Organic Solar Cell, Fill Factor, Ideality Factor, Open Circuit Voltage, HTL, ETL

Contents

1.   Introduction..335

2.   Simulation model and analysis of fill factor.................................336

2.1   Exponential current voltage relationship –...................................336

2.2   Polynomial current-voltage relationship-......................................337

2.3   **Effect of dark current on fill factor –** **338**

3.   **Results and conclusions**...**338**

Acknowledgement **342**

References **342**

1.   Introduction

Bilayer organic solar cell (OSC), as shown in Figure 1(a), is a device in which thin layer of organic material (donor and acceptor) is used between electrodes to convert light into electricity. This work is completely based on bilayer structure of organic solar cell as shown in Figure1(a) in which poly(9,9'-dioctylfluorene-co-bis-N,N'-(4-butylphenyl)-bis-N,N'-phenyl-1,4phenylenediamine) (PFB) is organic donor/HTL and poly(9,9'-dioctylfluorene-co-benzothiadiazole) (F8BT) is organic acceptor/ETL. Fig. 1(b) shows simplest conventional organic solar cell model without series and shunt resistances. Open circuit voltage, short circuit current, fill factor and efficiency are four important parameters of OSC.

$$FF = V_{max}\, I_{max} / V_{OC}\, I_{SC}$$

When $V_m = V_{OC}$ and $I_m = I_{SC}$ then $(FF)_{max} = 1$.

For a good photo-voltaic device, all three factors FF, V_{OC}, I_{SC} should be large so that it can deliver large output power for the same incidental optical power.

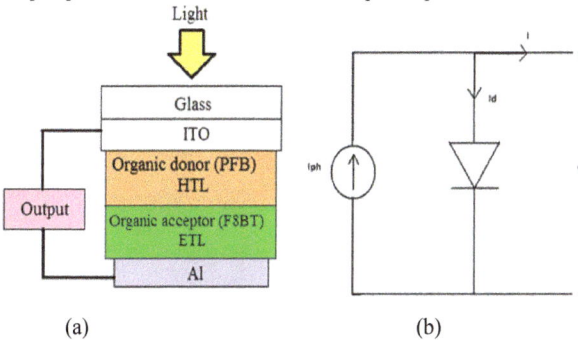

(a)   (b)

Figure 1 : (a) Bilayer organic solar cell structure. (b) Conventional organic solar cell model without series and shunt resistances.

2. Simulation model and analysis of fill factor

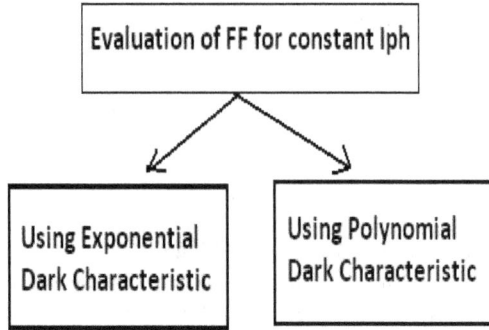

2.1 Exponential current voltage relationship –

In this model, dark characteristic is assumed to follow exponential current voltage relationship and I_{ph} is assumed to be constant.

$$I_d = I_0\left(exp\left(\frac{V}{nV_{th}}\right) - 1\right)$$

(1)

where n is ideality factor and V_{th} is thermal voltage, I_{ph} is light generated current, I_d is dark current and I is net output current.

Total output measured current can be written as a function of photo-generated current and dark current.

$$I = I_{ph} - I_d(V) = I_{ph} - I_0\left(exp\left(\frac{V}{nV_{th}}\right) - 1\right)$$

(2)

Output power of organic solar cell, when it is operating at voltage V and giving current I-

$$P = IV = I_{ph}V - I_0V\left(exp\left(\frac{V}{nV_{th}}\right) - 1\right)$$

If maximum power is obtained at voltage V_m,

$$\frac{\partial P}{\partial V} = I_{ph} + I_0 - I_0\left(\frac{V_m}{nV_{th}} + 1\right)exp\left(\frac{V_m}{nV_{th}}\right) = 0$$

$$\left(\frac{V_m}{nV_{th}} + 1\right)exp\left(\frac{V_m}{nV_{th}}\right) = 1 + \frac{I_{ph}}{I_0}, \text{ here assuming } y = \frac{V_m}{nV_{th}} + 1$$

$$yexp(y) = \left(1 + \frac{I_{ph}}{I_0}\right)2.718 \tag{3}$$

Here $y \exp(y)$ is Lambert's W function

$$y = lambertw\left[\left(1 + \frac{I_{ph}}{I_0}\right)2.718\right]$$

$$V_m = nV_{th}\left[lambertw\left[\left(1 + \frac{I_{ph}}{I_0}\right)2.718\right] - 1\right] \tag{4}$$

$$\text{and} \quad I_m = I_0\left(\frac{V_m}{nV_{th}}\right)exp\left(\frac{V_m}{nV_{th}}\right) \tag{5}$$

At V_{OC}, net output current will be zero. At this condition eq. (2) will give

$$V_{OC} = nV_{th}\ln\left(1 + \frac{I_{ph}}{I_0}\right) \tag{6}$$

2.2   Polynomial current-voltage relationship-

In this case it is assumed that dark current depends on the operating voltage in the following manner-

$$I_d = KV^m \tag{7}$$

Where K is constant and $m \geq 1$.

$$I = I_{ph} - I_d = I_{ph} - KV^m \tag{8}$$

If photovoltaic is operated at voltage V and output current is I, output power will be-

$$P = IV = I_{ph}V - KV^{m+1}$$

To calculate fill factor, one needs to find out the maximum power which photo-voltaic cell can supply. If maximum power is delivered at voltage V_m

$$\frac{\partial P}{\partial V} = I_{ph} - K(m + 1)V^m = 0$$

This will give, $V_m = \left(\frac{I_{ph}}{K(m+1)}\right)^{\frac{1}{m}}$ (9)

and $I_m = I_{ph}\left(\frac{m}{m+1}\right)$ (10)

At V_{OC}, net output current will be zero. At this condition eq. (8) will give

$V_{OC} = \left(\frac{I_{ph}}{K}\right)^{\frac{1}{m}}$ (11)

and $FF = \frac{m}{(m+1)^{(\frac{1}{m}+1)}}$ (12)

2.3 Effect of dark current on fill factor –

Simulation is done in MATLAB using 1D drift-diffusion electrical modeling of bilayer OSC. We obtained that the dependence of light generated current on the operating voltage, means that fill factor would depend on it as well, besides shape of dark characteristics. An estimate of variation of light current can be obtained by taking ratio of its value at short circuit and open circuit condition –

At 0 volt, $I(0) = I_{ph}(0) = I_{SC}$

At V_{OC}, $I(V_{OC}) = I_{ph}(V_{OC}) - I_d(V_{OC}) = 0$ i.e. $I_{ph}(V_{OC}) = I_d(V_{OC})$

The ratio $\frac{I_{ph}(V_{OC})}{I_{ph}(0)}$ is a measure of how drop in I_{ph} with the voltage. This ratio can be written as –

$\frac{I_{ph}(V_{OC})}{I_{ph}(0)} = \frac{I_d(V_{OC})}{I_{SC}}$

Thus $\frac{I_d(V_{OC})}{I_{SC}}$ shows an additional factor that would affect fill factor. As this factor increases or decreases, the fill factor should accordingly change too.

3. Results and conclusions

Eq. (3) suggests that as ideality factor n is changed, keeping reverse saturation current I_0 and photo-generated current I_{ph} constant, V_m changes in such a manner that (V_m/n) remains constant. So I_m will also be constant as it is a function of (V_m/n). From eq. (6)

open circuit voltage also changes with ideality factor n such that (V_{OC}/n) remains constant. It follows from the above reasoning that (I_m/I_{SC}) and (V_m/V_{OC}) will be unchanged if n will vary keeping the reverse saturation current constant. Hence as ideality factor n varies keeping the reverse saturation current I_0 constant, fill factor of the device will remain unchanged. Though if open circuit voltage (V_{OC}) assumed to be constant by varying reverse saturation current I_0 as ideality factor n changes, fill factor will change accordingly.

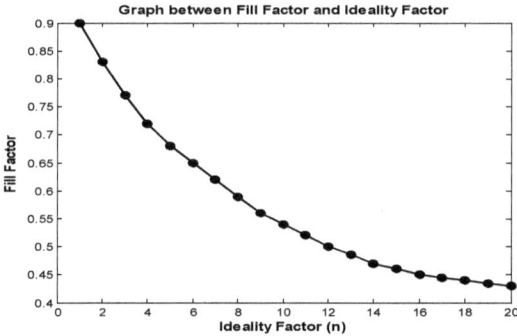

Figure 2 : Variation of fill factor with ideality constant n. open circuit voltage and light generated current are taken to be constant as 1.25 V and 1 mA-cm^{-2} respectively.

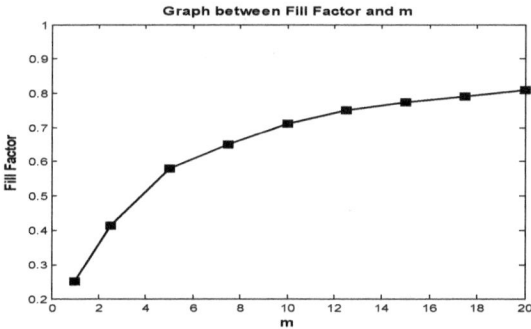

Figure 3 : Variation of fill factor with m. fill factor approaches to 1 as m becomes larger and larger.

Assuming I_{ph} to be 1 mA-cm^{-2}, I_0 to be 1×10^{-21} mA-cm^{-2} and ideality factor n to be 1, open circuit voltage and fill factor come out to be 1.25 volts and 0.9 respectively. Taking I_{ph} and V_{OC} constant, the variation of fill factor with ideality factor n is shown in Figure 2. We get that increase in the value of ideality factor n will reduce the value of fill factor.

Eq. (12) shows that fill factor is a function of m. Variation of fill factor with m is shown in Figure 3.

For m = 1, FF = 0.25. As m increases fill factor also increases and approaches to 1. However, FF will become only 1 when m is infinity. In this case also, m is a measure of the sharpness of the characteristic curve. As m increases, I-V curve becomes increasingly sharper resulting in a high fill factor. For polynomial dark characteristic with constant light generated current we get that increase in m will increase fill factor which approaches to 1.

Simulation results revealed in fig. 4 show that light generated current I_{ph} is a function of operating voltage, means FF would depend on it as well besides shape of dark characteristic. A large dependence of I_{ph} on increasing operating voltage would cause a significant drop in FF. Increase or decrease in an additional factor $\frac{I_d(V_{OC})}{I_{sc}}$ would accordingly change fill factor. Dark current can be varied in two ways, one by varying mobility and other by varying injection barrier heights. In both the cases fill factor increases with $\frac{I_d(V_{OC})}{I_{sc}}$ as shown in Figure 5 and Figure. 6.

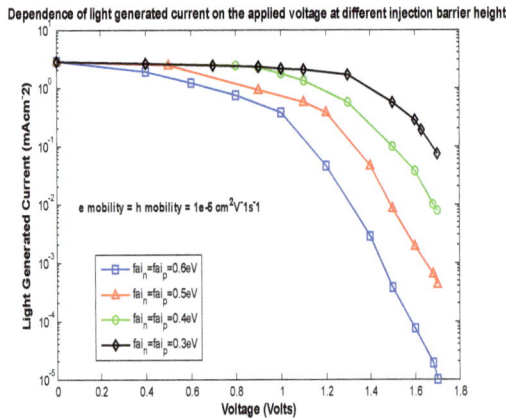

Figure 4 : Dependence of light generated current on the applied voltage. μ_p and μ_n are the hole and electron mobilities respectively. ϕ_p and ϕ_n are the injection barriers at anode and cathode respectively.

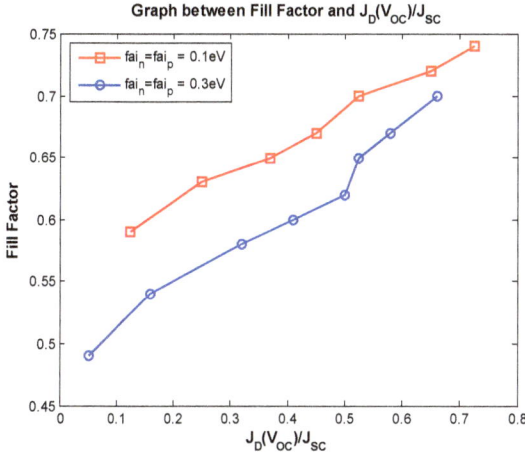

$$\frac{I_d(V_{OC})}{I_{SC}}$$

Figure 5 : Variation of fill factor with $\frac{I_d(V_{OC})}{I_{SC}}$ for 0.1eV and 0.3eV injection barrier heights. Different points have been obtained by changing mobility.

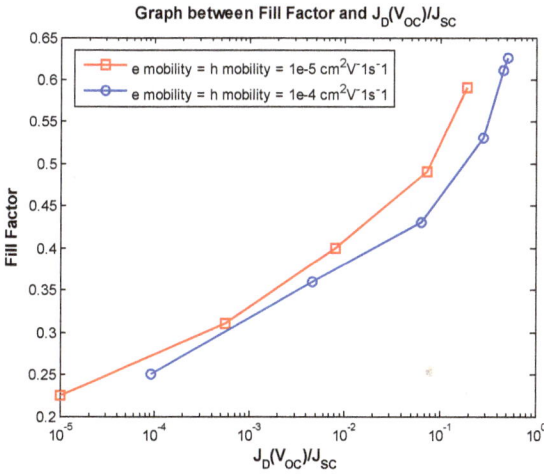

Figure 6 : Variation of fill factor with $\frac{I_d(V_{OC})}{I_{sc}}$ for carrier mobilities $1 \times 10^{-5}\ cm^2V^{-1}s^{-1}$ and $1 \times 10^{-4}\ cm^2V^{-1}s^{-1}$. Different points have been obtained by changing injection barrier height.

The analysis has indicated that for exponential V-I relationship, increase in ideality factor n reduces the fill factor and for polynomial V-I relationship, increase in m increases fill factor. Increase or decrease in an additional factor $\frac{I_d(V_{oc})}{I_{sc}}$ would accordingly change fill factor. Dark current $I_d(V_{oc})$ can be varied in two ways, one by varying mobility and other by varying injection barrier heights. In both the cases fill factor increases with $\frac{I_d(V_{oc})}{I_{sc}}$. A large dependence of light generated current I_{ph} on increasing operating voltage would cause a significant drop in fill factor.

Acknowledgement

I am extremely thankful to my organization Pt.Ravishankar Shukla University, Raipur (C.G.) for funding my research work.

References

[1]   B. K. Crone, P. S. Davids, I. H. Cambell and D. L. Smith, *"Device model investigation of bilayer organic light emitting doide"*, J. Appl. Phys., 84, (2000), 1974. https://doi.org/10.1063/1.372123

[2]   B. Mazhari, *"An improved solar cell circuit model for organic solar cells"*, Solar Energy Materials & Solar Cells, 90, (2002), 1021. https://doi.org/10.1016/j.solmat.2005.05.017

[3]   C. M. Ramsdale, J. A. Barker, A. C. Arias, J. D. MacKenzie, R. H. Friend and N. C. Greenham, *"The origin of open circuit voltage in polyfluorene-based photovoltaic device"*, J. Appl. Phys, 92, (2002),4266. https://doi.org/10.1063/1.1506385

[4]   D. P. Grubera, G. Meinhardtb and W. Papousekc, *"Modelling the light absorption in organic photovoltaic devices"*, Solar Energy Materials and Solar Cells, 87, (2005), 215-223. https://doi.org/10.1016/j.solmat.2004.08.011

[5]   J. A. Barker, C. M. Ramsdale, and N. C. Greenham, *"Modeling the current-voltage characteristics of bilayer polymer photovoltaic devices"*, Physical Review B 67, (2003), 075205. https://doi.org/10.1103/PhysRevB.67.075205

[6]   J. C. Scott and G. G. Malliaras, *"Charge injection and recombination at the metal-organic interface"*, Chem. Phys. Lett., 299, (1999), 115. https://doi.org/10.1016/S0009-2614(98)01277-9

[7]   J. Wagner, T. Fritz, and H. Bottcher, *"Computer modelling of organic thin film solar cells exciton model of photocurrent generation"*, Physica Status Solidi A, 136, (1993), 423. https://doi.org/10.1002/pssa.2211360215

[8] P. W. M. Blom, M. J. M. de Jong and S. Breedijk, "Temperature dependent electron hole recombination in polymer light emitting diodes", Appl. Phys. Lett., 71, (1997), 930. https://doi.org/10.1063/1.119692

[9] S. E. Shaheen, C. J. Brabec, N. S. Sariciftci, F. Padinger, T. Fromherz, and J. C. Hummelen," *2.5 % efficient organic plastic solar cells"*, Appl. Phys. Lett., 78, (2001), 841-843. https://doi.org/10.1063/1.1345834

[10] Y. Roichman and N. Tessler, "Generalized Einstein relation for disordered semiconductors implications for device performance", Appl. Phys. Lett., 80, (2002), 1948. https://doi.org/10.1063/1.1461419

Keywords

Bulk Hetero-junction 291

Cadmium Telluride Thin Films 185

CdSe Nanocrystals 241

Chemical Bath Deposition (CBD) 185

Copper Selenide Thin Films 134

Different Generation of Solar Cells 1

Drift Diffusion Model 310

ETL 334

Exciton 310

Fill Factor 334

Flexible Transparent Solar Cells 299

History of Solar Cells 1

HTL 334

Ideality Factor 334

Intermediate Bands 117

IPCE 291

Low-cost Self-assembled
Nanostructures 58

Multilayers 241

Multiple Electron Generation 117

Nanocrystals 117

Nanostructured Solar Cell Systems 58

Nanostructures 58

Non-uniform Generation 310

Open Circuit Voltage 334

Organic Electronics 269

Organic Solar Cell 291, 310, 334

PCBM 291

PCDTBT 291

Photo-active Polymer 310

Photoelectrochemical Cells 134, 185, 241

Photovoltaic Cells 1

Photovoltaic Devices 269

Photovoltaic Effect 37, 185, 241

Photovoltaics 58

Polymer Solar Cells 299

Polymeric Materials 269

PTE 291

Quantum Dot 117

Semiconductors 269

Solar Cell Characteristics 37

Solar Cell Materials 37

Solar Cell Parameters 37

Solar Cell 117, 269

Solar Power 58

Surface Morphology 134

Types of Solar Cells 1

UV-Vis Absorption 185

XRD 134, 185

About the Editor

Dr. Meera Ramrakhiani

Dr. Meera Ramrakhiani has been the Professor and Head, Department of Physics and Electronics and also Dean, Faculty of Science at Rani Durgavati University Jabalpur, India. She has done her graduation, M.Sc. and Ph.D from the University of Jabalpur (now Rani Durgavati University) and has 40 years of teaching and research experience at under graduate and post graduate levels. More than 25 students have completed their Ph.D. degree under her supervision and many more are working in the field of nanomaterials, luminescence and photovoltaic solar cells. She has participated in about 70 seminars/symposia etc. and presented her work at national/ international level and also organized a number of conferences/workshops etc. She has visited Italy, Hungary, Singapore, USA and China for the research work. Dr. Meera Ramrakhiani has authored or coauthored about 400 research papers/book chapters/articles and successfully carried out two research project. She has been reviewer to many national/international journals and has received many awards such as Vijaya Shree Award by India International Friendship Society in 1997, The 20th Century Award of Achievement by International biographical Center, Cambridge, England in 1998, Women of the Year 2005 Jeweler Issued by American Biographical Institute, Inc. etc. She is life member of many professional bodies and Fellow of Luminescence Society of India.

www.ingramcontent.com/pod-product-compliance
Lightning Source LLC
Chambersburg PA
CBHW071321210326
41597CB00015B/1298